Gender Issues
in Farming Systems
Research and Extension

Published in cooperation with the
Women in Agricultural Development Program,
University of Florida

Gender Issues
in Farming Systems
Research and Extension

EDITED BY

Susan V. Poats, Marianne Schmink, and Anita Spring

Routledge
Taylor & Francis Group

LONDON AND NEW YORK

First published 1988 by Westview Press, Inc.

Published 2018 by Routledge
52 Vanderbilt Avenue, New York, NY 10017
2 Park Square, Milton Park, Abingdon, Oxon OX14 4RN

Routledge is an imprint of the Taylor & Francis Group, an informa business

Library of Congress Cataloging-in-Publication Data
Gender issues in farming systems research and extension.
    (Westview special studies in agriculture science
and policy)
    "Based on the 1986 University of Florida
conference"--
    "Published in cooperation with the Women in
Agricultural Development Program, University of
Florida"--
    1. Agricultural systems--Research--Congresses.
2. Women in agricultural--Congresses. 3. Agricultural
extension work--Congresses. 4. Farms, Small--Congresses.
5. Agricultural productivity--Congresses. I. Poats,
Susan V. II. Schmink, Marianne. III. Spring, Anita.
IV. University of Florida. Women in agricultural
Development. V. Series.
S494.5.S95G46 1988 630'.88042   87-34315
ISBN 13: 978-0-367-01403-2 (hbk)

ISBN 13: 978-0-367-16390-7 (pbk)

Contents

Tables and Figures

FIGURES

Preface

This book is the product of an international
conference hosted by the Women in Agricultural
Development (WIAD) Program at the University of Florida
from February 26 to March 1, 1986. The WIAD Program's
general purpose is to promote an understanding of gender
and its relevance for agricultural development processes.
Women are critical to agricultural production in many
places, but access to resources and research technologies
may be constrained by gender and this may lead to detrimental effects on the design and implementation of sound
agricultural programs. The Program seeks to support and
develop expertise related to the roles of women and
intra-household dynamics in agricultural production,
research, and extension in order to improve the design
and implementation of agricultural programs. The
Program's primary audience is the faculty and students of
the University of Florida, but the Conference and this
book permitted us to reach out to a wider national and
international audience.

The Women in Agricultural Development Program began
as the Women in Agriculture (WIA) group at the University
of Florida (UF) in 1983 after five years of informal
activities related to women and development. It became
an official University of Florida program in 1984 and the
name was changed to Women in Agricultural Development in
1986. During early years, several initiatives emerged
from the International Programs office and from social
scientists, especially those in the Centers for African
and Latin American Studies. Momentum built up as a
critical mass of faculty came to the UF with experience
in the WID field. A group that formed in the fall of
1983, with Anita Spring as Director and Marianne Schmink
as Co-Director, served as an ad hoc committee to promote
awareness of issues related to WID and, more

specifically, to women's role in agricultural production. The WIAD group received support from the USAID Program Support Grant funds administered by the International Programs office and from other on-campus sources including the Centers for Latin American and African Studies and the Graduate School. Principal program activities include a bi-weekly speaker series and a bibliographic compilation of readings on women in agriculture, as well as work in curriculum development and in technical assistance.

From the beginning, the WIAD Program sought to focus its efforts specifically on the realm of agriculture, in recognition of the UF's strengths in this area. It took advantage of the opportunity to interface directly with a major USAID-funded project, the Farming Systems Support Project (FSSP) that had a worldwide mandate for technical assistance, training, and networking activities. As Associate Director of the FSSP based at the UF, Susan Poats was in a key position to facilitate that linkage. In addition to the institutional presence of the FSSP, the general philosophy and methodology of the "farming systems" approach provided promising avenues for attention to women. Yet gender had not been incorporated effectively and systematically as a variable in most farming systems work. The WIAD Program hoped to stimulate the development of conceptual and methodological approaches that could improve the incorporation of gender issues into farming systems activities. These ideas are discussed in more detail in Chapter 1.

In early 1984, the WIAD Program hosted a visit to the UF by several colleagues involved in an effort by the Population Council to produce useful case studies of gender issues in development projects. This meeting led to the creation of the FSSP/Population Council Intra-Household Dynamics Case Studies Project, co-managed by Susan Poats and Judith Bruce of the Population Council, with funding from USAID and from the Ford Foundation. The FSSP created a new task force to focus on the integration of household and family concerns into the farming systems perspective. In 1984, Cornelia Butler Flora (a member of both the FSSP's Technical Committee and of the task force on the family) visited UF. She also wrote a position paper on "Intra-Household Dynamics in Farming Systems Research: The Basis of Whole Farm Monitoring of Farming Systems Research and Extension." She and other task force members were instrumental in supporting the proposal to develop case study training materials through the FSSP/Population Council project.

The case study project, initially launched from a
WIAD activity, generated further momentum. In February
of 1985 the project's advisory committee contacted those
individuals and projects around the world who might be
interested in developing case study materials from their
own experiences in farming systems work. A survey
questionnaire requested more detailed information from
those who responded. The results of this questionnaire
provided some indication of the current state of knowledge on intra-household dynamics in on-going farming
systems projects (see Chapter 2). The FSSP/Population
Council project selected eight proposals to receive
support for the development of case study materials. Yet
the more than 75 respondents who expressed interest in
the project indicated that the incorporation of intra-
household variables was a greater concern to farming
systems practitioners than had been anticipated.

Parallel to the case studies project, the WIAD
Program proposed bringing together farming systems
practitioners to a conference on the UF campus. The
focus of the event would be purposely narrow in order to
maximize the potential for improving the integration of
gender into the farming systems approach, taking
advantage of the interest aroused by the case studies
initiative. Sessions would address specific issues of
theory, method, and policy related to Farming Systems
Research and Extension (FSR/E) in developing areas,
systematically comparing African, Latin American, and
Asian experiences. Marianne Schmink volunteered to
organize the conference and Susan Poats coordinated the
format of the program. WIAD's Steering Committee served
as an advisory body and a conference task force was
appointed. Funding came from the Center for Latin
American Studies, the Center for Tropical Agriculture,
and the Center for African Studies, as well as from the
Ford Foundation and the Rockefeller Foundation. A call
for papers circulated in July of 1985 defined specific
topical areas of concern and asked potential participants
to summarize the data they would present, its relation to
FSR/E, the geographic focus, and its relevance to the
designated themes.

The response was overwhelming. Over a hundred paper
proposals arrived from all over the world, expanding the
geographic focus of the conference. In an effort to
accommodate as many participants as possible, the organizers structured a comprehensive program of concurrent
sessions and explored as many sources as possible for
travel support, giving priority to visitors from developing countries. Over 40 participants received full or

partial funding to attend the Conference. There were 91 speakers in fifteen formal paper sessions, three round-table discussions, and an after-lunch lecturer. Carmen Diana Deere of the University of Massachusetts at Amherst delivered the keynote address on "Rural Women and State Policy: An Evaluation of the Decade" (later published in revised form in her book with Magdalena Leon, Rural Women and State Policy, Westview, 1987). Films and recent publications on FSR/E and gender were available for review during the Conference and several social events facilitated a high level of information exchange and networking. Attendance at conference sessions was consistently high. A total of 298 persons were register-ed or on the formal program and many others from the campus and local area attended without registering. The magnitude and complexity of the Conference demanded the assistance of many volunteer helpers. Special thanks are due to Carol Brown, Donna Epting, Jean Gearing, Janet Hickman, Patricia Kuntz, Cindy Lewis, Greg Moreland, Bill Reynolds, Barbara Rogers, Robin Sumner, and Darla Wilkes. Assistance was also provided by Wharton Williams Travel, Classique Cuisine, Renaissance Printing, Farming Systems Support Project, International Programs, International Food and Agricultural System (IFAS), and the College of Liberal Arts and Sciences.

Chris Andrew, Director of the FSSP, formally opened the Conference on behalf of the University of Florida. In his opening remarks he said:

> The intellectual capacity assembled in this room to address gender-related issues in on-farm research is second to none. We are fortunate this week to have presentations by people who have traveled from more than twenty countries, from forty universities (eleven non-U.S.), from three International Agricultural Research Cen-ters, and many other national, donor, private voluntary organizations, and private entities. The titles in the conference program mention twenty-two countries and we know that more will be discussed. Nearly every region of the world is represented.
>
> The intra-household concept, as an integral and dynamic part of the farm system, must con-sider gender issues. I emphasize integral and dynamic because we must take care not to esta-blish artificial boundaries to accommodate conventional simplicity. With a broad view of the household, third world food security can be

addressed in a national framework. We need a
conceptual focus that recognizes the farm household as a critical element in successful agricultural policy for research and development.
To enunciate and address the interdependent
needs of women and men in farming systems as
they interact with the bio-physical environment
is an important goal.

The critical importance of agriculture to the
vitality and strength of many of the world's
countries is widely recognized. At the same
time, the increasing diversity, complexity, and
intractability of the problems facing agricultural development make it imperative that agricultural research systems change and adapt to
address specific realities. One of these is the
role of women in agriculture. Another is that
gender issues are being considered together with
the farming systems approach, an agenda that
likely could not have found a platform ten years
ago. In the ebb and flow of agricultural
research, this Conference represents something
of a revolution in agriculture and in
communications.

The Conference offered an excellent opportunity for
exploring and testing new ideas and successful approaches
for incorporating gender sensitivity in agricultural
research and development. The FSSP/Population Council
project used the Conference to test a case study based on
FSR/E project activities in Zambia. The two authors of
the case, Charles Chabala and Robert Nguiru, were present
to view how the case worked in a training context.
Hilary Feldstein, Rosalie Norem, Kate Cloud, Susan Poats,
Nadine Horenstein, and Mary Rojas conducted the
abbreviated training sessions with approximtely sixty
conference participants. Their detailed evaluations and
suggestions were used to improve the case. Highlighted
repeatedly during the Conference was the need to
inventory and assess the usefulness of various methodologies used in dealing with gender issues. An ad hoc
special methodology workshop held on Saturday afternoon
following the close of the formal sessions was attended
by more than seventy-five people. Janice Jiggins and
Hilary Feldstein are using this session as the basis for
developing a methodologies handbook to accompany the
FSSP/Population Council case studies.

Aside from the high level of interest in the formal and ad hoc sessions, informal networking was intense among those present. These interactions underscored the remarkable level of interest in the practical aspects of integrating attention to gender issues into agricultural research and extension. The Conference provided a meeting place for people with diverse disciplinary and area expertise to learn from one another. The most exciting outcome of the Conference was the discovery of such a broad base of research already underway on gender issues in agriculture.

A total of sixty-four written papers were available in a three-volume set at the Conference. In order to make conference materials more widely available, the WIAD Program decided to organize a selection of the international papers for publication as a book. Although the quality of most of the papers presented was high, space considerations made it impossible to publish all of them. This volume includes less than half of the papers presented at the Conference. In making the difficult decision as to which papers to include, the editors used several criteria. Most important was adherence to the original thematic focus on farming systems work, as outlined in the call for papers. Secondly, preference was given to papers that presented new ideas or methodological approaches for the integration of gender into FSR/E projects. Finally, regional balance and the inclusion of non-U.S. authors who might not otherwise find an audience for their work in this country was sought. The overall goal was to produce a coherent reader with a comparative perspective, that would be both stimulating and helpful to people trying to implement gender-aware farming systems projects. We hope the many excellent papers that could not be included in this volume will find publication outlets elsewhere.

This book, and the WIAD Program's overall effort, have depended on the hard work and support of a dedicated group of faculty, students, and staff from the UF. A special thanks is due to Jean Gearing, who was responsible for much of the editing, correspondence, typing, and organizational tasks associated with the book. The editors are also grateful to Pamela Shaw for her editorial work, to Kathy Gladden for bibliographic and clerical assistance, to Sharon Leslie and Barbara Rogers for their assistance in artwork, and to Lana Bayles, Kenna Hughey, Shirlene Washington, Sabrina Byron, and Dana Whitaker who provided additional help with the word processing. The book and the Conference that produced it are, in turn, part of a larger effort in which the WIAD organizers have

been privileged to collaborate with many colleagues, some
of whom are mentioned above. It is hoped that the ideas
and experiences discussed by the authors of the following
chapters will help to stimulate and improve the consideration of gender as a crucial issue in agricultural
production and development.

Susan V. Poats
Marianne Schmink

1

Linking FSR/E and Gender:
An Introduction

Susan V. Poats, Marianne Schmink, and Anita Spring

The title of this book is like a code. The two terms
"gender issues" and "farming systems research and extension" are shorthand. Each represents an extensive field of
research and practice: women and development or WID, and
farming systems or FSR/E, respectively. The two fields
have much in common. Both emerged relatively recently in
response to dissatisfaction with the results of technological change in agriculture in developing countries.
Whereas in the 1950s and 1960s development theory and
practice emphasized growth in productivity, by the 1970s
there was a renewed concern to implement programs that
conceived of development more broadly, to mean the possibility of better lives for most people. This perspective
challenged a development field dominated by technical and
economic expertise. Efforts to develop more comprehensive
approaches that would bring together technical, economic,
and social considerations led to the two interdisciplinary
fields of WID and FSR/E.

In this brief introduction, justice cannot be done to
either field in its own terms. Rather, the historical and
practical considerations that favor their interaction and
the conceptual problems such a union can help to overcome
are reflected. The discussion will indicate how the
following chapters in this book contribute theoretical and
methodological insights that can help to make agricultural
development programs more efficient and equitable.

FARMING SYSTEMS RESEARCH AND EXTENSION

Most farms in developing countries are small scale,
with few resources other than family labor. Their
subsistence activities are multifaceted and their goals

complex, including both market and non-market considerations. Minimizing risk is especially important when family survival is at stake. Given the constraints they face, small farmers actively seek ways to improve their productivity and to maximize the few resources at their disposal. Agricultural technologies (including equipment, inputs such as fertilizers and pesticides, and management practices) are often designed for farmers with greater resources and market orientation based on an essentially economic calculation of costs and benefits. By the 1970s, development practitioners became concerned that the benefits of agricultural innovation accrued most easily to these wealthier farmers. Yet small farmers constituted the majority of producers and, ultimately, those most directly responsible for the welfare of rural families and communities. What were the social and economic costs of neglecting them? The design of technologies appropriate to the majority of low-resource farmers required an understanding of their particular constraints, goals, and practices that went beyond strictly technical and economic criteria. The farming systems concept emerged as a response to this challenge.

FSR/E is not a single approach, but an array of different perspectives and methods. This diversity is a source of debate and dialogue that continually enriches the field. In this book there is likewise no orthodoxy, but rather a collection of different points of view as to how to conceptualize and carry out farming systems work. The common elements that underly most versions of farming systems include: an explicit commitment to low-resource producers; a systems approach that recognizes the complexity of small farm enterprises; a focus on the farm family or household; and a recognition of the importance of including farmers in the research and extension process. The concept of "domains" is used to denote the specific client group (defined by environmental, ecological, and/or socioeconomic criteria) to whom the project is oriented.

WOMEN IN DEVELOPMENT

The WID field, similar to FSR/E, began with a concern for the distribution of development benefits. Like farming systems, women and development is far from a unified field of knowledge. Not only does it include many strands of research and practice, but the field has evolved rapidly over the approximately 20 years of its existence, since economist Ester Boserup published her ground-breaking work Women's Role in Economic Development in 1970. Boserup's work challenged the prevailing notion that economic development, or modernization, would automatically improve women's status by replacing traditional values and economic backwardness with new opportunities and an egalitarian ethos. She argued instead that economic innovations often replaced women's traditional economic activities with more efficient forms of production controlled by men. Examples included the decline of women's cottage industries due to competition from factories hiring predominantly men and, in some parts of the world, the growth of modern service and commerce sectors in which men predominated, in place of women's traditional marketing practices. The recognition that development, as practiced, might actually worsen women's position relative to men's crystallized the new field of women and development around a concern with equity.

By the late 1970s, however, a growing research base on women's economic activities showed that equity was intimately related to more technical problems of efficiency and productivity. If development undermined women's traditional economic contributions, was this loss compensated by the output of new forms of production? Were new economic opportunities opening up for women? What was the impact of these shifts on the welfare and productivity of the poor populations of the world? The new emphasis on the poor focused attention on women's importance as household producers and providers in addition to their domestic roles. No longer were they to be viewed simply as potential welfare beneficiaries whose needs might be neglected by development efforts. Instead, women were a mainstay of family and community welfare, active producers whose potential contributions were often overlooked or undermined. A clearer understanding of changes in women's role in production therefore was essential for the success of agricultural development projects.

A decade of theoretical experimentation and empirical research on women's role in development moved the field from the stage of raising awareness and clarifying issues to a search for practical applications. How could the WID insights be applied to development work? One solution was to create special projects or components devoted to women. While sometimes successful, these all too often emphasized women's domestic responsibilities rather than their productive work. They also distracted from the more general problem of improving the effectiveness of "mainstream" development projects by making them more responsive to gender differences among the client population. By what practical means could such a formidable task be undertaken?

The first attempts to answer this question produced an array of checklists of questions to be asked and data to be gathered in each project setting. A series of case studies were published as examples of how gender affected development projects. Various institutions compiled handbooks that specified how gender issues could be addressed at each step of the project cycle. But there were not enough experts trained in the analysis of such complex and variable matters as household division of labor, decision-making, and income management. Some of the basic issues could be specified in advance, but each setting required a unique assessment of their relevance and of the interaction with other important variables. While hiring more women as project staff members appeared to be a good idea, the gender of the researcher or practitioner turned out to be no guarantee of the requisite analytical skills.

In response to this dilemma, WID efforts in the 1980s sought to develop the tools of "gender analysis" and the methods by which development practitioners could learn and adopt them. USAID fostered a major effort to adapt the Harvard Business School's case study teaching method to training on gender issues in development projects. The Office of Women in Development sponsored the writing of several new analytical case studies that were compiled in a handbook that also provided a framework and set of basic concepts to be used in the case study analysis (Overholt et al. 1985). The cases and the training method have been widely used in training workshops that provide practice in tackling a set of questions that might otherwise seem hopelessly complicated. The strength of this approach is its emphasis on the link between project or development goals and gender differences in the client population. This focus helps to clarify the relevant issues and to indicate priorities for research and action.

GENDER ISSUES IN FARMING SYSTEMS RESEARCH AND EXTENSION (FSR/E)

The farming systems perspective is especially appropriate for such a process-oriented approach to gender analysis. The FSR/E methodology consists of a series of stages (diagnosis, planning and design of technology, experimentation and evaluation, and dissemination) that facilitate the specification of steps to be taken to address particular aspects of research and extension and how to make the best use of different kinds of data. But

FSR/E is also conceived to be an iterative, adaptive process in which, once the project is well underway, the various stages of research take place simultaneously. This philosophy is intended to maximize the potential impact of on-going farmer evaluations on the design and dissemination of future technological changes. The research process allows time to learn about the intricacies of farming systems and to incorporate new insights into more refined measures and project adaptations. Other characteristics of the farming systems approach especially important for gender analysis include its focus on small farm households and on the participation of farmers in the research and extension process.

Disaggregating Development Beneficiaries

The farming systems emphasis on reaching specific low-income groups helped to illuminate women's roles in agricultural development. Identifying small farmer constituencies required the disaggregation of society into "target" or "client" groups which brought women's activities into greater focus. The interaction of socioeconomic standing and gender was brought home by the growing recognition that women in poor families played essential economic roles that bore little resemblance to the activities of middle-class and elite women in the same societies. These observations were confirmed by mounting evidence from research that documented poor women's multiple economic activities, low earnings and long work hours, and restricted access to productive resources. Women played a central role in the low-resource farm households that were the focus of farming systems work.

The surprisingly high and growing proportions of female-headed households dramatized women's economic importance in poor populations and revealed the extra constraints under which they often labored to achieve family welfare (Buvinic and Youssef 1979). Rural out-migration of men was rising in many parts of the world as a result of development, leaving many women either temporarily or permanently in charge of their households (Palmer 1986). Their efforts were often undermined by labor constraints or by lack of access to productive resources, in part because research and extension services were primarily oriented to male farmers. The focus on female-headed households illustrated how disaggregation of beneficiary populations could more precisely delineate appropriate interventions for specific social groups. It also undermined the assumption that development projects focused on male farmers always would have the most effective impact on family welfare.

The Whole Farm System

The systems approach endorsed by the FSR/E constituency
lent itself well to illuminating women's economic importance. Small farm enterprises encompass multiple activities whose interaction is key to understanding management
decisions and practices. The configuration of a given
system changes readily over time in response to both internal and external factors. This holistic, dynamic perspective on small farming enterprises provided a framework
within which the family division of labor could be a key
focus. Social definitions of which tasks would be carried
out by men or women vary from one society, region, class,
or ethnic group to another. This variability indicates
that the division of labor is determined not by the physical difference between the sexes, but by the social
definitions of proper relations between women and men. The
concept of "gender" serves to distinguish the social character of these relationships, and the "sexual division of
labor" describes the allocation of tasks and responsibilities to men and women in a particular situation. In
practice, farming systems practitioners may disaggregate
only so far, stopping short at the analysis of the division
of labor within the household or family. Agricultural
research has historically focused on specific commodities
whose production is market-oriented. FSR/E recognized that
small farm enterprises combine crops and animals. Yet the
perspective still overlooked other essential activities
carried out by farm families, including off-farm work,
home-based production for use or exchange, and the work
required to maintain the home and its inhabitants. WID
research revealed that women were often predominant in
these activities, especially those based in the home that
tended to be overlooked or viewed as merely "domestic"
work. While men often specialized in income-generating
activities, women typically combined household management,
child care, and work to generate earnings (both on and off
the farm). These competing demands on their time could
serve as a significant constraint to the adoption of new
forms of production that relied on women's labor.

In small farm households, decisions reflected priorities and constraints related to a variety of activities and
goals, not just to those related to cash crop production.
The potential trade-offs between resources devoted to agricultural production and investments in improved family
nutrition were of particular concern to farming systems
practitioners whose objective was to stabilize or enhance
rural welfare. The systems approach adopted by FSR/E
practitioners provided a starting point for integrating the

diversity of farm and non-farm activities within a more complex model of the whole farm-household system. WID practitioners collaborated by focusing attention on women's importance in agricultural production, but also in focusing on activities not generally defined as "production" that are nevertheless essential to the well-being and economic livelihood of rural households and communities.

Intra-Household Dynamics

The focus on farm families brought development work much closer to the realities of poor families than was possible using the country level statistics. The concept of "household", sometimes used to denote a residential unit, sometimes synonymous with the nuclear family, was useful in the field of development and in the social sciences in general (Schmink 1984). It provided an intermediate level of analysis (between the individual and the aggregate society) and a convenient unit for the collection of empirical data. The existence of such primary domestic units in virtually all societies implied an attractive universality for the concept of household that was familiar to researchers and practitioners from their own personal existence. These perceived advantages, however, had hidden drawbacks. Development practitioners often generalized from their own experience, presuming that households elsewhere were similar to those in which they lived, when in fact household structure and functioning is highly variable. Whereas in advanced industrial society productive work is largely separated from the home, the same is not true for agrarian communities. Home-based food processing, handicrafts, care of animals, kitchen gardens, and manufacturing of such useful items as soap and clothing for a peasant family are not analagous to the domestic chores of a middle-class urban housewife. They are productive tasks essential to household welfare. Whereas a U.S. household typically depends on one or two monetary wages for its sustenance, rural families in the developing world rely on a diverse set of paid and non-paid activities for survival.

In many societies women and men have quite separate responsibilities, access to distinct resources, and differentiated control over returns from their own activities. In fact, households are themselves systems of resource allocation (Guyer 1981). The unitary neoclassical view of household income inherited from the advanced industrial nations is especially inappropriate for

such complex situations in which household members have
access to different resources and work opportunities, and
exercise differing degrees of control over separate income
streams that flow through the household. Household
decision-making is neither necessarily unitary nor
harmonious. Different members may decide about production
strategies, contribute labor to specific tasks, or bear
responsibility for the use of the commodities produced.
The complexity of intra-household dynamics implies that the
possibility of competing goals or priorities may require
negotiation among household members. Households are also
fluid; variability stems from responses to exogenous
changes (such as male out-migration), from internal
differentiation based on class, income, ethnicity, and
culture, and from demographic variables within the
household unit (that is, the pattern of family formation,
or the "life cycle" of the family).

Farmer Participation

The internal dynamics of small farm households affect
the process of client involvement in the research and
extension process. If work responsibilities, control over
resources, and decision-making are fragmented within the
family unit, who are the appropriate partners in the
research process and potential beneficiaries of the
proposed technologies? Since male household heads are
typically the public representatives of family groups, it
is often assumed that information and resources conveyed to
them will "trickle across" to others in their household.
But indirect communications strategies are inefficient and
may omit the actual "user" from the process of FSR/E. This
omission represents a loss of valuable indigenous knowledge
and may lead to inadequate or incomplete application of
technological innovations. Since women and men may know
about different factors relevant to agricultural production, the labor of one may not necessarily substitute for
the other. If farming systems projects are to succeed in
forging effective collaborative ties with their client
population, they must include both women and men farmers as
partners in the research process.

CONTENTS OF THIS BOOK

The chapters of this book are a selection of the papers
concerned with developing countries in Latin America, Asia,
Middle East, and Africa presented at the Conference in 1986
at the University of Florida (see Preface). The

Conference's primary objective was to bring together
scholars and practitioners with expertise and interest in
FSR/E to discuss state-of-the-art issues related to gender
in FSR/E. In order to maximize coherence, participants
were asked to prepare papers that would address specific
issues of theory, method, and policy related to FSR/E
across developing regions. The following questions were
posed under several topical areas:

The Whole Farming System. How can key components of a
farming system, including non-farm activities, livestock,
secondary crops, food storage, and food processing, be
identified? How does the division of labor by age and
gender constrain or facilitate specific economic or
productive goals? What is the potential impact of improved
agricultural technology on each of these goals and on
household members responsible for specific production
activities?

Intra-Household Dynamics. What are the key aspects of
internal heterogeneity of household units: differential
access and use of resources within households; multiple
enterprises and their interactions; substitutability and
specialization of labor in agricultural activities; marketing outlets and their relationship to differing or conflicting priorities and needs within farm units; and how
might proposed interventions alter the balance of power and
advantage?

Institutional and Policy Concerns. How does the surrounding environment beyond the farm gate at household,
community, and other social levels differ for men and
women? What specific constraints to production are posed
by these gender differences? How can FSR/E address
constraints such as legal status, restrictions on mobility,
domestic obligations, property rights, access to credit,
markets, and employment?

Definition of Research Domains. How can the key
components and actors within household and farming systems
be identified? What are key constraints to productive
activities and how does access to production inputs differ
by gender? What is the significance of, and interactions
between, multiple enterprises within the farm household
and how do they create different labor requirements, goals,
incentives, markets, and priorities for different family
members?

On-Farm Research and Extension. Who are the specific
audiences for direct involvement in on-farm research and
extension? How do labor constraints affect proposed
solutions and how do labor patterns impact on household
members who differ in their access to resources? How can

extension strategies be devised that are responsive to the
productive activities of both women and men? Which
extension mechanisms are most effective in reaching both
male and female farmers?

Monitoring and Evaluation. How can strategies be
designed for monitoring the differential impact of FSR/E
interventions on different individuals and enterprises
within the farming system? What are the unanticipated
effects of technological change? In what ways does
misunderstanding of gender issues lead to inadequate
planning and design or diminished returns to FSR/E
projects? How can these effects be minimized?

The papers contained in this book do not exhaust the
answers to these questions, but they do provide a
beginning. Authors were asked to include a common "minimum
data set" in their case study material to facilitate
comparison. The following chapters contain many innovative
approaches to conceptualizing and carrying out farming
systems projects that effectively take gender into account.
They highlight several features of the farming systems
approach that could be improved by more attention to gender, and they suggest practical ways that this could be
done. The book presents a comparative perspective on the
relevance of gender to farming systems work in the developing regions. Two dimensions run throughout the various
chapters: the presentation of site-specific data that will
permit in-depth analysis of specific cases and the search
for conceptual and methodological innovations.

Part I presents a set of articles that focus on key
theoretical and methodological issues relevant to the
farming systems approach. In Chapter 2, Rosalie Norem
summarizes the results of a survey of farming systems
projects that collected data on intra-household dynamics
and gender differences. Project staff expressed a need for
more intra-household data, especially on the factors
determining household variability (such as out-migration
and the family "life cycle"), on specific labor constraints
stemming from the gender division of labor, and on income
management within the household. Her findings also reveal
that different kinds of information are useful at each
phase of a project. In Chapter 3, Alison Evans discusses
some of the problems with FSR/E procedures that impede the
effective integration of gender considerations, including
the emphasis on market criteria and measure, and the
assumed homogeneity of the farm household. She presents a
framework of ideas to help in constructing a broader, more
dynamic model of farm-household systems. She also discusses
the relevance of gender at different points in the FSR/E

process, and institutional constraints that must be overcome to improve attention to gender.

Janice Jiggins, in Chapter 4, continues the discussion by focusing on the problems of communication between researchers, farmers, and extension workers, using examples from Zambia and Lesotho that show the rationality and flexibility within the domestic domain. She explores the difficulties of reconciling scientific knowledge systems with those of indigenous people in the course of conducting on-farm research. Lessons drawn from her examples point to the need for establishing key field-household interactions at an early stage of the diagnostic process and to developing methods for mutual communication of key concepts across researchers and female producers' distinct knowledge systems. She proposes the use of situation-analysis based on critical incident technique and peer group workshops as appropriate methodologies for improving communication and diagnosis.

Amalia Alberti focuses in Chapter 5 on the problem of generating data sufficiently sensitive to gender differences to guide the definition of client groups during the initial phases of a project. Echoing Evans' point that a priority for wealthier, more market-oriented farmers will tend to exclude women, she discusses the pros and cons of different sources of techniques for data collection in FSR/E projects. The following chapter, by Peter Wotowiec, Jr., Susan Poats, and Peter Hildebrand, explores in more detail how definitions of client populations need to be modified in accordance with the problems posed at different stages of the project cycle. They offer a refinement of the conventional FSR/E concept of "domains" to distinguish between "research domains" (that maintain variability), "recommendation domains" (homogeneous for technology testing), and "diffusion domains" (for disseminating new technologies). In Chapter 7, Jonice Louden summarizes the compelling reasons for incorporating gender issues into FSR/E monitoring and evaluation systems, especially in a country such as Jamaica where women play key roles in agricultural production. The FSR/E process presents an opportunity to collect systematically valuable information that can help to inform project implementation and refine gender-sensitive measures of key indicators of development.

Chapters 8, 9, and 10 take up conceptual and methodological issues beyond the farming systems universe. Dissatisfied with the market bias of standard economic models of the household, authors Lila Engberg, Jean Sabry and Susan Beckerson propose an alternative production activity model based on measures of time allocated to

income-generation, subsistence, and home production. The
more integrated model suggests trade-offs between labor
allocated to cash and to subsistence activities in Malawi
that could have nutritional implications. Also concerned
with the concept of household Art Hansen presents data, in
Chapter 9, from Africa that suggests caution in conceptualizing and measuring the frequency of female-headed
households. His findings show that static surveys may
underestimate the probability that a woman will be a head
of household at some moment or moments in her life, thereby
reinforcing the importance of involving both men and women
in development efforts. Eva Wollenberg's Chapter 10
discusses the strengths and weaknesses of various time
allocation methodologies and their relevance to farming
systems work. She explores how four different approaches
to the collection of time use data were used in a Philip-
pine project. Her discussion emphasizes the dynamic nature
of the FSR/E process, a theme common to all of the chapters
in this section. Gender patterns and intra-household
relationships become relevant to different degrees and in
different ways at each point in the project cycle. These
chapters, and others following, provide concrete suggestions as to how an unfolding farming systems project team
can collect and analyze the information that will enable it
to develop and adapt production technologies to the needs
of different users.

Diane Rocheleau, in Chapter 11, draws upon experience
from a broad range of countries; she details a land user
perspective as an appropriate method for incorporating
women as clients and active participants in agroforestry
projects. Her paper and that of Owusu-Bempah (Chapter 29)
play an important role in expanding the horizon of FSR/E to
consider the rural landscape as the context and focus for
projects in order to address the gap between natural
resource management and farming systems research. The role
and domain of women in the interface of these two areas is
clearly laid out as the next critical frontier for
expansion of household research and gender-based analysis.

Beginning with Part II, the Chapters explore gender
issues in FSR/E on a regional basis. Women's roles in
agricultural production are less visible in Latin America
than in other developing regions. The same is not true in
the Caribbean nations such as St. Lucia, described in
Chapter 12 by Vasantha Chase. While island women play a
significant role in farm work and decision-making, they
receive less income and fewer extension services than do
male farmers. Informal data collection methods reveal that
female-headed households face particular labor and input
constraints that limit their output, choice of crops, and

amount of land planted. A concern with integrating food consumption into the farming systems approach led the Caribbean Agricultural Research and Development Institute (CARDI) to recommend labor-saving methods of backyard garden production, oriented to improvement of family nutrition. The link between nutrition and agricultural change is also the focus of Eunice McCulloch and Mary Futrell's Chapter 13. Their measures of the nutritional output of cropping activities reveal the "low level steady state" farming system that maintains Honduran families at risk of persistent malnutrition.

In Chapter 14, Patricia Garrett and Patricio Espinosa describe the steps taken by the Bean/Cowpea Collaborative Research Project in Ecuador to adapt project activities to gender and social class differentiation. Their rich discussion of the FSR/E process demonstrates the importance of women's participation in production and decision-making, even in Andean Latin America where farming commonly is assumed to be the domain of men. The same is true in the Peruvian highland community described in Chapter 15 by Maria Fernandez, were women are responsible for most tasks associated with livestock production. Recognizing women as a key source of knowledge on traditional livestock production practices, the project team experimented with a variety of strategies to draw them into active participation. Fernandez' argument reflects the emphasis in Rocheleau's and Owusu-Bempah's papers on agroforestry of the importance of women's knowledge about traditional resource management practices.

Part III presents five case studies from Asia and the Middle East. Chapter 16 by Rita Gallin and Anne Ferguson and Chapter 17 by Jane Gleason present case studies from Taiwan. Gallin and Ferguson use longitudinal data from one village to show that a limited focus on farming activities ignored off-farm work and failed to analyze the interactions between the agricultural and industrial sectors of society. The authors propose the term "household enterprise" as a way of dealing with interrelated farming and off-farm work, and encourage researchers to view off-farm activities as "central rather than tangential to FSR/E analysis." Gallin and Ferguson discuss farm mechanization and note that it did not displace women, but rather concentrated certain tasks among some male specialists and caused some women to assume managerial positions previously restricted to men. Older women also took over tasks of younger women who then sought off-farm employment.

Gleason's detailed labor study in Southern Taiwan complements the previous work, and together they present a good example of why generalizations about gender and

agriculture should not be made for an entire country.
Gleason argues that in Southern Taiwan, as agricultural
mechanization increased, more women than men were displaced
and forced to other sectors of the economy. In her study,
availability of female labor increased the variety of crops
grown and the level of diversification, indicating that
women will be the users of modern vegetable technology and
will be most affected by changes in vegetable production.

Chapter 18 by Bahnisikha Ghosh and Sudhin Mukhopadhyay
studies time allocation by men and women in a rice-based
farming system in West Bengal, India. Though female labor
is subject to sociological constants, the contribution of
women is often larger than men's. A change to new rice
technology increased female labor, however the authors show
that the increased workload falls within the home production sector and was largely unaccounted for in the traditional economic literature.

Chapters 19 and 20 are based on work conducted at two
of the International Agricultural Research Centers. Thelma
Paris, in Chapter 19, describes how women were successfully
integrated into a crop-livestock project of the International Rice Research Institute (IRRI), in the Philippines.
Beginning as observers and slowly integrating themselves by
collecting and disaggregating data on household and production activities, the section members of the Women in Rice
Farming Systems were able to produce useful information and
become fully participatory members of the project. As a
result of their efforts, the whole team began to recognize
that specific production activities are the responsibility
of women and that on-farm research needed to target them.
Women's livestock, particularly swine, rootcrops, and
vegetables that had not been previously addressed by the
project, were proposed as new areas of research as a result
of the incorporation of women's concerns. In addition,
subsequent training courses included a significant number
of women participants, and the course addressed women's
production problems.

Andree Rassam and Dennis Tully in Chapter 20, discuss
research at the International Centre for Agricultural
Research in Dry Areas (ICARDA) on gender and agricultural
labor in Syria. They find that though male and female time
contributions to crop production are similar, males are
more often involved in new technologies, especially mechanization, while females are more involved in more traditional ones such as hand labor. The authors expect continuing

mechanization to further reduce female agricultural activities and they propose additional research to determine the impact of these changes.

Part IV contains a number of case studies from Africa that includes the descriptions of cropping systems, labor patterns, and work in on-farm, farmer-managed trials. Some themes that emerge from the papers are: the separate economies of men and women within households; the variability of the sexual division of labor in farm tasks; and increasing numbers of households headed by women and the concomitant increase in work burden due to male migration and divorce. Concerning the sexual division of labor, tasks may be the same or different for both sexes; in female-headed households the so-called "male tasks" are performed by women out of necessity. Concomitantly, the authors report that researchers and extensionists have failed to recognize women's roles in farming, ignored gender in the design of FSR/E projects, and not included women much as trial cooperators. Women, especially female heads of households, are often low-resource farmers who may have special problems that research and extension need to address. (Indeed, there are male low-resource farmers who have many of the same constraints.) The question as to whether or not gender accounts for separate recommendation domains finds different answers in the papers due to differential ecologies, social organization, and cropping systems.

Chapter 21 by Margaret Norem, Sandra Russo, Marie Sambou, and Melanie Marlett provides an example from The Gambia of how a women's component was formulated and operationalized as part of a larger, FSR/E project. Existing women's societies were used as a basis for organizing women and a maize-cowpea intercropping package was developed. The women experienced difficulties with the package related to pests, seed varieties, and labor patterns. The project was able to use women's participation in the trials and the problems they encountered to argue for the need to include women in subsequent research and extension efforts.

Jeane Henn in chapter 22, examines how government policies and environmental constraints impact on intra-household dynamics in Cameroon. Labor patterns, proximity to roads and urban areas, and farm gate prices affected incomes, differently in two villages. Food sales were very important to women's incomes but marginal to men's incomes that were mostly derived from cash crops. In one village, men withdrew labor from food crops resulting in an increase in women's work. However, the women close to roads and urban markets were able to increase their labor output and foodstuffs produced, and double their income while women in the other village were not.

Jean Due provides data in Chapter 23 from Tanzania, Zambia, and Malawi on how gender is important to FSR/E work arguing that unless there are person(s) on the FSR/E team who are sensitive to the issues, important information will be missed. In Tanzania, a diagnostic survey for bean/-cowpea research revealed that women select seed and contribute more labor on the crop than men. FSR/E work in Zambia would be hindered by not knowing the extensive labor contribution of women on the one hand, and the extent of off-farm and non-crop income, some of which is generated by women, on the other. Data on extension agents contacts with farmers in Tanzania and farmers' farm income are correlated and show that male contact farmers have seven times the income of female heads of households, who are rarely contacted by extensionists.

In Chapter 24, Timothy Mtoi discusses labor patterns in one region of Tanzania and uses a model to test the significance of female labor on expected risk and productivity under two cropping systems connected to a FSR project. The analysis shows that farm income would increase if the new alternative farming system had female labor transferred to it. However, policy decisions affect whether or not women can participate in the new technology (i.e., be trial cooperators) and obtain extension advice on the packages.

Chapters 25, 26, and 27 focus on Zambia and provide a more detailed set on the farming systems there and on the results of FSR/E diagnostic surveys. Mary Tembo and Elizabeth Chola Phiri discuss the traditional chitemene system of shifting agriculture and its sexual division of labor, the results of the colonial period that drained off male labor, and the lack of extension credit services to women. The result of these conditions has affected the diet and nutritional status of the population because women have taken to growing cassava (a crop that is less labor intensive, but also less nutritious than millet or maize), and farmers neglect food crops for household consumption because of growing cash crops.

Chapters by Robert Hudgens and Alistar Sutherland examine FSR/E diagnostic survey work of Adaptive Research Planning Teams that became sensitized to the need to target women farmers in research activities and in the determination of recommendation domains. Hudgens details the charactaristics of male and female-headed households in terms of land holdings, draft power, source of inputs, and cash sales. The comparisons show that there are both similarities and differences between the two household types and that the female-headed households experience labor shortages. Women tend to be isolated from government services and their production is constrained by lack of exposure to new ideas, inputs, and capital.

As part of the diagnostic phase of FSR work, Sutherland compares women's and men's roles in three regions of Zambia and argues that even within one country, gender roles are influenced by cultural, economic, political, and ecological factors. Labor, cash availability, and draft power tend to divide households into recommendation domains. Gender is a distinguishing factor in one region, but not in the others.

In Chapter 28, Anita Spring reports on two different on-farm, farmer-managed trials in Malawi. In the first, the inclusion of low-resource female farmers along with high-resource male farmers generated two recommendation domains. Improved maize cultivars and use of fertilizer worked well in the better environments and with high resource farmers, but would be disastrous in the low-resource environments and with low-resource farmers. On the other hand, the traditional cultivar was better in these situations. The second set of trials involved all female cooperators to solve a technical problem of inoculating soybeans as well as the issue of whether or not male extensionists could work with female farmers. It was found that the women could do trials with precision and that male extension and research staff could work with women farmers in terms of training and credit programs.

Kofi Owusu-Bempah, in the final paper, argues for inclusion of farmers in the planning and design of projects, and particularly calls for the involvement of women in the selection of species to be included in agroforestry projects in Ghana. His work represents a largely private sector effort and, like Rocheleau, calls for expanding the framework of analysis to include the landscape perspective and the intersection of crops, livestock and forest enterprizes.

In conclusion, the papers in this volume contribute to an understanding of how gender affects farming systems and the way that FSR/E operates. The papers demonstrate that by linking the two codes – gender and FSR/E – the agricultural research and extension system can become more efficient and effective in dealing with different groups of farmers. The papers provide details of specific cases and the methods used to incorporate gender perspectives and analysis.

There is no single recipe for action. Instead, these studies from an array of ecological, social, and political contexts demonstrate that it is both possible and practical to use gender analysis as a tool in the work of agricultural development.

REFERENCES

Boserup, E.
    1970 Women's Role in Economic Development. New York:
        St. Martin's Press.
Guyer, J.
    1981   Household and Community in African Studies.
        African Studies Review 24: 2/3.
Overholt, C., M. Anderson, K. Cloud, and J. Austin
    1985   Gender Roles in Development Projects: A
        Casebook. W. Hartford, CN: Kumarian Press.
Palmer, I.
    1985   The Impact of Male Out-Migration on Women in
        Farming. W. Hartford, CN: Kumarian Press.
Schmink, M.
    1984   Household Economic Strategies: Review and
        Research Agenda. Latin American Research Review
        19:3:87-101.

2
Integration of Intra-Household Dynamics into Farming Systems Research and Extension: A Survey of Existing Projects

Rosalie Huisinga Norem

This paper reports the results of a survey designed to study farming systems projects that include an intra-household focus in data collection, design, or implementation. Farming systems models (Shaner et al. 1982) have recognized the importance of the household as a component of the farming system, but until recently little has been done to systematically "open the black box" of the household component in those systems models. Projects responding to the survey being reported here are among those attempting to gain a more systematic understanding of the inter and intra-household factors influencing farming systems.

The primary purposes of the present survey are to assess the types of information collected and used by projects, to evaluate the methods used for obtaining the information, and to gain some insight into how and why the intra-household information is helpful. In addition, project managers were asked to identity types of information they would like to have, but are not available and the constraints affecting various phases of their projects.

RATIONALE

This study evolved from on-going work relating household concerns to farming systems work. When the Farming Systems Support Project (University of Florida/USAID) was first initiated, a family task force was organized to focus on the integration of household and family concerns into the farming systems perspective. One of the recommendations of this group was to develop case studies and training materials that would promote such integration. In a position paper on "Intra-household Dynamics in Farming Systems Research: The Basis of Whole Farm Monitoring of

Farming Systems Research and Extension," Cornelia Butler-Flora (1984) set the stage for an intra-household dynamics and farming systems case studies project that was subsequently funded and implemented.

Concurrent with the effort to develop training materials that will sensitize people to the importance of intra-household factors in farming systems work as well as in other development efforts, there is a need for more knowledge about the kinds of data being collected and the methods used by existing projects that are attempting to focus on intra-household factors. As an attempt is made to understand the complexity of household dynamics, ways to obtain and analyze information within reasonable time and other resource limits also must be found. Questions about how much information on aspects of the household should be obtained from whom have yet to be answered (Norem 1983).

This is not to suggest that one "right" way of focusing on household dynamics and farming systems can be identified. Rather it is to suggest that by examining what is being done and how effective researchers and practitioners involved find current efforts, some guidelines can be identified that will be helpful in future planning.

It may be helpful to consider how information relating to households can be broken into different units depending on the purposes of the individual study. Overall, the unit of interest in intra-household dynamics is the household. The unit of data collection might be one or more household members, other informants, or other existing information. The unit of analysis can be an individual, the household or subsystems thereof, a work group, or the farming system, among other possiblities. Designing parsimonious data collection and analysis procedures requires an understanding of how these units relate in various situations.

For example, it may be possible to obtain good data on the unit of interest from only one person if what is required is basic demographic information such as age, gender, and education of household members. The household is also the unit of analysis in this example. Information about the tasks performed in the household is more likely to require data collection from more than one person, or extensive observation or record keeping to permit the collection of enough information to focus on the household as a unit of analysis. As a clearer picture of the state of the art develops, it is hoped a clearer set of guidelines will evolve. The survey summary presented in this paper is a first step.

DESIGN OF THE STUDY

The Farming Systems Support Project and Population Council Intra-household Dynamics and Farming Systems Case Studies Project was initiated with support from USAID and the Ford Foundation in late 1984. In February of 1985, a request for expressions of interest was sent to projects and individuals on a variety of international mailing lists. Over seventy-five expressions of interest were received in response to the request. These expressions of interest were used to develop initial lists of types of data and data collection methods being used in projects. The lists were used in conjuction with the case studies project conceptual framework to draft a survey questionnaire. The questionnaire was reviewed by the case studies project advisory committee and revised using the committee's suggestion.

The questionnaire was mailed to all projects that had responded to the original request for expressions of interest in the case study project, since those projects had self-selected themselves in terms of interest in intra-household concerns. A few other projects were also included in the survey. All questionnaires received (n=19) are included in the summary, regardless of type of project. Most are farming systems oriented, with one project being specifically focused on women in farming systems. Seven projects from Asia, seven from Africa, two from the Middle East, and three from Latin America are included. The titles and identifying information about the projects are presented in Table 2.1.

Each project has a different specific target group, but all projects have target groups of farms with multiple crop systems. Fifteen projects report farms in their project also have livestock, most for multiple use, including cash income, food, traction, wealth, and prestige. The average land holdings for farmers in the projects ranged from .89 hectares to 30 hectares, with an overall mean of 6.34 hectares.

RESULTS

Types of Intra-household Data

Each project was asked to indicate whether or not it has data about five general categories: (1) demographic information, (2) household members' participation in activities, (3) household members' access to production resources, (4) household members' participation in decision making, and (5) income and expenditure data, benefits from

TABLE 2.1

PROJECTS RESPONDING TO SURVEY

COUNTRY	PROJECT TITLE	CONTRACTOR AND UNIT IN CHARGE (SOURCE OF FUNDS)
ASIA		
Indonesia	TROPSOILS Soil Management CRSP	University of Hawaii with University of N. Carolina & Center for Soils Research (USAID)
Philippines	Farming Systems Development Project Eastern Visayas Non-Farm & Resource Management Institute	Ministry of Agriculture and Food and the Virginia State University (USAID)
Nepal	Agricultural Research & Production Project Farming System Development Division	Winrock International & Ministry of Agriculture (USAID)
Bangladesh	Women in Farming Systems	Bangladesh Agricultural University (Bangladesh Agricultural Research Council)
India	Role of Farm Women in Decision Making Related to Farm Business	Haryana Agricultural University (Haryana)
Philippines	Balinsasayao Agroforestry Project	Silliman University Research Center (Ford Foundation)
Philippines	Farming Systems Development Project Eastern Visayas Non — Farm & Resource Management Institute	Cornell University, Ministry of Agriculture & Food, & the Visayas State College of Agriculture (USAID)

AFRICA

Burkina Faso	Fulbe Agropastoral Production in Southern Burkina Faso — for USAID Agric. Sector Grant	Frederick Sowers, Univ. of California, Berkeley (USAID)
Burkina Faso	Income & Agric. Investment in a Bobo Village	University of Illinois, (Wenner Gren Foundation, NSF)
Sierra Leone Extension	Adaptive Crop Research of Sierra Leone	Southern Ill. University, Louisiana State University & Ministry of Agriculture & Natural Resources (USAID, and Government of Sierra Leon)
Ghana	REDECASH/BIRD	Bureau of Integrated Rural Development (BIRD, REDECASH)
Botswana	Agricultural Technology Improvement Project	Midwest Int'l Agricultural Consortium (MIAC) (USAID, Kansas State University)
Kenya	Dryland Farming Research	Ministry of Agric. National Dryland Farming Research Station, (FAO/UNDP, Kenya Gov't.)
Sierre Leone	Phase I: Land Resources Survey Project Phase II: Strengthening Land & Water Division	FAO/Ministry of Agriculture (UNDP)

(continued)

TABLE 2.1 (continued)

COUNTRY	PROJECT TITLE	CONTRACTOR AND UNIT IN CHARGE (SOURCE OF FUNDS)
MIDDLE EAST		
Israel	Irrigation Innovation and Family Farming Strategies in Israel	City Univ. of New York, (Faculty Research Grant)
Syria	Syrian Households: Women's Labor & Impact of Technologies	Andree Rassam, (ICARDA & MEAWARDS)
LATIN AMERICA		
Mexico	Livestock Production Systems in Central State of Veracruz	Centro de Investigacion Ensenanza en Granaderia Tropical, CIEEGT, Facultad de Medicina & Zootechnia (Universidad Nacional Autonomo de Mexico, UNAM)
Honduras	Honduras Agricultural Research Project	Consortium for International Development, New Mexico State Univ. (USAID)
Panama	EMPARAD: Socio-economic (Agricultural Development Bank) Case Study	Miriam Reyes, Martinez Elba, Rosa Hernandes

farm production, food consumption and nutrition information. Each of these categories include several specific
kinds of data. Table 2.2 shows the types of data collected
and the number of projects that collected each type.
Table 2.3 summarizes the ways each type of information was
or is used by the projects that responded to the survey.

TABLE 2.2

TYPES OF INTRA-HOUSEHOLD DATA COLLECTED BY
PROJECTS RESPONDING TO SURVEY

Type of Information (N=19)	Number of projects
Demographic	
Household Structure, Membership, and Size	19
Education	17
Ethnic Identity	16
Migration Patterns	9
Variation in Household Structure Over the Life Cycle	7
Household Members' Participation in Activities	
Cash Crops by Crop	13
Subsistence Crops by Crop	12
Livestock Production	11
Other Primary Income Generating Activities	8
Major Tasks of Household Reproduction	11
Household Member's Access to Production Resources	
Land	
In General	13
By Tenure Category	9
By Production Potential (Irrigated, Non-Irrigated)	8
Labor	
Family	12
Hired	15
Exchanged	11
Capital	
Seeds	14
Tools	15
Equipment	15
Animals	14

(Continued)

TABLE 2.2 (cont'd.)
TYPES OF INTRA-HOUSEHOLD DATA COLLECTED BY
PROJECTS RESPONDING TO SURVEY

Type of Information (N=19)	Number of projects
Household Members' Access to Production Resources	
Innovations or Improved Production Inputs	
Information (Extension Contacts, Training, Etc.)	14
Technology Inputs Requiring Cash or Credit	10
Credit	
Informal	12
Formal	12
Other	1
Household Member's Participation in Decision-Making	
Land Use	13
Use of Family Labor	13
Use of Hired Labor	11
Use of Exchange Labor	9
Use of Technology Inputs	14
Use of Credit	12
Cropping Choices	13
Cultivation Practices	13
Uses of Harvested Crop and Residue	13
Marketing	12
Income and Expenditure Data	
Each Household Member's Sources of Income	7
Each Household Member's Expenditures	7
Benefits from Farm Production	
Use of End Products From Crop Production	11
Desirable Characteristics of Each Crop or Crop Product	8
Each Household Member's Access to or Control of End Products	6
Food Consumption and Nutrition Information	
Diet Survey	4
Nutritional Adequacy Analysis	4
Food Preparation Practices	6
Food Preferences	6
On-Farm Household Food Production	7

TABLE 2.3

USES OF TYPES OF INTRA-HOUSEHOLD DATA
BY PROJECTS RESPONDING TO SURVEY

Use	Number of Projects Reporting Use of Information by Type of Information*					
	1	2	3	4	5	6
Initial Project Design	11	5	5	3	3	–
Selection of a Target Group	11	4	8	3	4	–
Identification of Recommendation Domains	4	5	5	1	4	–
Choice of Research Topic	7	6	8	3	6	–
Designing Trials	6	4	6	3	4	–
Selection of Participating Farmers for Field Trials	6	5	6	2	4	3
Evaluation of Field Trials	2	3	6	2	4	2
Redesign of Trials	3	6	6	6	4	2
Technology Recommendations	8	5	5	4	6	2
Extension Efforts	7	4	5	6	4	–
Project Evaluation Design	5	2	4	1	3	1
Assessing Time and Labor Constraints	13	11	7	8	1	1
Assessing Opportunity Costs for Innovation	7	5	5	3	3	–

Note: *Type of Information:

1 Demographic Information
2 Household Members' Participation in Activities
3 Household Members' Access to Production Resources
4 Household Members' Participation in Decision-Making
5 Income and Expenditure Data, Benefits from Farm Production, Food Consumption, and Nutrition
6 Other kinds of information collected include: religious affiliation, inheritance data, and information gathered from husband and wife together.

Demographic Information

The most frequently used methods for obtaining demographic information are pre-existing national surveys,
formal project surveys, participant observation, and son-
deos (rapid reconnaissance surveys). This information is
summarized in Table 2.4 for all types of data. Demographic
information is also available through pre-existing anthropological studies and local village records for some
projects. Other projects collected information through
farmer records, community informants, time allocation
studies, team members' personal knowledge, and indepth case
studies.

Nine projects collected demographic data before the
project began, five during the diagnosis stage, and seven
parallel with on-farm testing. Ten projects collected
demographic data on an on-going basis.

All projects collected demographic data on household
structure, membership, and size. Most also have information about education and ethnic identity. Nine projects
have data on migration patterns and seven have them on
variation in household structure over the life cycle.
Table 2.3 shows that demographic information was most
useful in the early planning stages of projects.

Respondents were asked to identify the most helpful
information for each type and to give an explanation of how
or why the information was helpful to their project. They
were also asked to indicate any information in each category that they did not have but wished they did. These
open-ended questions provide more detail related to intra-
household concerns than the tabulated results shown in the
tables.

The specific demographic information identified as most
helpful to a project varies by project, however, some
generalizations can be made. Gender and age structure of
the household is mentioned by several respondents,
sometimes alone and sometimes in conjunction with other
information such as labor and income. This information was
useful in identifying target groups and in designing trials
that ascertain labor bottlenecks and total household
activity patterns. Household structure is also reported as
an important consideration in designing extension efforts.
Ethnic information is mentioned second most frequently as
the most helpful demographic information, because farming
practices and values about female participation vary
according to ethnicity.

TABLE 2.4

MOST FREQUENTLY USED METHODS OF DATA COLLECTION
BY TYPE OF DATA

Type of Data	Data Collection Method
Demographic Information	National Surveys Formal Survey Participant Observation Sondeo
Household Members' Participation in Activities	Formal Survey Participant Observation Community Informants
Household Members' Access to Production Resources	National Surveys Formal Surveys Participant Observation
Household Members' Participation in Decision-Making	Formal Survey Team Member's Personal Knowledge; Participant Observation
Income and Expenditure Data, Benefits from Farm Production, Food Consumption, and Nutrition	Formal Surveys Participant Observation

The two kinds of demographic information least often available, migration patterns and variation in household structure over the life cycle, are the ones most frequently named in response to the question, "Are there demographic data you do not have that you wish you had collected?."

Household Members' Participation in Activities

Formal surveys, participant observation, and community informants are the methods most frequently used to obtain information about the participation of household members in various activities (see Table 2.4). As with demographic information, projects obtained this information in a variety of ways. Each of the methods listed above for demographic information is used by at least one project to obtain activity data, group meetings provide an additional source of information about household members' activities.

Three projects collected activity data before the project began, six during initial diagnosis, and six parallel with on-farm trials. Nine projects collect activity data on an on-going basis.

Specific questions were asked about type of activity data collected and method of disaggregation. Twelve projects collected task assignment data disaggregated by gender and age. Nine projects disaggregate by position in the household as well. Five projects have information about time allocation.

Thirteen projects report collecting some information about the participation of household members in various activities. Most frequently collected information (n=13) is about activities related to production of cash crops, with subsistence crops and livestock production information available for twelve out of nineteen projects. Other activities within the household receive less attention as indicated in Table 2.2. Table 2.3 indicated that projects use activity data less often than demographic data. Activity data are used most often to assess time and labor constraints.

Information on household members' activities was most helpful in designing research and targeting interventions. Respondents would like more detailed information about non-production activities and several express a desire for activity data covering a period of time up to a year. The complexity of activity data is pointed out and difficulties with processing such data are mentioned.

Household Members' Access to Production Resources

This study breaks production resources into sub-categories of land, labor, capital, innovations, and credit. The projects represented use a variety of methods to obtain resource information; the most frequent are pre-existing national surveys, project-conducted formal surveys, participant observation, and team members' personal knowledge.

Six projects collected data on access to resources before the project began, seven during initial diagnosis, and seven during on-farm trials. Five projects collect these data on an on-going basis.

As indicated in Table 2.2, fourteen of the seventeen projects have some resource information. But examination of Table 2.3 suggests that fewer projects overall use this information than use demographic data. More projects use resource information than activity data for most purposes, but more projects use activity data to assess time and labor constraints.

The answers to questions about the most useful resource data and why and how it is useful indicate that land resource information is perceived as most helpful for more projects than other kinds of resource data. The responses also indicate that the usefulness of resource access data is very project specific. Data on access to resources is likely to be helpful in research design and in selection of field trial locations. There is a pattern among responses about the kind of resource information respondents would like to have had but that was not available. More information about monetary income, including gifts and remittances, is mentioned in several contexts, including credit, opportunity costs for innovations, and access to capital.

Household Members' Participation in Decision-Making

Fourteen projects in the survey have some data about decision-making within households. These data are collected most frequently through formal surveys, team members' personal knowledge, and participant observation. Other methods are used, but in a project-specific manner. Only three projects report having decision-making data to use in initial project design. One project collected data during the initial diagnosis and four collected data parallel with on-farm trials. Six projects are continuing to collect decision-making data.

Table 2.2 indicates that projects which have decision-making data also have information about most of the subcategories identified, including land use, labor use, technology use, cropping and cultivation practices, and use of production outputs. Table 2.3 suggests that projects are not using decision-making data extensively. Eight projects use decision-making data to assess time and labor constraints, and this is the most frequent use reported.

Responses to open-ended questions about the usefulness of decision-making data express the theme that a better understanding of household dynamics permits more knowledgeable identification of target groups. Eight respondents indicate their projects could use more detailed decisionmaking data that would reveal more about the effect of decision patterns.

Income and Expenditure Data, Benefits From Production, and Food Consumption and Nutrition Information

Seven projects have information about these categories of data. Formal surveys and participant observation are

the most common methods of obtaining the information.
There are some differences among the categories. Participant observation is most likely to be the source of
information about food consumption and nutrition information, and is not as likely to be a source of production
benefits data. Formal surveys are used by several projects
for all three categories.

Only one project had data in these categories before
the project began. Three projects collected the data
during the initial diagnosis, four collected the data
parallel with on-farm testing, and eight collect the
information on an on-going basis.

Table 2.2 details that eleven projects have information
in at least one of these three categories. Income and
expenditure data are least frequently available. Table 2.3
shows a fairly equal distribution of the use of specific
kinds of data available in these categories over the various phases of the projects, especially in the design and
implementation of field trials.

Since this section of the survey combines three
categories, the answers to questions about which information was most helpful, why so, and how, are somewhat
complex, but they also point out the need to integrate
information about overall production and consumption
patterns in the household. For example, respondents noted
the importance of looking at off-farm income and cash
income from food crops and of understanding the households'
reliance on local markets both for food and income. They
also mention the need to assess the opportunity costs of
innovations based on total inputs and total income
generating possibilities.

Respondents answering the question about information
they wish they had collected, mention primarily better
income data for individual household members monitored over
time. Several note the difficulty of obtaining reliable
income data and state the importance of finding better ways
of obtaining such information.

Most Effective Methodologies

Respondents were asked to select the study or activity
of their project that was most effective in collecting
information about intra-/inter-household variables relevant
to farm production and the ones that were most useful in
determining project decisions concerning research priorities, cooperating farmers, technology acceptance, etc.
Nine respondents name the formal survey as most helpful.
This is usually done at the beginning of the project.
Eight respondents identify participant observation as the

most useful activity for obtaining household information. This tends to be on-going. Four respondents state that the sondeo as most useful. The sondeo took place anywhere from the beginning to the third year of the project. Ten respondents report the head of household as the primary informant, whether male or female. Seven projects tried to include at least one other adult household member. Three relied on whoever was at home with a preference for the head of household. One case study involved all members of the household.

Constraints to Projects

Respondents were asked to identify constraints that affected the study design, sample selection, conduct of the study, data analysis, or applications of the data to their project or activity. These responses are summarized in Table 2.5. Ten respondents report physical, logistical or resource contraints on sample selection for their projects. Detailed answers to open-ended questions about these constraints show the most common one is transportation, either because of the lack of means or because of the difficulties of the terrain.

TABLE 2.5

CONSTRAINTS INFLUENCING PROJECTS

Phase of Project (N=19)	Number of Projects Reporting	
	Physical/ Logistical/ Resource	Cultural/ Social Political
Study Design	9	5
Sample Selection	11	4
Conduct of Study/Activity	7	6
Data Analysis	7	1
Application of Data to Project/Activity	3	1

In order of descending frequency, other constraints mentioned are funds, language, personnel, political situations, and ethnic group considerations. In many instances the constraints are named in conjunction with one another, such as ethnic concerns and language difficulties.

SUMMARY

Several summary points can be made from the information provided in the survey. First, there is a wide variation in the kind of data being collected about households with a common focus on the household as a unit of interest. The data are most often collected from heads of household, so for some kinds of data there may be difficulty in using the household as a unit of analysis. For example, data about decision-making describe a dynamic intra-household process, but data involving several household members require complex data collection procedures. It is important to examine alternatives in determining which information is important for which stage of a project and how to obtain it as efficiently as possible. One respondent pointed out that designing more standardized methods of data collection and analysis is difficult because of the unique aspects of each project, but the same respondent also emphasized that efforts to move in this direction will save significant resources and hopefully eliminate the need for each future project to make the same mistakes.

There was general agreement among respondents that more information about intra-household concerns was important. One project is collecting additional information in most of the categories in the final stages of the project with the hope that it can be used in future planning.

This survey gives us one measure of what projects are currently doing and assesses perceptions of the importance of intra-household data. The next step is to suggest a household minimum data set that projects could collect. This would not eliminate the need for specific data for specific projects, but would provide the potential for comparative analysis and the beginning of theoretical generalization.

REFERENCES

Flora, C. B.
   1984   Intra-Household Dynamics in Farming Systems
      Research: The Basis of Whole Farm Monitoring of
      Farming Systems Research and Extension. A Position
      Paper. Manhattan, KS: Department of Sociology,
      Kansas State University.
Norem, R. H.
   1983   The Integration of a Family Systems Perspective
      into Farming Systems Projects. Conference
      Proceedings, Family Systems and Farming Systems
      Conference. Blacksburg, VA: Virginia Polytechnic
      Institute and State University.
Shaner, W. W., P. F. Philipp and W.R. Schmehl
   1982   Farming Systems Research and Development:
      Guidelines for Developing Countries. Boulder, CO:
      Westview Press.

3
Gender Relations and Technological Change: The Need for an Integrative Framework of Analysis

Alison Evans

BACKGROUND

Over the past decade, agricultural research has focused on the rural poor and the need to strengthen the production systems of small farms. In sub-Saharan Africa a substantial proportion of agricultural activity and food production continues to be organized at the household-farm level, and with concern increasing over the region's food producing capabilities, researchers and policy makers have sought to identify ways of raising the productivity and incomes of self-provisioning farming households (Henn 1983; World Food Council 1982).

Farming systems in sub-Saharan Africa are extremely diverse, but a number of shared conditions and characteristics permit some generalizations. First, household members typically comprise the main source of on-farm labor, but households also require large amounts of labor for non-farm and reproductive tasks. Second, rural women, especially poor rural women, make a substantial contribution to all types of agricultural activity, and in most cases have responsibility for the production, preparation, and distribution of food for the household and for sale, and for the reproduction of family labor (FAO 1984). Thus, if agricultural research and policy is to have a positive impact on food production and security in sub-Saharan Africa, researchers and policy makers must link new technology and extension advice more directly to the needs and priorities of smallholder producers in general and women producers in particular.

The record of technological intervention in sub-Saharan Africa has been very variable and the literature contains many examples of the contradictory and unintended effects of technology change for traditional farming households (Dey 1984; Bryceson 1984; Agarwal 1981). It is always the poorest and smallest farming households that are most vulnerable to the "unplanned" effects of "modernization," but it is the effectiveness and well-being of women producers in poor households that appears to be most at risk (Agarwal 1981; Kisseka 1984; Whitehead 1981).

The fact that technological changes have complex, differential, or even perverse effects on the productivity of small farm systems indicates that agricultural research and extension must be reoriented to identify and deliver solutions to on-farm problems that are reliable and predictable within the whole farm system and above all consistent with the needs and interests of both women and men (CGIAR 1985; Hahn 1985; Fresco 1985; Maxwell 1984).

AGRICULTURAL RESEARCH AND DEVELOPMENT

The driving force behind recent research and development in small farm agriculture has been Farming Systems Research and Extension (FSR/E). Although FSR/E does not follow one model it does tend to pursue a single aim: to develop technology that helps small farmers improve output (yield) and increase marketed food supplies while protecting the natural environment (FSSP 1984). A major advantage distinguishing FSR/E from earlier research and development approaches is its potential for integrating comprehensive, interdisciplinary knowledge about small farmers' technical needs with the wider systemic implications of technological change (Shaner et al. 1982).

The FSR/E approach tries to identify the distinctive components and parameters of the "total" farm system in which a small farm-household functions. In most FSR/E models, the family labor unit or household is at the center of the conceptual framework and the farming system is defined as:

a unique and reasonably stable arrangement of farming enterprises that the household manages according to well defined practices in response to the physical, biological and social environments and in accordance with the household's goals, preferences and resources (Shaner et al. 1982).

In their functionalist approach to farm systems analysis, FSR/E practitioners try to understand how factors under the control of household members (endogenous factors) like family labor and the techniques of production, interact with physical, biological, and socioeconomic factors beyond household control (exogenous factors) like climate, prices, and the availablity of credit, affect the agricultural welfare of the farm household. By combining

knowledge of the whole farm system with extensive on-farm
research and experimentation, systems researchers should be
able to recommend technological alternatives that meet the
priority needs and preferences of small farmers and evaluate which technologies fit where and why, and monitor who
benefits.

In practice, the FSR/E approach is used primarily to
reduce the degree of error in finding the appropriate technical "fix" for market-based problems of agricultural production. Thus, FSR/E practitioners have restricted themselves to finding technological solutions to a limited
range of crop and livestock production problems, using
mainly market criteria to measure the costs and benefits to
farm-households. The methodology used to identify farming
households' needs has been implicitly restricted and the
choice of recommendations has been limited to suit a nar-
rowly defined, technocratic approach to on-farm development. FSR/E practitioners have not yet built a broad
understanding of the diverse needs and preferences of small
farms or of the differential effects of technological
change for members of farm-households. The FSR/E approach
is particularly limiting when considering gender issues in
agricultural development. In particular, FSR/E is ill
equipped to:

(1) understand the distribution of costs and benefits of
    alternative technological choices and extension pro-
    jects between men and women within farm households and
    under what conditions technological change benefits men
    but not women, or in fact harms women, individually or
    as a group or
(2) identify and meet the technical needs of women as
    farmers, particularly as producers and consumers of
    food.

FSR/E has not accomplished what it set out to achieve
at the socio-economic level. Researchers have failed to
consider how farming systems are constituted along gender
lines and to analyze the influence that gender differentiation has in shaping technological and economic outcomes in
agricultural development. The methodology contains procedures and assumptions that have tended to reinforce gender
blindness.

First, FSR/E uses market or monetary criteria to identify productive components in the farm system and to
measure their contribution to household income and welfare.
Many of the value-adding or service activities that are not
directly comparable in monetary terms are left out or are
considered costless in the economic equation. For example,

FSR/E does not value non-monetized, household-based activities such as food production (especially secondary foods), processing and preparation for direct consumption, fuel and water collection, and house maintenance and repair.

From micro-level research, it is shown that preferences for market criteria have important gender implications because while men, women, and children all perform some of these non-market tasks, it is women who contribute most of the labor, energy, and resources (as well as exchanging labor and resources with other households) to get this work done. By treating non-market, home-based work as complementary to the farm unit rather than an essential component of it (Fernandez 1985; Lele 1985), production economists have neglected important links between market and non-market enterprises. For example, the linkages ignored include those between water and fuel collection, that in sub-Saharan Africa is almost always non-monetized, with crop and livestock production, food production, and food processing. More importantly, economists have neglected the significance of gender differentiation, the division of labor, and resources by gender in shaping relations between food and non-food production, market and household-based enterprises, and consequent responses to economic incentives and new technology.

Second, production economists and technicians have concentrated on ways of improving yields and the volume of marketed output to increase household income and welfare. In this process they tend to focus on specific sub-systems like the cropping system, and within that, particular crops with obvious market potential. This selective approach has often been pursued without careful consideration of the possibility that production constraints stem from other parts of the farming system or that farmer decision-making has to do with the operation of the farming system as a whole and not with isolated parts of it.

This implies that planned changes within the cropping system, for example, that should increase aggregate output and income, may have unanticipated and differential effects on individual household members' production. The welfare effects of planned changes can be predicted only if the full extent of household members' involvement in the farming system is known, and more importantly, if it is known how the activities of men, women, and children fit together within the prevailing technical and social relations of production.

Systems researchers have often overlooked appropriate indirect or non-farm solutions to farm-based problems. For instance, farm output may increase if the productivity of household-based activities, many of which are women's responsibility, were to improve. Higher productivity in these tasks may have a positive impact on the amount or quality of labor available for market-oriented production. It has been demonstrated empirically that women's household-based activities will feed directly into the satisfaction of household basic needs whereas the link between market-based activities and income and household welfare is more spurious (Hart and Popkin 1975). If women can collect water or fuel more easily, they could expend more time and energy in directly productive activities. Or by raising the productivity of subsistence food crop production and processing women could improve the health and efficiency of labor, especially during the low energy or "hungry" seasons.

Third, the FSR/E approach is based on a stereotyped model of "the household" as a homogeneous unit of production and consumption that is at variance with the reality in many rural areas.

Furthermore, it is assumed that the distributional effect of technological change on farm household members can be read off from shifts in the demand curve for family labor within a specific sub-system. Technical change, however, does not effect the division of labor and resources in any systematic way (Whitehead 1981). The relations between women and men in farming households are not solidified within the technical division of labor nor are their interests and needs necessarily complementary. Households are not homogeneous units and the needs and preferences of women and men are not easily defined with respect to a single household production function. Even more importantly, the household decision-making process can be only partially explained by economic factors. The choices of men and women in the household and beyond are the outcome of an intricate bargaining process mediated by normative relations of power and control.

In many West African countries, men and women occupy different positions within the farm economy, not only in terms of task allocation, but also in terms of their different access to factors of production — inputs, credit, others' labor time, off-farm employment opportunities, and markets. In polygynous households, senior wives also have structurally different economic positions from junior wives. In a number of decisions, men and women have a joint economic interest; in many others, often those associated with direct welfare activities, their interests conflict or their economic and technical needs are different. For example in Sierra Leone, women's interests in improving swamp rice farming and intercropping to produce a marketable surplus have been in conflict with

men's interests in upland farming and their use of work
gangs and household labor in the production of market
crops. Polygyny also compels wives to have independent
access to cash income that involves management of non-farm
income generating enterprises such as palm oil processing,
soap and pottery making, and dyeing cloth. These enterprises are kept separate from men's and require women to
maintain separate budgets and resources even when living
with men in the same extended household (Richards et al.
1981).

THE SEARCH FOR A NEW FRAMEWORK

Most FSR/E models take a selective and comparatively
static view of the farming system. There is a need for
more searching concepts and methods to look at gender in
farm-household systems, to understand the gender
distribution of costs and benefits of technological
choices, and to identify women's technical needs.
Unfortunately, due to a lack of empirical evidence this
paper can only offer some analytical suggestions on how
such a framework might be built.

Conceptual Issues

First, a conceptual model that treats the small farming
unit as an interlocking system of market production,
subsistence, and reproductive activities performed at farm
and household levels should be developed. Reproductive
activities include primarily the domestic services
necessary for family survival. Central to this
inter-related model are the multi-activity household and
the different sets of economic and social conditions that
shape the economic participation and welfare of women, men,
and children. These sets of conditions are not immutable.
Depending on the total matrix of productive and reproductive activities that comprise any one farm-household
system, they are likely to alter with seasonal shifts, the
life cycle development of the household, and emerging
employment and market opportunities in the local economy.
Such a framework should be able to generate data on the
key sets of conditions that shape the differential participation of household members and give explicit attention to
the most important interactions between enterprises that
vary by gender.
The anthropological literature for sub-Saharan Africa
suggests that data needed center on intrahousehold resource
allocation and on the relative weights of market, non-
market, social, and economic forces in shaping the

differential distribution of women's and men's labor, resources, management, and control within the total farm-household system. Such information could be used to investigate the possibility for productivity gains in non-monetarized enterprises that have either a direct or an indirect productive impact on household members. Agricultural researchers and extension personnel also need such information to understand why planned changes in the technological and economic balance of farming systems can have unanticipated and sometimes detrimental effects on other enterprises outside of the production arena.

In Africa, a commonly cited problem for women farmers, particularly in the poorest households (Dey 1984; Guyer 1980; Whitehead 1981), is that technological changes have intensified women's workloads without adequate compensation or have eroded their access to land and non-farm income. Without understanding exactly how women's multiple activities interlock with each other and with other enterprises in the system, neither the aggregate net effect of technological change upon the total workload or returns to labor of individual household members, nor the differential productivity and welfare effects on women and men can be established clearly.

Methodological Issues

A broader conceptual model raises a number of important methodological considerations for farming systems research.
Recommendation domains. While agronomic, topographic, and general socioeconomic data will continue to be crucial in selecting recommendation domains, different household forms, family structures and composition must be increasingly incorporated. In Africa, it is essential to consider polygynous households, consensual family structures, female-headed households, and the complex variation in authority and power structures that determine the location of decision-making and access to resources in household units (Allison 1985). At this stage it is important that researchers be aware of the different economic and social positions that women and men occupy in African households and how these are manifested in different decision-making processes, separate budget and production functions, and different access to inputs and markets, incentives and off--farm employment opportunities.
Problem identification. Information about the different positions of women and men in the farming household must be linked to precise data about patterns of resource allocation, labor utilization, and the multiple fit of women's and men's responsibilities in order to identify

major constraints and bottlenecks as they vary by gender.
The degree of flexibility and substitutability of the
organization of farm-household activity must be taken into
account as well as the potential trade-offs or conflicts
that might emerge between the genders when a constraint
within one sub-system creates another elsewhere in the
system. Further research should focus on:

(1) the degree of flexibility and substitutability between
    the role of women's labor in farm production and
    household-based production relative to men's;
(2) the substitutability between women's labor and capital
    (productivity enhancing technologies) relative to other
    household members; and
(3) the flexibility and substitutability between women's
    labor and that of other family or household members,
    including husbands, children, neighbors, or hired labor
    (Lele 1985).

Much of the information that can be gathered about
these points will be qualitative rather than quantitative,
but is nonetheless valuable. Researchers and extension
personnel should resist the temptation to devalue the
importance of qualitative data in planning on-farm research
and in evaluating responses by farm-households.

On-Farm Research, Extension, and Evaluation. Gender
differences must be an integral criterion woven into the
analysis of technological constraints and the recommendation of solutions and must also be considered when choosing
target farmers with whom to conduct on-farm trials. It is
particularly important that researchers recognize the mixed
strategies that women employ to meet basic household needs
and consider the potential impact that trade-offs or cur-
tailments within these strategies have for individual and
household welfare.

It is also important to trace the distribution of male
and female labor time and resources within total production
cycles. Thus, crop production from cultivation to processing, consumption, or sale should be disaggregated to
understand the possible welfare implications of changes
within the cycle. For example, changes in cropping patterns
may intensify women's unpaid labor during weeding,
harvesting, and processing stages and conflict with their
labor needs elsewhere in the farm-household system.
Despite creating a greater workload, a new crop variety may
offer women potential income gains through the processing
and sale of crop by-products over which they have marketing
control. The dominant effect cannot be established a
priori.

Finally, it is necessary to acknowledge that household services play an economic role in maintaining the productive function of the farming system and to identify the non-pecuniary costs borne by individual household members (usually women) in providing such low productivity services under time-consuming and arduous conditions.

CONCLUSIONS

If gender issues are to be given explicit attention within agricultural research and technology development, then an alternative analytical framework is required. Such a framework should identify the different sets of conditions that characterize the economic participation of women and men in small farm systems and view the integrated farming systems as a gendered system of production and reproduction. The purpose of such a framework is not only to examine existing evidence with greater rigor, but also to generate specific hypotheses, data, and methods that break down gender biases within agricultural research and extension. A number of problems that must be addressed to build an alternative framework include:

(1) Overcoming the methodological division between technocratic and social science research that challenges the integration of quantitative and qualitative methods and data;

(2) Resolving time conflicts, such as the time scale that agricultural research institutes or donors demand for FSR/E but that cannot accommodate the amount of time that sociological or socioeconomic research usually requires; and

(3) Channelling research funds and personnel into more extensive data collection and methodological approaches. (Funding for scientific research into minor food crops and technologies for non-farm activities is extremely scarce.)

Despite these problems, to have a clear understanding of the relationship between gender relations and the economics of farming systems and technological choice it is important to question the conceptual basis of existing methodological approaches in agricultural research and development (see Figure 3.1) and to build alternative frameworks for gathering empirical data and testing hypotheses in which gender relations are an explicit variable.

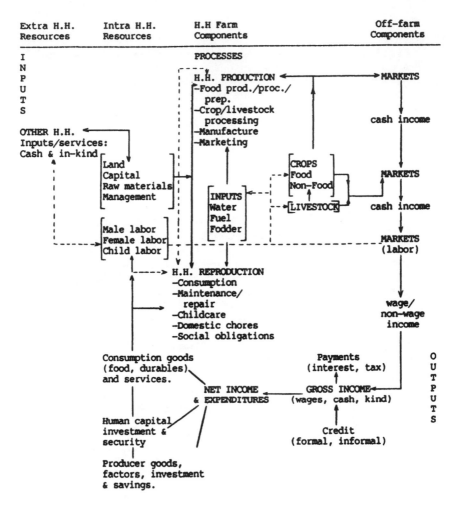

| Extra H.H.
Resources | Intra H.H.
Resources | H.H Farm
Components | | Off-farm
Components |

I
N
P
U
T
S

PROCESSES

H.H. PRODUCTION
-Food prod./proc./
  prep.
-Crop/livestock
  processing
-Manufacture
-Marketing

MARKETS

cash income

OTHER H.H.
Inputs/services:
Cash & in-kind

Land
Capital
Raw materials
Management

CROPS
Food
Non-Food

LIVESTOCK

MARKETS

cash income

INPUTS
Water
Fuel
Fodder

Male labor
Female labor
Child labor

MARKETS
(labor)

H.H. REPRODUCTION
-Consumption
-Maintenance/
  repair
-Childcare
-Domestic chores
-Social obligations

wage/
non-wage
income

Consumption goods
(food, durables)
and services.

Payments
(interest, tax)

O
U
T
P
U
T
S

NET INCOME
& EXPENDITURES

GROSS INCOME
(wages, cash, kind)

Human capital
investment &
security

Credit
(formal, informal)

Producer goods,
factors, investment
& savings.

KEY: - - - - = flows of family labor in/outside the F-H.H. system.
      _____ = flows of non-labor inputs & outputs (& payments).

PLEASE NOTE: For clarity, the elements of the system have not been disaggregated along gender and age lines. These variations are captured when the
framework is applied to the situation of each family member in turn and then
compared.

Acknowlegement: Many of the ideas for this diagram come from C.D. Deere and
A. de Janvry "A Conceptual Framework for the Empirical Analysis of Peasants"
American Journal of Agricultural Economics, 1979.

FIGURE 3.1

A CONCEPTUAL MAP FOR LOOKING AT THE FARM-HOUSEHOLD SYSTEM
FROM A GENDER PERSPECTIVE

REFERENCES

Agarwal, B.
  1981   Agricultural Modernization and Third World
        Women: Pointers From the Literature: An Empirical
        Analysis. WEP Research Working Paper WEP/10/WP21.
        Geneva: International Labor Organization.
Allison, C.
  1985   Women, Land, Labour and Survival: Getting Some
        Basic Facts Right. IDS Bulletin 16:3:24-29.
Bryceson, D.
  1984   Women and Technology in Developing Countries:
        Women's Capabilities and Bargaining Positions.
        Prepared for INSTRAW, United Nations.
CGIAR
  1985   Women and Agricultural Technology. Bellagio,
        Italy: Rockefeller Foundation and International
        Service for National Agricultural Research.
Dey, J.
  1984   Women in Food Production and Food Security in
        Africa. Women in Agriculture 3. FAO: Rome.
FAO
  1984   Women in Agricultural Production. Women in
        Agriculture 1.
Fernandez, M.
  1985   Social Components in Peasant Farming Systems: a
        Research Proposal. Unpublished.
Flora, C. B.
  1984   Intra-Household Dynamics in Farming Systems
        Research: The Basis of Whole Farm Monitoring of
        Farming Systems Research and Extension. A Position
        Paper. Manhattan, KS: Department of Sociology,
        Kansas State University.
Fresco, L.
  1985   Food Security and Women: Implications for Agri-
        cultural Research. Presented at the International
        Workshop on Women's Role in Food Self-Sufficiency
        and Food Security Strategies. Paris: ORSTOM/CIE,
        January.
FSSP
  1984   Newsletter 2:3.
Guyer, J.
  1980   Female Farming and the Evolution of Food
        Production Patterns amongst the Beti of
        South-Central Cameroon. Africa 50:4:341-356.

48

Hart, G. and B. Popkin
    1985   A Note on the Interdependence between Economic
        and Welfare Factors in Rural Filipino Households.
        University of the Philippines, Institute of Economic
        Development and Research. Pp. 75-5.
Henn, J.
    1983   Feeding the Cities and Feeding the Peasants:
        What Role for Africa's Women Farmers? World
        Development 11:12:1043-1055.
Kisseka, N.
    1984   Implications of Structural Changes for African
        Women's Economic Participation. Unpublished.
Lele, Uma
    1985   Women and Structural Transformation. Economic
        Development and Cultural Change 34:2:192-221.
Maxwell, S.
    1984   I. Farming Systems Research: Hitting a Moving
        Target. II. The Social Scientist in Farming Systems
        Research. Institute of Development Studies
        Discussion Paper 199.
Richards, P., J. Karimu, et al.
    1981   Upland and Swamp Land Rice Farming in Sierra
        Leone: The Social Context of Technological Change.
        Africa 51:2
Shaner, W.W., P.F. Philipp and W.R. Schmehl
    1982   Farming Systems Research and Development
        Consortium for International Development. Boulder,
        CO: Westview Press.
Whitehead, A.
    1981   A Conceptual Framework for the Analysis of the
        Effects of Technological Change on Rural Women.
        Technology and Employment Program Working Papers WEP
        2-22/WP79. Geneva: International Labor
        Organization.
World Food Council
    1982   National Strategies to Eradicate Hunger. New
        York, NY: World Food Council, United Nations.

4
Problems of Understanding and Communication at the Interface of Knowledge Systems

Janice Jiggins

One of the fundamental justifications for the practice of FSR/E is that its methodologies promote a desirably close relationship between farmers, researchers, and extensionists in the determination of research criteria and the design and choice of interventions. Considerable effort is going into persuading researchers and extensionists to understand that:

> The goals and motivations of farmers, which will affect the degree and type of effort they will be willing to devote to improving the productivity of their farming systems, are essential inputs to the process of identifying or designing potentially appropriate improved technologies (Norman et al. 1982:25).

Much less effort is being devoted to helping farmers to understand the values, rationales, and objectives that lie behind research and extension behaviors. As FSR/E becomes more involved with on-farm trials, particularly farmer-designed and managed ones, the more important it is for FSR/E practitioners to find ways of articulating their own rationality and making it accessible to farmers.

Some of the FSR/E practitioner rationality is conveyed implicitly in the course of working closely with farmers. A 1985 circular on the Agricultural Technology Improvement Project (ATIP) program, distributed to local agricultural staff in Mahalapaye, Botswana, notes:

> In explaining these meetings, it is important the farmers understand this is a new approach to research in which we want to work together with them to discuss and evaluate alternatives, rather than just rely on collecting information from them (ATIP 1985).

But conversations with farmers in areas where FSR/E teams have been working, reveal that farmers remain greatly puzzled by such things as to why researchers insist that fields or plots be measured in certain ways and what those measurements mean. Or, what it is in the logic of the reseachers' world that makes them value, for example, certain livelihood activities such as field cropping above other activities that seem to the farmers themselves equally necessary components of their livelihood. The breakdown of communication and understanding seems greater between women -- as farmers, food processors, traders and consumers -- and male researchers. This is due not only to sociocultural distances between them. Male researchers may understand little of the rationality of the domestic domain in their own worlds. The researchers' lack of an explicit frame of reference in this sphere -- or use of a partial or biased one -- influences their own set of mental constructs by which they perceive and interpret the world of women within farming systems. Communication difficulties thus are compounded.

There are a number of threads that might be disentangled here. This paper will pull out and untwist only one: the logic of flexibility within the domestic domain, illustrated by examples from Lesotho and northern Zambia (1). The data are not complete from the FSR/E point of view, having been collected for other purposes, but they do highlight a number of points that FSR/E theory and practice needs to take into account. The paper concludes with suggestions about how this might be done.

The data are drawn from two areas of acute seasonal stress. In the Lesotho site, a longitudinal study of energy flows suggests a bimodal pattern of stress (Huss-Ashmore 1982) while the second area in northern Zambia has a short period of moderately erratic, within-season rainfall and a long, dry, cool period of six to seven or even seven and a half months, with the time of acute hunger falling in the January-February period after weeks of heavy labor and declining food stocks (author's unpublished field notes, 1979-80).

Both are areas of high male out-migration and income insecurity. Risk and loss minimization figure highly in farming system strategies. Women, as household heads and as farmers within male-headed enterprises, respond to climatic and income uncertainties by trying to maintain flexibility, typically in four spheres of the domestic economy:

(1) in production, by maintaining reserve crops and
    varieties in household gardens and in wild habitats;

(2) in the timing of operations, volume of product handled, and technique used, in the spheres of food processing, storage and food preparation;

(3) by altering the mix, timing, and quality of performance of their multiple roles; and

(4) by manipulating whatever room there might be for substitutabilty of labor and obligation between men and women.

For reasons of space, the following illustrations are taken from only the first two spheres.

LESOTHO

In the peculiarly distorted economic situation of Lesotho, the day to day survival of rural households is largely a matter of how women maintain themselves and their children. The rationality of the farming system is not determined by physical and climatic features -- these only set limits to what is possible. It is determined by the rationality of women's lives, that is centered in (though by no means confined to) the domestic domain. Within that domain, there is one resource which is critical: fuel supply. It could be said to be the key both to cropping choices and household food availability. Huss-Ashmore writes:

> Because fuel is essential for processing almost all foods, it can be considered a critical resource for the maintenance of health and nutritional status. In Mokhotlong the type of fuel used and the time spent to procure it, vary according to the seasonal availability of dung (1982:156).

The preferred fuel is compacted dung, readily available during the winter from the kraal close to the homestead, that, when dried in uniform slabs, burn with the slow, even heat necessary for the long cooking of dried grains and pulses. In households without a kraal, women have to purchase the dung or manipulate kin relations to obtain it. When the cattle are moved in the summer to the high pastures, women must use horse and cattle dung, picked up from the fields and trails, that is less dense and takes more time to gather. Both sorts are kindled with resinous, woody shrubs that become scarcer as the summer passes but may be the only source of summer fuel if insufficient dung can be gathered. For a short period, kraal dung may be kindled with maize cobs as they are threshed in the winter.

It is fuel availability rather than food availability that determines which foods are eaten at different seasons:

> The supply of slow-cooking protein sources is not used equally throughout the year but is depleted during the cold season when appropriate fuels are available. During the summer the population relies heavily on wild vegetable protein sources, which require more time to locate and gather but which can be rapidly cooked (Huss-Ashmore 1982:157).

One might think that these interactions -- and their further entwining with the water/fuel/grain seasonalities of sorghum beer-brewing, a source of income and wage work that is important for being one of the few available to women -- are not so terribly difficult for a researcher to discover. However, single visits to households during an exploratory survey may fail to discover seasonalities that are both interdependent across disciplinary boundaries (fuel-forestry; cropping; post-harvest domestic food technology) as well as across gender boundaries (cattle are men's business). The fact that a critical key to the functioning of the farming system operates within the domestic domain may continue to conceal its significance during a verification phase of FSR/E work as well.

A further difficulty arises when researchers try to measure the quantities involved such as the cooking time of various foodstuffs using different fuels. An anthropological study of Sesotho measurement concepts points out:

> A woman knows how long to cook vegetables because she knows when they are ready. One woman, preparing bread, was asked how she would cook it: Until it is ready (bo butson). Pressed for precision, she thought carefully and then said: Five or six hours (li-hora - again the English word). In fact, she cooked the bread for an hour and a quarter and saw that it was perfect when she took it from the steam oven. It was ready both in English terms and her own. At no time did she refer to any kind of time, not even in the sun. It was not the time that made it ready. It was the cooking (Wallman 1965:240).

The researcher is concerned with the measurement of time but the woman is concerned with the measurement of "readiness" and there is no reason why the measurement process should not begin with readiness (What does it look like? Is it hard or stiff or does it run? etc.). Instruments such as time allocation studies, useful as they are

as indicators of the range of activities and claims on
labor, make invisible whatever it is that women themselves
see themselves as allocating or conserving. In Huss-
Ashmore's case, women collect wild vegetable proteins not
because they are a preference food nor because there is
nothing else available, but because women primarily wish to
conserve fuel.

The difficulty does not lie in using the measurement
units that make sense within the rationality of the user.
Researchers and extensionists alike are trained in the concepts of scientific agriculture and these concepts may have
no equivalents in the knowledge system of the woman while
the woman is trained in the concepts of her indigenous
knowledge system and may have no way of knowing the significance of the concepts used by the researcher and exten-
sionist. This has little to do with any differences in
ethnic background of the actors and a great deal to do with
the difficulty of articulating the rationality of one system in the terms of the rationality of another.

NORTHERN ZAMBIA

In the northern Zambia location, the fact that local
vegetables and fruits form an important part of the diet is
well-established and there are even a few research programs
investigating the more important species (MAWD 1983). What
is not so readily accepted is that these may have characteristics that yield benefits not provided by modern varieties of the main food crops, however abundant they might
be. Interventions that make their production more difficult by switching labor or land use, for example, may also
make the seasonal management of diets more difficult,
unless the market provides substitutes at affordable
prices.

The question of the timing of agricultural innovations
with respect to the role of market provision of those goods
and services previously supplied within domestic and local
economies merits a short digression here. In agriculture
in industrial countries and albeit to a lesser degree in
irrigated agriculture in developing countries, research
organizations work within and for production and knowledge
systems that are well-defined, well-organised, and highly
interactive. There are at least four main components:

(1) farmers, who are organised and able to contribute to
    research programming through a variety of channels;
(2) powerful industrial organizations engaged in the
    business of transforming primary production into a
    range of consumer and industrial goods, who are well

able to signal to researchers their own technical
requirements or even, by paying for research, to
determine what crop characteristics meet the needs of
their own technical processes;
(3) powerful commercial organizations engaged in the business of wholesaling and retailing produce and processed
foodstuffs, that are able to insist on high quality
standards in defence of existing and new sales; and
(4) consumers, who, either through their purchasing power
or through consumer organizations and lobbies, also
signal their preferences to researchers.

The case is quite different in most dryland areas in
developing countries. Except perhaps for the richest
farmers, producers are not organized nor politically
powerful and have few if any links with researchers. The
range of transformation processes occurs largely within the
domestic domain using local technologies. Wholesalers and
retailers operate in fragmented and often non-competitive
arenas in which the overall level of sales is depressed and
quality carries no premium. Finally, consumers have weak
purchasing power and few, if any, organized channels for
expressing their preferences.
If the inherent yield potential of many dryland areas
is judged to be low, with scant chance that the value of
the marketed output will ever pay for or induce the kind of
infrastructural developments witnessed in irrigated environments, then presumably, it will be necessary to preserve
a continuing capacity to derive benefits from the goods and
services presently obtained from the biomass through transformation within the local community and the household
economy. The challenge becomes that of raising capacity
without displacing too many of the benefits presently
obtained from within the micro-economy rather than in
raising capacity by concentrating on only a few benefits
(higher yield) and externalizing the provision of the rest.
Varietal characteristics must continue to an unknown extent
to meet the demands of domestic transformation processes,
technologies, and end uses. The following example describes just such a situation.
The local vegetables (fruits not included here for the
sake of simplicity) being produced on one farm during
February 1980 at Sambwa in Mpika District of Northern
Zambia include: four cassava varieties (masanga uko,
matutumushi, muntulunga, ucongo); three finger millet
varieties (mwaangwe, mutubila, mwambe); two varieties of
groundnuts; and local maize. Each has a very specific
place in seasonal production and food management. For
example, mwaangwe is a sweet, very early maturing finger

millet that provides one of the first new food crops in the
year and a sweet beer for working parties as the main
harvesting period approaches. Two of the cassava varieties
have palatable leaves (masanga uko and matutumushi - the
latter much sought after by the wild pig) but these fall in
the cold season (June, July), so some are dried early in
the season for later consumption.

In addition to these main food crops, there were five
distinct production sites around the compound (and one further away in the dambo or wide valley bottom) tended by the
two adult resident women, on which were grown a mix of wild
and semi-wild plants and "crops" promoted and officially
marketed by the government. A number of the wild plants,
such as busoshi (Sesamum alatum) and chimamba (Sphenostylis
erecta), also occur naturally (respectively on disturbed
soil and around anthills) but on this farm could be considered as true crops, for they were deliberately planted
on chosen sites, protected from chickens and weeds, fertil-
ized with household rubbish, and the product traded in
Mpika market. Their utility is partly a reflection of the
low and erratic yield of groundnuts but, they have a
utility as snack foods at a time of the year when women may
cook only once a day or even once in two days. The
perennial chimamba ensures that some kind of snack is
always going to be available. It would only lose its
utility if alternative snack foods were to become available
at the critical time of the year when women are busiest or
the preparation and cooking time of cassva and millet were
to become less or women's cultivation labor were to be
reduced.

Another example is provided by the great care the women
took to maintain the balance between the availability of
the staples (millet, cassava) and the availability of oily
or slippery foods for the relish that are added to the
nsima or thick, coarse porridge. The nsima is almost
inedible in sufficiently large quantities without such a
relish to ease it down. In conditions of scarce and expensive commercial cooking oils, unreliable groundnut harvests
in the face of erratic rains, and the time required for
shelling and pounding groundnuts, the softness and slipper-
iness of some local vegetables were highly desired characteristics. Pupwe (Fagara chalydea) is an important dry
season resource in this respect. Slippery local vegetables
have the additional advantage of needing no blanching or
treatment with potash when they are dried for preservation.

Both the men and the women had been experimenting with
vegetable production, as the following examples illustrate.

The male head of the compound had been trying white cabbage, tomatoes, and cucumbers from seeds supplied through the Horticultural Marketing Board in Mpika, in a dambo garden at the end of the rainy season. He found that the cucumbers grew best but were the least needed for domestic use as they already grew a satisfactory range of cucurbits. The tomatoes were well-liked for their flavor and the softness of the flesh but were tiresome to eat because of their tough skins, the very characteristic that made them suitable for the rough marketing conditions. The women had been experimenting for many years with lubanga (Cleome gynandra) selecting for larger leaf size without sacrificing any of the tenderness. They reported, too, that they could get a higher price in the local market for the larger-leaf variety.

The men in the compound scored consistently lower than the women on the following tests using vegetable sources: identification by sight; recall of the main physical descriptors and husbandry, processing techniques and length of storage; and preparation for eating. Zambian and expatriate members of the nearby agricultural college who were engaged in conducting and supervising trainee extension workers in farm surveys, were asked to share their views of the role of local vegetables in the farming system. They all referred the question to the home economic staff, who were trained to work with "western" vegetables, and with few resources to work in the field, knew only those local vegetables that had been used in the household in their own home areas.

There are further problems of communicating knowledge between distinct knowledge systems. The production sites where the local vegetables were grown changed shape and size as women took advantage of rainfall patterns in the advancing season to make additional sowings. At the same time, neither the market value of the product traded nor the opportunity cost of female labor (based on market wage rate) would seem adequate measures of the value of either women's labor time or of the local vegetables to the farming systems. The women themselves used a notion of convenience that appeared to be a combination of characteristics such as: ease of growing near the house; availability (fresh or dried) at moments critical from the point of view of diet management; ease of processing and preparation; timing of labor inputs; and substitutability for other crops. The notion encompassed the principle of flexibility. In this respect, they were reluctant to choose paramount characteristics either for any one crop or between the range of crops. Instead, local vegetables were viewed as a bundle of biomass that enabled them to manage their resources and responsibilities to the best advantage.

THE IMPLICATIONS FOR FSR/E

There are a number of important "lessons" that can be drawn from this brief review. To summarize, these are reduced to two:

(1) the need to develop methodologies for establishing the key field-household interactions at an early stage of the diagnostic process, and
(2) the need to develop methodologies for mutual communication of key concepts across the boundaries of researchers' and female producers' distinct knowledge systems.

Two techniques that might prove to be useful diagnostic instruments for researchers of any background are situation analysis based on the critical incident technique and peer group workshops.

The former is widely used in diagnostic sessions between researchers and carefully drawn panels of users in industrial and commercial practice. It involves informal but structured interviewing which, as users identify problem areas and describe the boundary conditions, focuses on a "critical incident" that exemplifies one of the problems. The incident is then analyzed in depth, leading into discussion of desirable ways to deal with it. Each of the problems is similary treated in turn.

Peer group workshops are widely used throughout the ESCAP region in the development of local, self-managing groups and income-generating projects and by the Food and Agriculture Organization's (FAO) Marketing and Credit division in the promotion of female entrepreneurship. They are based on the understanding that knowledge and expertise exists among local communities, together with a diagnostic capacity attuned to local realities. They draw on the expertise of those who are locally recognized as knowledgeable within the subject problem area by facilitating the preparation of case studies of their successes. These cases are exchanged and analyzed at workshop sessions, leading to identification of interventions that would allow these successes to be replicated. A great deal of experience now exists to guide the preparation and implementation of workshops with those who have little formal education and to facilitate the participation of service officers (agricultural researchers or extension workers).

Both these techniques have the added advantage that they eliminate some of the stages of "translation" of knowledge concepts and, with careful preparation, it is not too difficult to identify those items which, though denoted differently, refer to a standard unit. Returning for example, to the case of the cake that is "ready", the researcher can measure the hours it takes to cook and the baker the cooking that is needed to make it "ready". Both are referring to a standard referent, although the baker might be interested in the number of mouths it feeds and the researcher in its unit weight and composition. The difficulty comes when one is using a knowledge concept that has no referent in the knowledge system of the other. The difficulty is, in a sense, one-sided. Researchers are often keen to learn about and understand the concepts of producers but have little awareness of the constructs and values inherent in their own knowledge system. Where the knowledge system of male agricultural researchers and extension officers does not encompass either an experiential nor trained understanding of the domestic economy, the problem seriously undermines FSR/E practitioners' claims to be conducting systems based technology development.

NOTES AND ACKNOWLEDGEMENTS

(1) The research was carried out between January 1979 and September 1980 in the Central, Northern and Luapula Provinces of Zambia by members of the Rural Development Studies Bureau, University of Zambia, Lusaka.

REFERENCES

Huss-Ashmore, R.
  1982   Seasonality in Rural Highland Lesotho: Method and Policy. In A Report on the Regional Workshop on Seasonal Variations in the Provisioning, Nutrition and Health of Rural Families, pp. 147-161. Nairobi: AMREF.
Ministry of Agriculture and Water Development (MAWD)
  1983   Handbooks for Agricultural Field Workers, Zambian Local Vegetables and Fruits. Lusaka: Department of Agriculture, Ministry of Agriculture and Water Development.

Norman, D., E. Simmons, H. Hays
    1982   Farming Systems in the Nigerian Savanna,
        Research and Strategies for Development. Boulder,
        CO: Westview Press.
Wallman, S.
    1965   The Communication of Measurement in Basutoland.
        Human Organization 3:4:236-243.

5
From Recommendation Domains to Intra-Household Dynamics and Back: Attempts at Bridging the Gender Gap

Amalia M. Alberti

One of the concepts to emerge from the Farming Systems approach to agricultural research and extension is that of the recommendation domain. Defined as "a group of roughly homogeneous farmers with similar circumstances for whom we can make more or less the same recommendation" (Byerlee, et al. 1980), the underlying assumption is that farmers of households within the same recommendation domain will have similar responses to proposed technology (Shaner, et al. 1982: 44). Recommendation domains are used to focus the research process and expedite dissemination of the technology thereby facilitating the extension phase.

The debate continues in the farming systems literature and among farming systems practitioners about both the more relevant criteria and the preferred timetable to identify and to elaborate recommendation domains. The position of those who maintain that the early delineation of recommendation domains precludes considerations that are not readily evident or initially salient (Cornick and Alberti 1985), is countered by others who maintain that the early identification of recommendation domains permits their progressive refinement (Franzel 1984). Still others (Norman and Baker 1984) point out that in the last analysis both the target groups identified and the nature of the technology recommended tend to reflect the expertise of the team members in a particular Farming Systems Research/ Extension (FSR/E) project.

This paper argues that, first, recommendation domains sensitive to gender issues are difficult to develop due to scant documentation of women's participation in agricultural and farm-related activities in local areas, and second, if developed, they are difficult to implement due to several features common to many FSR/E projects. Indeed, it seems that the greater the pressure for prompt elaboration of recommendation domains, the greater is the likelihood that women's roles, as well as their concerns within the FSR/E context, will be overlooked because of insufficient time to draw them out. The long term solution to satisfactorily addressing gender issues in farming systems, however, lies less in attempts to develop appropriate recommendation domains and more in efforts to revise the FSR/E framework so that gender issues are deliberately and self-consciously entertained. Until these changes occur, several key questions are proposed to assist FSR/E practitioners in assessing what gender related issues are potentially relevant in a particular FSR/E site and whether or not they can be addressed feasibly within the existing project framework.

OBSTACLES TO DEVELOPING
GENDER SENSITIVE RECOMMENDATION DOMAINS

Among the more common techniques suggested for the initial stages of problem diagnosis leading to the formation of recommendation domains are reviews of secondary data, informal interviews with persons such as local officials, residents, and extension workers, and an exploratory survey of farmers sometimes combined with or followed by a formal survey (Harrington and Tripp 1984; Shaner, et al. 1982). The obstacles to uncovering the extent of women's involvement in the total or select phases of a farming system embedded in each of these techniques are discussed briefly.

Secondary Data Reviews

Much of the literature on women in agriculture published within the last ten years underscores the extent to which the involvement and contributions of women in this area have been underrepresented (Deere and Leon 1981; Lewis 1981). Nevertheless, secondary sources such as census data and local agricultural reports that continue to ignore or underestimate female contributions abound. When FSR/E staff consult these materials they are likely to accept the data as factual unless they are aware of the possibility that female participation in agriculture may be masked or otherwise distorted. Only when sensitized to this bias may they be persuaded to seek additional corroboration before dismissing gender as a potentially relevant variable.

Informal Interviews and Exploratory Surveys

Local officials and extension agents can often provide site-specific information that a FSR/E project staff member would be hard-pressed to obtain so efficiently otherwise. Information on female involvement in agriculture, however, is less likely to be obtained for several reasons. First, cultural values may intervene. When female agricultural activity is associated with poverty, not only are male officials unlikely to discuss such activity on the part of female members in their own household, but they may well be reluctant to discuss such activity on the part of female residents in general presuming that it would reflect negatively on the socioeconomic status of the community.

Assuming that these local officials and extension agents are almost always exclusively male, attempts to adjust for these gender-related "blind-spots" by speaking with their wives or other female household members may not yield substantially different results (Alberti 1980). To the extent that these women partake of the elevated social status of their households, they are unlikely to make public their own involvement in farm-related tasks or to imply difficult socioeconomic conditions within the community by referring to such activity on the part of other women unless it is to demean them.

Second, it has been found that male farmers routinely underestimate the degree and undervalue the importance of female involvement in farm-related activities in which they too participate (Bourque and Warren 1981; Deere and Leon 1981; Alberti 1986) and ignore or are unaware of the extent of female involvement in farming activities in which they do not share. Hence asking male farmers about the participation of females in agriculture will not necessarily elicit accurate information.

Finally, the reluctance of the national male FSR/E staff members, especially if they are from the local area, to ask questions that are deemed inappropriate by local standards must be considered . Moreover, cultural norms may restrict male field staff members' access to women for interviews. As yet another possibility, male FSR/E staff members may resist interviewing women because of their own attitudes about female participation in agriculture.

Formal Surveys

The advantages and limitations of formal surveys have been widely discussed. Within the farming systems literature, Chambers (1981, 1983) Chambers and Ghildyal (1985) is

perhaps the most outspoken and graphic critic as he
conjures up visions of "30 pages of questionnaire ... which
if asked are never coded, or if coded never punched, or if
punched never processed ... examined ... or analyzed..."
that a number of us have also seen (1980: 4).

Two points about formal surveys on women's involvement
in agriculture must be raised. The first is that preparing
a questionnaire assumes knowing what is necessary and how
to design the queries. While this ought eventually to be
true for initial surveys conducted during project development, this is not always the case for women's issues precisely as a consequence of some of the limitations just
discussed. Secondly, many formal surveys are designed to
be administered to either the male or female head of household, but not to both. Generally, the household member
available when the interviewer arrives responds. However,
the survey form frequently lacks an item to indicate who
was actually interviewed and whether or not that person was
male or female. Hence, even if relevant questions about
women's involvement in agriculture and farm-related
activities are included, it is impossible to disaggregate
male and female responses and to analyze them for
consistency and comparability.

LOCATING WOMEN IN THE FARMING SYSTEMS CONTEXT

Given these constraints what site-specific information
might be readily available that would expedite developing
recommendation domains sensitive to gender differences?
Readily available refers to information that could be
elicited over a few days through informal conversations
with local residents, teachers, and other persons working
in the area. Rapid Rural Appraisal procedures recommend
doing this in conjunction with a field trip around the
project area (Chambers 1980; Beebe 1985). The field trip
is essential to provide visual information to accompany
verbal accounts. Lines of inquiry otherwise not considered
may be opened when the information from these two sources
does not concur.

The information obtained from responses to the following questions ought to enable FSR/E practitioners to
contextualize the situation of women in the FSR/E setting
in broad strokes. At the same time, it would facilitate a
quick assessment of whether or not the FSR/E project, as it
exists or could feasibly be modified, can viably address
the gender issues relevant to that site. Where addressing
those issues is possible,

collecting the kind of information needed should follow
ideally to inform analyses of intra-household dynamics in
FRS/E (See for example, Flora 1984; Feldstein 1987).

What Are the Local Cultural Norms Regarding Female Agricultural Activity? Is More Than One Culture Represented in the Project Area?

In many parts of Latin America, particularly indigenous
regions of the Andes, women work side by side with men in
the fields. In other areas, such as Honduras, women are
seldom seen working in the fields beneath the direct rays
of the sun and may well be embarrassed if they are seen.
Asian women such as those from Bangladesh are rarely field
workers while many of their Indian counterparts assume the
major role in most if not all phases of rice production.
The differences are largely the result of cultural
variations whose dominant mode of expression may be
religious or ethnic or some combination of the two. What
is important is that when a certain portion or subportion
of the population of an area shares a particular cultural
orientation, it is possible to make certain assumptions
about the kinds of roles women are likely to assume within
an agricultural setting and to be forthcoming with
information about those roles. For example, if visible
agriculturally productive activity on the part of women is
highly circumscribed, it can be expected that even when
women do engage in such endeavors, they will be extremely
difficult to document.
When more than one cultural group is represented in an
area, additional factors may come to bear on the situation.
Is one group dominant and the other subordinate? Is the
participation of women in agricultural and farm-related
tasks the same for both groups? Are the norms regarding
such involvement the same? If the norms vary, which norms
do agricultural extensionists and field workers represent?
In culturally complex settings, it is important to specify the cultural group or groups to which a recommendation
domain applies. This should help clarify and explain what
would otherwise be unanticipated responses to a recommended
technology. Factors that might be involved include differential access to extra-household labor by ethnic group or
different production objectives despite use of the same
traditional technologies.

Does Women's Participation in Agriculture Vary by Social Class? If so, in What Ways?

There is an ever-growing consensus that participation by households, and by women within those households, in the farming system is highly contingent on social class. Women from land-poor households who engage in farming tasks tend to work longer hours at those tasks and generate proportionately lower returns than other women. Often they are the women who have been left behind while their male partners migrate in search of wage employment. Women from landless households are clearly the most vulnerable as they are increasingly dependent on an ever more tenuous agricultural wage labor market that relegates them to more restricted and marginal employment opportunities even as it expands commercially (Hart 1978; Stoler 1977; Sen 1985; Stolcke n.d.; Young 1985; Horn and Nkambule-Kanyima, 1984; Chaney and Lewis 1980).

While these trends may be widespread, they are not universal. Knowing whether or not they are valid for a particular setting should provide some clue as to how candid men or women are likely to be about female involvement in agricultural and farm-related activities.

Do Women Specialize in Food Production and Subsistence Agriculture?

Despite broad variations in patterns, the preeminent role of women in the production of food for home consumption appears to cross continental bounds (Chaney and Lewis 1980).

In Latin America the evidence is widespread that the majority of women who directly engage in agricultural production at the household level primarily raise basic crops intended for home consumption though they may also market small portions of those crops. If the household also raises a cash crop, it is likely to be under the control of the male head of household, even when women contribute labor to its cultivation. The more the household's agricultural activities are commercially oriented, the less likely it is that women of the household will be directly involved in agricultural production. However, when hired laborers are present, women of the household are usually expected to provide support services such as food preparation and are occasionally called upon for managerial activities (Deere and Leon 1981; Bourque and Warren 1981; Alberti 1986).

In Asia, the scenario is distinctly different. Despite broad variations in the extent of women's direct involvement in rice-based agricultural economies due to ethnic and religious differences, women are always involved in the processing of rice and frequently bear major responsibility for its transplanting, weeding, and harvesting. When the household's access to rice fields is insufficient to meet its own consumption needs, women as well as men are likely to seek work as agricultural laborers with rice as the preferred medium of payment. Participation in the harvest of kin and neighbors, if not in the planting as well, is another strategy geared to insure a ration of rice (Hart 1978; Sen 1985; Dey 1985). In each instance, the overarching objective is to obtain food that can be immediately used by the household.

In contrast with rice cultivating areas, in the Philippines for example, the cultivation of cash crops such as coconut, tobacco, and commercial varieties of root crops such as cassava and sweet potato, is dominated by men. Root crops grown for home use, however, are often under the immediate control of women (Cornick and Alberti 1985).

Until recently, the situation in Africa presented what had probably been the most consistent association between crops and gender. Even now, food crops are grown almost exclusively by women, though some women, particularly those near urban areas, have begun to cultivate cash crops as well (Ferguson and Horn 1985). In contrast, men continue to concentrate their efforts in cash crop production. Getting a sense of the pattern that predominates for a given FSR/E project should help us identify the crops and animals that women tend to work with as well as to assess the FSR/E project's capabilities in those areas.

UTILIZING THE INFORMATION
WITHIN A FARMING SYSTEMS FRAMEWORK

After a discussion of how this information may be fruitfully used, the following areas must be considered. First, the knowledge gained should enable researchers to better identify the variables that are particularly relevant in relation to women in farming systems in the project area. Second, it can provide guidelines to estimate the validity of the information and data that does exist. Third, it highlights the kind of information that is available while giving some indication of what is lacking. This should help to assess what additional information is needed and to appraise how sensitive its collection may be.

For example, the knowledge that there are two ethnic groups within the FSR/E project area should immediately prompt the question as to whether or not their attitudes toward agriculture and women's involvement in agriculture are the same. If they differ, ways of systematically distinguishing responses by ethnic group becomes important.

The shortcoming of these illustrations is that real life situations rarely fall into compartments that vary so neatly along a single dimension. Rather, multiple variables combine and fuse, whether systematically or erratically, resulting in ever more complex relationships. Their salience is heightened as they interact with some of the more common features of farming systems projects. Let us examine some of these characteristics and the way they interact with gender concerns.

FARMING SYSTEMS CONSTRAINTS
AND THEIR IMPLICATIONS FOR GENDER ISSUES

Site selection for farming systems projects often results from political and economic decisions that occur outside project bounds (Shaner et al. 1982; Harrington and Tripp 1984). Marginal areas are less likely to be selected. Not only do they tend to lack political leverage, but projects in such areas are more prone to failure due to their residents' restricted access to resources. Women who engage in agricultural and farm-related activities, however, are frequently concentrated among the resource poor who are commonly located in more marginal areas.

Despite occasional efforts to the contrary, farming systems projects are frequently commodity-oriented either as the result of project mandate, team member expertise, or a combination of these factors (Norman and Baker 1984). Furthermore, a commodity orientation is frequently aligned with a commercial orientation. As has been discussed, women are more likely to cultivate food crops with a view to household consumption. Hence, when a FSR/E project has a commodity orientation it may implicitly ignore women by excluding crops of most concern to them.

Another constraint is that FSR/E projects tend to adapt already existing technology, or "shelf technology," to a particular situation, rather than to develop new technology for a specific situation. They justify their approach on the basis of insufficient resources and a time frame inadequate to allow for additional research. However, existing technologies have tended to be capital intensive, and until recently, to give demonstrated results only when adopted as an entire package, rather than in steps over time. To the extent that the women who engage in agricultural and farm-related activities are concentrated among the resource poor, they may be unable to adopt the new technology because of insufficient cash resources, or if they have the resources, they may be unwilling to adopt the new techno-logy because it is inappropriate to their goals when they are subsistence rather than commercially oriented.

Finally,to paraphrase Chambers, these factors interlock (1980:3). As Harrington and Tripp note: "domains are formed so that researchers can effectively deal with the majority of farmers in a particular area" (1984: 14). However, the only majority that women tend to constitute as household level agriculturalists is that of the rural poor. Nevertheless, even among these, some women have partners, others are single, some are the only farmer in the household, and still others are the only sources of labor. Though women who directly engage in farming and farm-related activities are unlikely to be wealthy, it is likely that there is considerable variation in their access to resources, even among those broadly labelled as "poor."

Women in agriculture tend to share a disadvantaged position in male-oriented agricultural research and devel-opment programs. The way they experience that disadvan-tage, however, is mediated by their culture, resources, and civil status. It is difficult for recommendation domains that depend on homogeneous circumstances in key variables to locate issues that relate to "women" equally despite their diversity. What a true incorporation of gender issues in farming systems implies is a revision of the farming systems unit of analysis from the household to the male and female members within the farming systems household for the stages of problem diagnosis and design. The information thus provided would enable farming systems practitioners to make conscious though difficult choices about where the FSR/E resources will be channelled, knowing full well and in advance how those choices are likely to affect men and women differentially.

REFERENCES

Alberti, A. M.
  1979   Metodologia Apropriada para el Estudio de la
         Mujer Rural en los Andes del Ecuador. Presented at
         the workshop Metodologia Apropriada para Estudiar a
         la Mujer Rural. Quito: CEPLAES/Ford Foundation.
  1982   Observations on the Productive Role of Women and
         Development Efforts in the Andes. Proceedings of
         the First Women in Development Workshop sponsored by
         HIID/MIT. Cambridge, MA: Harvard University Press.
  1986   Gender, Ethnicity, and Resource Control in the
         Andean Highlands of Ecuador. Ph.D. Dissertation.
         Stanford, CA: Stanford University.

Beebe, J.
  1985   Rapid Rural Appraisal: The Critical First Step
         in a Farming Systems Approach to Research. FSSP
         Networking Paper No. 5.
  n.d.   Factoring New Information into Decisions About
         On-Farm Trials. USAID Mission, The Philippines.

Bourque, S. C. and K. B. Warren
  1981   Women of the Andes: Patriarchy and Social
         Change in Two Peruvian Towns. Ann Arbor, MI:
         University of Michigan Press.

Chambers, R.
  1981   Rapid Rural Appraisal: Rational and Repertoire.
         Public Administration and Development 1:95-106.
  1983   Rural Development: Putting the Last First.
         London: Longman.

Chambers, R. and B.P. Ghildyal
  1985   Agricultural Research for Resource-Poor Farmers:
         The "Farmer-First-And-Last" Model. Agricultural
         Administration 20:1-30.

Chaney, E. M. and M. W. Lewis
  1980   Women, Migration, and the Decline of Smallholder
         Agriculture. Washington, D.C.: USAID Office of
         Women in Development.

Cornick, T. R. and A. M. Alberti
  1985   Recommendation Domains Reconsidered. Presented
         at the Farming Systems Symposium: Farming Systems
         Research and Extension: Management and Methodology,
         Kansas State University, Kansas.

Deere, C. D. and M. Leon de Leal.
    1981   Women in Agriculture: Peasant Production and
            Proletarianization in Three Andean Regions. ILO
            Rural Employment Policy Research Program:
            Unpublished Working Paper.
Dey, J. M.
    1985   The Concept and Organization of a Research
            Network on Women in Rice Farming Systems. In the
            Report of the Project Design Workshop on Women in
            Rice Farming Systems. Los Banos, Philippines:
            IRRI.
Feldstein, H., S. Poats, K. Cloud, R. Norem
    1987   Intra-Household Dynamics and Farming Systems
            Research and Extension Conceptual Framework.
            Population Council/FSSP-University of Florida
            Working Document.
Ferguson, A. and N. Horn
    1985   Situating Agricultural Research in a Class and
            Gender Context: The Bean/Cowpea Collaborative
            Research Support Program. Culture and Agriculture
            26:1-10.
Flora, C. B.
    1984   Intra-Household Dynamics in Farming systems
            Research: The Basis of Whole Farm Monitoring of
            Farming Systems Research and Extension: A Position
            Paper. Mimeograph.
Frankenberger, T. R.
    1985   Adding a Food Consumption Perspective to Farming
            Systems Research. Prepared for the Nutrition
            Economics Group, Technical Assistance Division,
            Office of International Cooperation and Development,
            United States Department of Agriculture.
Franzel, S. C.
    1984   Modeling Farmers' Decisions in a Farming Systems
            Research Exercise: The Adoption of an Improved
            Maize Variety in Kirinyaga District, Kenya. Human
            Organization 43:3:199-207.
Harrington, L. W. and R. Tripp
    1984   Recommendation Domains: A Framework for On-Farm
            Research. CIMMYT Economics Program Working Paper.
            CIMMYT: Mexico.
    1985   Formulating Recommendations for Farmers,
            Researchers and Policy-Makers: Issues in the Design
            and Use of Socio-Economics Research. Presented at
            the ESCAP CGRPT Center Conference "Towards
            Recommendations for Research, Policy and Extension:
            Methodological Issues in Socioeconomics Analysis of
            Coarse Grains and Food Legumes," Bandung, 18-23
            November.

Hart, G.
    1978   Labor Allocation Strategies in Rural Javanese
        Households. Ph.D. Dissertation, NY: Cornell
        University.
Horn, N. and B. Nkambule-Kanyima
    1984   Resource Guide: Women in Agriculture in
        Botswana. Michigan: Bean/Cowpea Collaborative
        Research Support Program.
Lewis, B. C.
    1981   Invisible Farmers: Women and The Crisis in
        Agriculture. Washington, D.C.: USAID Office of
        Women in Development.
Norman, D. W. and D. C. Baker
    1984   Components of Farming Systems Research FSR
        Credibility and Experience in Botswana.
        Agricultural Technology Improvement Project,
        Department of Agricultural Research. Botswana,
        Africa: Ministry of Agriculture.
Sen, G.
    1985   Women Workers and the Green Revolution. In
        Lourdes Beneria, ed., Women and Development: The
        Sexual Division of Labor in Rural Societies. NY:
        Praeger Publishers.
Shaner, W. W., P. F. Philipp, and W. R. Schmehl
    1982   Farming Systems Research and Development.
        Boulder, CO: Westview Press.
Stolcke, V.
    n.d.   Social Inequality and Gender Hierarchy in Latin
        America. Position paper presented at the SSRC
        Workshop.
Stoler, A.
    1977   Class Structure and Female Autonomy in Rural
        Java. Signs 3:74-89.
Young, K.
    1985   The Creation of a Relative Surplus Population:
        A Case Study from Mexico. In Lourdes Beneria, ed.,
        Women and Development: The Sexual Division of Labor
        in Rural Societies. NY, NY: Praeger Publishers.

6
Research, Recommendation and Diffusion Domains: A Farming Systems Approach to Targeting

Peter Wotowiec, Jr., Susan V. Poats, and Peter E. Hildebrand

TARGETING FARMING SYSTEMS ACTIVITIES: HOMOGENIZING VARIABILITY?

Inherent in the farming systems approach is the recognition of the variability of the complex circumstances farmers face while managing farms that are comprised of inter-related crop, animal, household, and off-farm enterprises. Diversity in farming systems must be recognized in developing appropriate technologies for the farmers that manage those systems. However, it is not practical to conduct research tailored specifically to a few individual farmers. Targeting entails the grouping together of similiar clientele so efforts can be sufficiently focused. Although the concept of targeting might seem contrary to the recognition of heterogeneity among farms, it is an essential component of the farming systems approach. When Farming Systems Research and Extension (FSR/E) practitioners target a group of farming systems as relatively homogeneous based on a few simple factors, the existing variability among farms is often not sufficiently considered. How can FSR/E teams define and target homogeneous groups of farming systems without losing sight of the heterogeneity among them? Farming systems practitioners take different positions on this issue (Cornick and Alberti 1985).

One perspective stresses the early definition of homogeneous groups of farmers using the recommendation domain concept to guide subsequent research activities. Collinson (1979, 1980), Gilbert et al. (1980), and Franzel (1985) advocate ex ante delineation of recommendation domains based on secondary data and preliminary surveys, followed by a formal survey to refine the domain boundaries. Both Collinson and Franzel describe a technique of defining recommendation domains through interviews with extension agents and local authorities before actually initiating activities with farmers. Early definition of recommendation domains is usually based upon a few relatively easily identifiable factors such as soil type, agroecological zones, crop type, and management (Harrington and Tripp 1985). These authors note the importance of continuing the refinement of domain boundaries throughout the sequence of on-farm adaptive research, but the subsequent reassessment of recommendation domains is often not vigorously pursued.

A more recent view states that grouping farming systems should not take place until the researchers have an adequate understanding of the variability inherent in local farming systems, usually not accomplished early in the work in an area. Cornick and Alberti argue that recommendation domains established early are frequently poorly conceived and lead to a premature assumption of homogeneity. The failure to consider potential variability from factors such as long-term climate induced trends in cropping patterns, household decision-making and labor allocation, or relationships between on- and off-farm activities, may bias subsequent technology development. For example, Cornick and Alberti (1985:1) note:

> ...the roles of women and children that can be critical factors in the development and subsequent adoption of technologies are often explicitly excluded from consideration in recommendation domains. This occurs because the usual time frame for development of recommendation domains is inadequate to the task of understanding intra-household dynamics and the importance they hold in the system.

In particular, socioeconomic factors are often not fully integrated into domains defined early, either because of the longer period of time necessary to gather this information, or because of the absence of trained social scientists as part of farming systems teams. One area often poorly covered in early definitions of domains is the different agricultural roles of men and women. Proceeding with on-farm research and other activities on the basis of a hastily achieved assumption of homogeneity could result in inefficient subsequent research and the promotion of solutions that are not appropriate to farmers (Cornick and Alberti 1985:25) or technologies that may favor some farmers (male) while causing disadvantages for others (female).

In this paper the issue of variability versus homogeneity in the targeting of farming systems research and extension activities is explored. After a brief review

of targeting in FSR/E using recommendation domains, problems in the conventional use domains in FSR/E are described
in an attempt to bring together the two differing view-
points and to begin to resolve the question. The refined
concept of targeting allows for better inclusion of gender
variables in the definition of domains.

OVERVIEW OF TARGETING AND RECOMMENDATION DOMAINS

Targeting for Efficiency and Social Equity

FSR/E must differentiate between various potential
farmer-client groups and determine the particular needs of
each, if technologies are to be developed that clearly meet
those needs (Byerlee and Hesse de Polanco 1982). Most
literature on the subject of targeting in farming systems
has stressed the increase in efficiency of FSR/E activities
made possible through focusing upon specific, relatively
homogeneous farmer groups.

Efficiency in allocation of research resources is
essential if a program is to reach and benefit a maximum
number of farmers. By focusing scarce resources upon
roughly similiar groups of farmers, research programs are
able to carry out investigations on a selected number of
representative farms and later can transfer the findings to
the comparable situations faced by other farmers.

Targeting is also important in justifying the farming
systems approach to institutional policy makers who are
concerned about social equity in the distribution of
resulting benefits. Farming systems practitioners use targeting concepts to assist them in making decisions which
increase the likelihood of an optimal distribution of
research results among the members of a community.

Conventional Concept of Recommendation Domains

The concept of "recommendation domains" has been widely
used in targeting farming systems research since Perrin et
al. (1976) first introduced the idea. It is described and
defined by Byerlee et al. (1980:899) in the following
manner:

> ... a recommendation domain (RD) is a group of farmers
> with roughly similiar practices and circumstances for
> whom a given recommendation will be broadly appropri-
> ate. It is a stratification of farmers, not area:
> farmers, not fields, make decisions on technology.

Socioeconomic criteria may be just as important as agroclimatic variables in delineating domains. Thus resulting domains are often not amenable to geographical mapping because farmers of different domains may be interspersed in a given area.

Using this definition, neighboring farm households might be placed in different recommendation domains because of differences in availability of family labor. In societies where women cultivate different crops than those of the men, female farmers could comprise a recommendation domain separate from male farmers even if they are from the same household.

Expanding Upon the Definition of Recommendation Domain

Perrin et al. (1976) originally conceived of the notion of recommendation domains as an aid to researchers for targeting the development of technologies to specific audiences. The concept has been expanded since then to include a number of additional situations and purposes. Some of the most common applications of recommendation domains include the following gleaned from current literature on the topic:

(1) making policy decisions;
(2) identifying priority issues for research;
(3) specifying clientele for developing recommendations;
(4) selecting representative sites and farmer-cooperators;
(5) focusing analysis of surveys and on-farm trials;
(6) orienting extensionists to groups of similar farmers;
(7) transferring adapted technology to appropriate farmers; and,
(8) enhancing equitable distribution of FSR/E benefits.

As Harrington and Tripp (1985) point out, the domain concept is vital to every stage of the on-farm research process. However, it is apparent from reviewing the literature on the subject that the definition of "recommendation domain" not only changes at each stage, but also varies according to the individual who applies it as well as to the end result. The wide variability among farmers and farms, and the dynamic nature of the farming systems development sequence, contribute to the confusion that exists among FSR/E practitioners as to the general meaning and use of the term recommendation domain.

On-Farm Variability and Conventional Recommendation Domains

The emphasis by Byerlee et al. (1980) upon "farmers, not fields" as the sole basis for the delineation of recommendation domains is not always warranted because of the variability found in some field situations. Cornick and Alberti (1985) cite the case of farmers in the community of Quimiaq in the mountains of Ecuador who manage different cropping patterns in different agro-ecological zones, a product of altitude, temperature, and rainfall variation on the mountain slopes. Not only does each farm cross agro-ecological zones, but the cropping patterns found in each field vary greatly from year to year. For example, depending upon a farmer's perception of trends and yearly changes in climatic conditions, bean or fava bean intercrops will be assigned to maize fields located at varying elevations along the slope.

Gender and intra-household variables are often neglected in the process of defining a recommendation domain because of the relatively more difficult and time consuming task of collecting and analyzing data on these variables. Existing information on gender and household variables often offers few useful insights for defining recommendation domains when compared to the secondary data available on agroecological characteristics. In addition, the gender and household data that may exist may be unobtainable locally. Nevertheless, superficial understanding of these variables or the transfer of erroneous assumptions without continued investigation can hamper design and delivery of appropriate technology.

Refining the Concept of Domains

The argument here is that the issue of targeting in FSR/E has become confusing because the definition of the term "recommendation domain" has been stretched to cover too many situations and too many different purposes. Farming systems practitioners must develop a common understanding of how the use and definition of "domains" change as the farming systems sequence progresses from initial characterization through problem diagnosis, testing, adaptation, evaluation, and finally, to the delivery of the new technology to farmers.

It is essential to account for the heterogeneity in farming systems, even while delineating relatively

homogeneous groups. Refinement and expantion of the use of
domains in targeting will enable researchers to distinguish
applications of the domain concept, while still recognizing
the diversity among farm households and farming systems.

This can be accomplished by recognizing a problem focus
in the definition of the domains, by tying the changing
concept of domain more closely to the farming systems
sequence, and by stressing a greater inclusion of socioeconomic considerations into the targeting process. The
refinements outlined below are a sharpening of focus not a
changing of terminology, that will lead to increased
utility of this method of targeting in the field.

Any of the three types of domains described below may
be defined within specific geographic boundaries for ease
in conceptualization, but it is imperative to realize that
domains do not necessarily include all the area within the
boundaries prescribed. Because domains are based upon a
specified problem focus and upon socioeconomic considerations in addition to the more geographically mappable factors of climate, altitude, and soil, they are actually
interspersed intermittently in a discontinuous pattern
throughout a geographic area.

The examples here will emphasize gender as a key factor
in delineating domains; other factors, such as class,
education, language use, or food preferences, could also be
used.

Research Domains: Targeting for Variability

A "research domain" is a problem-focused environmental
(agro-ecological and socioeconomic) range throughout which
it is expected that hypothesized solutions to a defined
problem could have potential applicability and therefore
should be tested. Research domains are determined during
the initiation of research activities, largely by consideration of biophysical (agro-ecological) factors, with some
attention to socioeconomic and gender issues.

Recommendation Domains: Targeting for Homogeneity

Research domains are comprised of one or more agro-
socioeconomic "recommendation domains", that are tentatively defined based upon the response of a specific technology to the actual agro-socioeconomic conditions found on
farms. A "recommendation domain" is a group of farmers (or
farmers and their fields) with a common problem for whom a
tested solution meets their biophysical and socioeconomic
requirements for adoption.

In the Ecuadorian case cited by Cornick and Alberti, recommendation domains would be based not only upon farm households, but also upon their separate fields that are not contiguous but widely dispersed in location and altitude. Each household might fall into several recommendation domains depending upon: (1) where their fields are located along the agroecological gradient of the mountainside; (2) the climate-related crop management decisions made for each of those fields; and, (3) the particular problem solutions to be tested.

Other examples from West Africa demonstrate how gender can be used to differentiate recommendation domains. In many areas, men and women have separate fields, often inherited from their same sex parent, that are not managed communally by the household. Women traditionally grow rice on their lands while men produce upland crops such as groundnuts or sorghum on their own fields. In this system, fields managed by a household pertain to different recommendation domains depending upon both the cropping system and the gender of the farmer manager. In one area of the Ivory Coast, men plant yams in a cleared field. Women will often care for the yam plants by weeding them while they plant their vegetable crops in the space between the yam plants. In this system, fields are neither men's nor women's, nor would entire fields fall into a single problem-focused recommendation domain. Rather, domains would be determined by crops and their managers, male or female, and contain pieces of many fields.

Recommendation domains are seen as tentative in nature throughout the on-farm adaptive research process. Recommendation domains are initially hypothesized by the FSR/E team on the basis of on-farm exploratory and refinement trials, information collected through directed surveys, and subsequent on-farm verification trials. Over time, as more information is gathered, the recommendation domains are refined and redefined to closer approach reality.

Diffusion Domains: Targeting for Communication

"Diffusion domains" are interpersonal communication networks through which newly acquired knowledge of agricultural technologies naturally flows (Hildebrand 1985). Informal flow of information through a community grapevine is substantial (Rogers 1983). From farmer to farmer, neighbor to neighbor, store operator to patron, information about new ideas moves through a farming community. Awareness of a new technology being verified in on-farm trials takes place among farmers and their families who are not directly involved in the on-farm research.

A farming systems team can enhance the informational effect of on-farm research activities in a community. By understanding the local communication networks in an area, the FSR/E team can strategically locate on-farm verification trials in each diffusion domain to enhance the diffusion of information about a new technology among potential users. This ensures a broader, more equitable distribution of information because it has the potential of reaching farmers who are difficult to reach through conventional extension methods and who rely greatly upon localized interpersonal communication to acquire information.

Frequently, information about new technologies developed in agricultural programs tends to be communicated only through male information networks. In some societies information about technologies is diffused only slowly, if at all, from men to women even within a household. Female farmers are clearly disadvantaged in learning about new technologies if they cannot participate in male-oriented dissemination programs. Definition and use of diffusion domains in the FSR/E testing process allows practitioners to recognize and plan for the fact that men and women often have different communication networks. For example, if men gather and exchange information about agricultural technology at certain locales (cooperatives, local seed and feed stores, bars) where women are usually not permitted by custom to enter, women may effectively be excluded from the process of dissemination.

FIELD USE OF THE DOMAIN CONCEPT

In practice, farming systems teams work in a project area located on the basis of geographical and political considerations rather than with biological conditions or socioeconomic concerns. Within a project area, project focus can be based on a specific priority commodity commonly produced by farmers in the area or may be based on socioeconomic considerations such as an emphasis upon small farmers or women farmers. The farming systems team working in the area may have responsibility for determining project focus. Seldom will the team have input into defining the project area. Even though it is of great importance in targeting farming systems efforts, the process of selecting the project area and project focus lies beyond the scope of this paper. This discussion will commence with subsequent stages of the targeting process. For the sake of brevity and clarity, a relatively simple example will be used.

A Case of Targeting in the Farming Systems Approach

The following example is drawn from farming systems activities in Central America (Ruano 1977; Hildebrand and Cardona 1977; Reiche Caal et al. 1976). Although based on actual experiences and cases, some liberty has been taken with its portrayal here to show how this refined concept of domains might have been advantageously applied.

A farming systems team from the national research institute composed of three agricultural technicians, one economist, and one anthropologist (all men) was assigned to a certain hilly section of the country. In accordance with national agricultural production objectives, the team's mandate was to work on improving the production of basic grains among small, resource-limited farmers in the project area (a commodity and socioeconomic based project focus).

Initial informal reconnaissance of the area and a review of secondary information revealed that the area was comprised of relatively flat, fertile lands in the valley bottoms and poorer, rocky soils on the slopes. The larger, fertile farms in the valley bottoms were owned by wealthier farmers who were able to employ mechanization in their cultivation systems. Tractors were used in their mono-cultural stands of maize and short, improved sorghum varieties. In contrast, the hillsides were largely devoted to small farmer cultivation, with farms averaging about 3.5 hectares. Sorghum and maize were interplanted using mostly traditional, taller sorghum varieties. A few farmers employed bullocks and plows on their farms, but most cultivated their crops by hand.

Unfortunately, little secondary information existed concerning the socioeconomic conditions of the area. However, generally for this region, people say that men plant and tend the crops while women manage the household, food processing and preparation, and the marketing. Little was known about the role of women in production. The team assumed that this was generally true for the project area. The team did not at this point have any female members.

In keeping with their project focus, the team decided their attention should be targeted on the smaller hillside farmers and farms. A sondeo (Hildebrand 1982), or diag-nostic survey, conducted in the hillside region revealed that farmers in the hillside areas used similar systems of intercropping maize and sorghum. They complained that the scarcity and irregularity of rainfall had made maize cul-tivation an increasingly risky endeavor. Farmers were unable to grow enough maize to meet their consumption needs. Only the male heads of households were targeted for the sondeo.

Since irregular rainfall frequently caused the failure of the maize crop, the more drought-tolerant sorghum was being grown to supplement it. However, farmers expressed a dislike for eating sorghum and indicated they only grew it to sell for animal feed, using the proceeds to purchase maize. In this sense, substituting cultivation of sorghum for maize reduced the risk of crop failure yet provided for the household subsistence needs.

Sorghum production in the area was higher per unit planted than maize, but still below production levels achieved elsewhere in similar environments with improved varieties. As one facet of their farming systems program, the team hypothesized that selected improved sorghum varieties within the traditional cropping system could lead to a partial solution to the identified production problem.

Based on these findings, the team considered the hillside maize and sorghum farmers and their fields, with declining maize yields as a single "research domain" (problem-focused, agroecological range). A series of exploratory trials were designed for placement throughout the research domain.

At harvest, the team collected production data as well as information on farmer opinions about the new varieties. Even though the new, earlier varieties performed well on all test sites, there were sharp differences among farmers as to their acceptability. Some farmers were planning to keep seed and plant the new varieties again the following season. Others were quite disinterested in the varieties, but their reasons were unclear to the team. Based on farmer evaluations, the team partitioned the "research domain" into two groups of farmers; those interested in planting the sorghums again and those not interested. The former group became a tentative "recommendation domain" and more precisely refined trials were designed to continue testing the varieties under farm conditions, while further determining the reasons for the farmers' acceptance of the new sorghums. For the other group, more information was needed by the team to determine why the new sorghums were unacceptable. Thus, this group continued to constitute a "research domain."

Information had been collected to characterize the farming systems of the area while monitoring the exploratory trials. Continuous contact of the team members with farmers during this period had yielded much additional socioeconomic information not apparent from the initial sondeo activity. All hillside farmers and their farming systems no longer appeared alike.

Some farmers at slightly lower elevations had soils with better water retention characteristics than other farmers on higher slopes. These farmers could plant maize with a greater assurance of obtaining a harvest than those at higher locations with poorer soils. Through additional directed interviews, it was found that lower elevation farmers tended to grow sorghum primarily as a cash crop. Because of their favored soil conditions, they possessed enough cash from crop sales to ensure a continuous supply of maize in the household. These farmers did not consume sorghum.

Over time the team came to realize that even though most people claimed they did not eat sorghum, many were actually using it as a substitute for maize. The team hypothesized that sorghum consumption increased among less well-off households farming the poorer, higher elevation fields. It was apparent that farmers of this group also were not interested in the new higher yielding sorghums. As one aspect of their attempt to resolve this seeming contradiction, the team initiated informal surveys with women of the households within this group. Unfortunately, owing to socio-cultural and linguistic barriers, the male team members were unable to obtain adequate information.

This was corrected by temporarily adding a female social scientist from the institute headquarters to the team to conduct the interviews. She found that these families did consume sorghum, although they had not always done so. Decreasing maize harvests and lack of resources for the purchase of maize had forced them to consume sorghum. Women interviewed indicated that consumption of sorghum implied a certain social degradation, a "shame" in the eyes of neighbors. In many cases, a farmer whose family consumed sorghum was considered a poor provider. To the casual observer, sorghum consumption was not apparent among the farmers; but as the team moved deeper into the community, they found that sorghum was an important part of the diet among families lacking maize.

Further study of sorghum preparation, cooking and taste preferences revealed that sorghum, like maize, is primarily eaten in the form of tortillas, either prepared with maize or alone. Women said some of the new varieties tasted bitter and were not fit for consumption. One of the new varieties was not bitter-tasting, but due to purple glumes, it left telltale dark spots when made into tortilllas. Although the purple glumes could be removed after many washings, this was an unacceptable alternative for most families because of a scarcity of readily available water in the higher elevation areas of the research domain.

Using this information, the team partitioned the original research domain into two "recommendation domains." For the earlier tentatively defined recommendation domain, consisting of farmers who produced sorghum destined for the animal feed market, on-farm testing of the previously introduced new sorghum varieties was continued. For the second recommendation domain, composed of relatively poorer farmers producing sorghum for home consumption under less favored soil conditions, the team recommended that the research institute acquire or develop varieties with less coloring and no bitter taste that could then be tested on-farm with the farmers in this group.

Through this experience the FSR/E team and the research institute began to realize that while women were not directly involved in sorghum production, they did have considerable influence in making cropping decisions that affect household concerns such as consumption. Newly cognizant of the need for an augmented social perspective in their development activities, the team began a second phase of on-farm experimentation targeted towards the two separate recommendation domains.

At the same time, they began to work with local extension personnel to study the flow of agricultural information among the farmers and households in the region. Recognizing the role that household consumption preferences play in the adoption or rejection of sorghum technologies, female team members and interviewers were added to the farming systems program to ensure a balanced gender perspective.

Among the many local information pathways, it was found that women exchanged much information about sorghum and other agricultural crops with other women at the weekly markets. Among men, interpersonal communication concerning farming and crops took place on Sundays when farmers from the surrounding countryside congregated in the town plaza to converse and visit. By the close of the second season of farming systems activities, the team had tentatively defined several local "diffusion domains" based on gender, religious affiliation, locality groups, and other factors. On-farm trials and extension efforts were managed to ensure information flow to each diffusion domain.

CONCLUSION

This greatly simplified case provides an example of how the refined domain concept allows grouping of roughly

homogeneous farmers while not losing sight of the
heterogeneity inherent among them. This conception of
domains is not a static one, but one that recognizes the
changing nature of the targeting process as a result of
on-going information gathering through surveys, participant
observation, and on-farm experimentation. Maintaining a
flexible determination of domains allows for a greater
understanding of the diversity of local farming systems, of
the rationale behind the behavior of farmers, and of the
effect of gender and social factors upon the local practice
of agriculture.

REFERENCES

Byerlee, D., M. Collinson, et al.
   1980   Planning Technologies Appropriate to Farmers:
      Concepts and Procedures. Mexico: CIMMYT.
Byerlee, D. and E. Hesse de Polanco
   1982   The Rate and Sequence of Adoption of Cereal
      Technologies: The Case of Rainfed Barley in the
      Mexican Altiplano. Mexico: CIMMYT.
Collinson, M. P.
   1979   Deriving recommendation domains for Central
      province, Zambia. Report no. 4. Nairobi, Kenya:
      CIMMYUT.
   1980   A Farming Systems Contribution to Improved
      Relevancy in Agricultural Research: Concepts and
      Procedures and Their Promotion by CIMMYT in Eastern
      Africa. Nairobi, Kenya: CIMMYT East Africa
      Economics Program.
Cornick, T. and A. Alberti
   1985   Recommendation Domains Reconsidered. Paper
      presented at the 1985 Farming Systems Research and
      Methodology Symposium. Manhattan, KS: Kansas State
      University.
Franzel, S.
   1985   Evaluating a Method for Defining Recommendation
      Domains: A Case Study from Kenya. Paper presented
      at the 1985 Farming Systems Research and Methodology
      Symposium. Manhattan, KS: Kansas State University.
Gilbert, E. H., D.W. Norman and F.E. Winch
   1980   Farming Systems Research: A Critical Appraisal.
      Michigan Sate University Rural Development Paper No.
      6. East Lansing, MI: Michigan State University.
Harrington, L. W. and R. Tripp
   1985   Recommendation Domains: A Framework for
      On-Farm Research Working Paper 02/84. Mexico:
      CIMMYT.

Hildebrand, P.E.
    1982   Combining Disciplines in Rapid Appraisal: The
            Sondeo Approach. Agricultural Administration 8:
            423-432.
    1985   On-Farm Research: Organized Community
            Adaptation, Learning and Diffusion for Efficient
            Agricultural Technology Innovation. Farming Systems
            Support Project Newsletter 3:4:6-7.
Hildebrand, P.E. and D. Cardona
    1977   Sistemas de Cultivos de Ladera para Pequenos y
            Medianos Agricultores: La Barranca. Guatemala:
            ICTA.
Perrin, R. K., D. L. Winkelmann, E. R. Moscardi and J. R.
        Anderson
    1976   From Agronomic Data to Farmer Recommendations:
            An Economics Training Manual. Mexico: CIMMYT.
Reiche Caal, E.C., P. E. Hildebrand, S. R. Ruano and J. Wyld
    1976   El Pequeno Agricultor y Sus Sistemas de Cultivos
            en Ladera. Guatemala: ICTA.
Rogers, E.M.
    1983   Diffusion of Innovations. New York, NY: Free
            Press.
Ruano, S.R.
    1977   El Uso del Sorgo para Consumo Humano: Caracter-
            isticas y Limitaciones. Guatemala: ICTA.

7
Incorporating Women into
Monitoring and Evaluation in
Farming Systems Research and Extension

Jonice Louden

OVERVIEW OF FARMING SYSTEMS RESEARCH AND EXTENSION

During the past few decades, a number of programs designed and financed by national and international agencies tried to improve the productivity of rural populations. Among the strategies adopted were the "Green Revolution," extension, research and development, credit, irrigation, and soil conservation. While these innovations have made technological advances, a number of limitations have also been detected. The plant breeding breakthroughs of the "Green Revolution" of the 1960s, that produced high-yielding grain varieties favored more progressive farmers. Most of this research concentrated largely on plantations and export crops and provided little technical assistance to the small farmer.

The strategy employed by extension services took results generated on research stations to the farmer. Implicit in this approach was the assumption that farmers have inadequate knowledge about agriculture and must depend on information from professional groups. Farmers often rejected advice based on what they perceived as "book learning" rather than practical experience about farming. Due to the limited success of the extension approach, it was imperative that new strategies be employed to improve agricultural production and to correct the food deficit situation now becoming acute in most developing countries.

The farming systems approach was introduced during the 1970s to work more effectively with the problem of increased agricultural production through improved technology. Farming Systems Research and Extension (FSR/E) aims at improving the effectiveness of national research and extension services in generating and disseminating technologies appropriate to farmers. A farming system may be broadly defined as the way in which a farm family manages the resources it controls to meet its objectives within a particular ecological, social and economic setting.

FSR/E is an applied approach with the following characteristics:

(1)   Potential technologies are evaluated from the farmers' point of view taking into consideration their environment, objectives and priorities, resource endowment, constraints, and present management strategies.

(2)   Researchers consider the farming system as a whole; when focusing on a single commodity or operation, they must consider how it relates to other system components.

(3)   Research is location or farmer specific; experiments are planned for roughly homogenous groups of farmers called recommendation domains.

(4)   The research team is interdisciplinary, including both social, biological, and physical scientists, and works together on a commonly defined agenda.

(5)   Experimentation is carried out on farmers' fields to develop technologies relevant to the farmers' environment and to facilitate farmers' participation and evaluation.

(6)   FSR/E crosses the division between research and extension.   Extension staff play an active role in technology generation and researchers participate in dissemination activities.

(7)   A FSR/E Program is consistent with national policy guidelines and the long term interests of society (Shaner et al. 1982; Franzel 1984).

The promise of farming systems research is that small farmers will enjoy a higher standard of living through increased agricultural production provided with appropriate technologies designed with sensitivity to their needs. Placed in its proper context, it is a second generation "Green Revolution". The purpose of this paper is to address the topic of how to incorporate gender issues into monitoring and evaluation systems within the FSR/E perspective.

WOMEN'S ISSUES AND THE DEVELOPMENT PROCESS

The issue of incorporating gender into farming systems research and extension arises from the growing concern during the last two decades about the participation of women in the development process. Women have played a crucial role in national food security, yet this role has largely been "invisible" as agricultural statistics do not adequately reflect their presence.

Conventional methods of data collection are based on an inadequate conceptualization of the role of women that underrepresents their contributions to agriculture. Conceptually, the identification of the farm as the unit of observation is problematic because it isolates crops and livestock decisions and activities from other productive and social activities. Operationally, it leads to gathering information from the "farmer," typically, the man with social authority over the household.

Recognizing the need to utilize all resources in the development process, a number of agencies have placed the role and status of women as a critical issue on their agendas. The incorporation of gender issues into the development agenda recognizes women's participation in agricultural development at various levels: as farmers and producers of crops and livestock and users of technology; as active agents in the marketing, processing, and storage of food; and as agricultural laborers.

Assessment of Women's Participation in Rural Development

Two approaches to policy-oriented research evolved to assess the participation and status of women. The first, the equity-oriented approach, developed in the early focus on women's issues. This approach argues that women lose ground economically as the development process proceeds. Because of the lack of an adequate data base, this hypothesis about the negative impact of development on women could not be empirically tested.

The poverty-oriented approach evolved out of the equity approach. It links womens' issues with poverty and tries to quantify the positive effects that may result from incorporating women's concerns into economic development programs. Women are perceived as participants in, rather than beneficiaries of, development programs. This approach restricts the women studied to those in economic need.

This shift in emphasis from an equity-oriented to a poverty-oriented approach substantially changes research questions and methods. There was a shift from description to analysis of women's conditions, from the definition of women's economic problems to the quantitative documentation of their existence, and from anthropological to sociological and economic methodologies.

This shift in approaches offers a challenge to evaluation research to be innovative in developing a methodological and theoretical framework for analyzing social phenomena (Buvinic 1983; Fortmann 1981). More empirical research requires the development of specific definitions

and concepts that will provide for more rational data collection and analysis based on deductive logic.

Monitoring and evaluating systems involves social scientists in evaluative research and adds to our knowledge
about existing social problems. The problem of poverty
among women is a major problem within the society. To
offer solutions to these social problems, it is necessary
to evaluate the extent of the problem, and also to develop
problem solving mechanisms. FSR/E aims at solving the
problem of poverty among agricultural producers, many of
whom are women, by improving the way in which they manage
their resources.

Incorporating women into monitoring and evaluation
systems brings into the foreground two relevant issues:
(1) How does selecting gender as a variable for research
affect scientific objectivity? and (2) How does one analyze
individual orientations within a holistic farming systems
perspective? Integrating women into monitoring and evaluation systems may require selecting out women for special
attention. This presents a degree of bias particularly
within a farming systems perspective.

Women in Agriculture – Some Evidence from Jamaica

In Jamaica, women have a long history of involvement in
work away from home, particularly in agriculture. During
the pre-emancipation period, women worked side by side with
men (Mathurin 1975). More recent statistics on women in
agriculture are drawn from special studies carried out in
specific localities. While the data base is not large
enough to make generalizations, the findings throw some
light on women's role in the farming systems and food
production in Jamaica.

A survey of higglers (market vendors) in 1977 estimated
that women traders handle approximately 80 percent of the
marketing of fruits, vegetables, and staples in the island.
According to this survey, about 30 percent of post-harvest
losses to women farmers are caused by conditions associated
with marketing (Smikle and Taylor 1977).

In the central region of Jamaica, the results of a survey showed that women farmers manage about 22 percent of
farm holdings. Even when women were not principle farm
operators, 47 percent of male spouses said that women
assisted in farming operations, while 21 percent reported
collaboration in at least planting and harvesting (USAID
1979). Women were also involved in farm management and
decision-making; 65 percent of male respondents said they

usually consulted their wives on changing cropping patterns.

Women play an important role in food production, as small producers and as agricultural laborers. The 1978 agricultural census revealed that 35,188 or 19 percent of the total 182,169 single holder farms island wide were operated by women, but the amount of land owned by women represented only 12 percent of the total. Women are less likely to own land in their own right, and those who have land are likely to have smaller holdings than men. Consequently, they are less likely to have access to credit and extension workers (Jamaica Department of Statistics 1979).

Although agricultural statistics and social and agroeconomic data on farmers in banana production are abundant, little of this information is disaggregated by sex. The data presented here were drawn from the limited studies available in an attempt to highlight the socioeconomic situation of women in banana production. An agrosocioeconomic survey in six parishes interviewed 6,269 farmers, 45 percent of whom were women (Jamaican Ministry of Agriculture 1982).

An exploratory study of a banana plantation revealed that women constituted 52 percent of employees. All laboring activities were carried out by women employed as farm hands. Their duties included planting, fertilizer application and weeding, as well as caring for benches, pruning, sleeving, dehanding, etc.. Women were also extensively involved in the boxing plant, making up 75 percent of the workers involved in this activity (Louden 1985).

TOWARD A MONITORING AND EVALUATION SYSTEM FOR INCORPORATING WOMEN INTO FSR/E

Monitoring and evaluation has been accepted as an essential feature of the process of planning, organizing, and maintaining rural development projects. The two components are necessarily interrelated and form the basis on which on-going policy decisions are taken to improve or modify the project to suit changing conditions or requirements. In agricultural development, monitoring focuses on the operation, performance, and impact of agricultural projects. A monitoring system is an information system for management decision-making (FAO 1983).

As an applied science, evaluation research can be defined as "the application of scientific principles, methods, and theories to identify, describe, conceptualize and measure, predict, change or control those factors or variables important to the development of effective human service delivery systems" (Struening and Guttentag 1975).

The following critical steps are involved in completing evaluation studies: conceptualizing the problem; reviewing relevant literature; developing a research strategy; determining a research design; selecting and maintaining a sample; choosing measures and assessing their sociometric properties; maintaining data collection standards; analyzing the data; and communicating results.

In order to be effective and useful, evaluation like other research must satisfy the following criteria:

(1) Validity - it should provide evidence about the problem the evaluator wants to solve.

(2) Reliability - the sample should be large enough and sufficiently representative to permit conclusions about the total population.

(3) Objectivity - the instruments and job devices should be free from personal prejudice or bias.

(4) Practicability - the instruments and procedures should be easy to use and readily understood by respondents and easily tabulated, summarized, and reported (Agarwal 1960).

Evaluation connotes measurement, appraisal, or assessment of progress made in any project. It is a process of analysis that turns the "search light" on the relative merits and deficiences of persons, groups, programs, situations, methods, and processes with a view to improving the operational deficiences of the undertaking and determining how far it has progressed and how much farther, and in what ways it should proceed (Agarwal 1960).

Academic and Practical Difficulties in Evaluation Research

Monitoring and evaluation systems do distinguish between men and women, but there is usually no in-depth analysis as to whether or not there is unequal access to resources or if women are the primary producers and users of technology. The fundamental problems seem to be the choice of models or the unit of analysis and the methods of data collection used.

Buvinic (1983) identifies three problems associated with using the household as a unit of analysis. The first derives from the assumption that physical household boundaries define the unit of social and economic organization.

However, individuals sharing a house do not always constitute families; this is particularly true among female-
headed households.

The second problem arises from the economic principle
that defines the household as the basic decision-making
unit behaving according to the rule of maximization of
household utility. This assumption does not allow for
recognition of the different preference schedules of
individual members and masks any sex or age discrimination
in the household allocation of production and consumption.

The third problem derives from the implicit assumption
based on the model of the industrialised world that only
non-market production and consumption taking place within
the household are in the woman's hands. Therefore, when
market consumption and production are investigated, farms
and firms rather than households are chosen as units of
analysis and household (women's) contribution to market
(farm) production is ignored.

Few studies recognize the interdependence of production
and consumption activities in most farm households. The
failure to see this interdependence can confound the analysis of the effect of development programs that are dependent on farm/household behavior and decision-making.

A more fundamental issue is that conventional methods
used in collecting information on rural producers tend to
underestimate their actual participation, particularly in
developing countries. The characteristics of small farmers
and their farming systems present serious problems in data
collection. The major difficulty arises from the fact that
small farmers and research workers tend to differ in their
definitions and thought processes. For example, small farmers tend to make only limited use of established official
measures of weight, volume, distance, area and time. Conceptual problems are often encountered in relation to
rights of ownership and control of land, crops, and livestock (Edwards and Morgan-Rees 1964; Alleyne and Benn 1979;
FAO 1982).

SOCIOECONOMIC INDICATORS FOR MONITORING
AND EVALUATION SYSTEMS

With the exception of total fertility rate and mean age
at marriage, indicators with particular significance for
women, there are no other census indicators that can be
considered relevant only for women. For the most part, it
is possible to monitor and evaluate the impact of rural
development on the role of women only in terms of the corresponding role of men. In most cases the same indicators
apply to men and all that is needed is disaggregation by

sex when the data are collected for individuals or disaggregation by sex and type of head of household when data
are collected for households. A household is defined as
"one or more persons voluntarily living together and
sharing at least one meal in general, father, mother,
children and other relatives, as well as other persons
sharing their household arrangements (UNESCO 1983)."

A monitoring and evaluation system with a farming
systems perspective should examine the following:

(1) How are the resources of the households (land,
    capital, labor) utilized?
(2) What activities does each member of the household
    perform in crop production (land preparation,
    planting, weeding, fertilizing, harvesting,
    marketing, and processing)?
(3) Who makes decisions about what to produce,
    selection of planting material, employment of farm
    labor, and the use of technology?
(4) The use of technology and credit: How does the lo-
    cation of farm affect the use of technology? What
    are the constraints to the adoption of available
    technology? How does the size of farm and inade-
    quate collateral affect access to credit? Due to
    their low income, are women farmers unable to use
    available technology (fertilizer, pesticides,
    insectides, farm machinery, and implements)?
(5) Livestock: Who cares for livestock? How does this
    interact with the cropping system? What are the
    constraints to introducing livestock into the
    cropping system?
(6) Farm Mangement: Who has knowledge about crops/
    livestock, crop care, and record keeping? How
    does household member's participation in community
    organizations influence their activities?
(7) Sociological characteristics of the farmer: age,
    sex, educational attainment.
(8) Structural variables: social organization of house-
    hold, land tenure and fragmentation of holdings,
    employment of labor, and wages paid for agricul-
    tural labor.

CONCLUSION

While micro-level data on women's situations might be
useful, the feasibility of this approach in developing
countries where men and women experience similar economic
conditions must be considered carefully. We argue, therefore, that we need instead to develop monitoring and

evaluation systems that record gender specific information on men's and women's activities; that identify benefits, costs, and constraints by gender; and that determine men's and women's relative access to resources (land, capital, labor, and technology). There is a need to generate country by country statistics, as evidence from one country in Latin America cannot be used to make generalizations about another, especially countries in the Caribbean. The conceptualization and measurement of key indicators with policy implications, such as the activity rates of men and women in agriculture, must be appropriate and comprehensive to produce valid and complete enumeration of all of men's and women's activities. Our research instruments should be designed to generate valid data and correct analyses, placing production in its broad micro- and macro-socio-economic context.

While women scientists tend to recommend "women-oriented research", the long run aim of any program is to promote sustained agricultural development with benefits to all regardless of gender or social status. What is requir-ed is that evaluators develop indicators that measure accurately the status of men and women and to make recom-mendations that will benefit both as participants in national development. These mechanisms should not be used to further sexual dualism in the market place but to reduce inequalities and create new opportunities (Powell 1984).

NOTES AND ACKNOWLEDGEMENTS

The author thanks John Campbell, Paulette Lewis, Faith Innerarity, and Dorian Powell for their constructive com-ments on an earlier draft of this paper.

REFERENCES

Agarwal, B.L.
    1960   Techniques for Evaluating Rural Development
        Programmes. The Indian Journal of Agricultural
        Economics 15:1:103-110.
Alleyne, S. and S. Benn
    1979   Data Collection and Presentation in Social
        Surveys. University of the West Indies: Institute
        of Social and Economic Research.

Buvinic, M.
   1983   Women's Issues in Third World Poverty. In Women
           and Poverty in the Third World. M. Buvinic, M. A.
           Lycett, and W. P. McGreevy, eds., New York, NY:
           Johns Hopkins Press.
Edwards, D.T. and Morgan-Reis, A.M.
   1964   The Agricultural Economist and Peasant Farming
           in Tropical Conditions. In International
           Explorations of Agricultural Economics. Roger Dixey,
           ed., Pp. 73-85. Ames, IO: Iowa State University
           Press.
Food and Agricultural Organization
   1982   Monitoring Systems for Agriculture and Rural
           Development Projects. Rome: FAO.
Fortmann, L.
   1981   The Plight of Invisible Farmer: The Effect of
           National Agricultural Policy on Women in Africa. In
           Women and Technological Change in Developing
           Countries. Roslyn Dauber and Melinda Cain, eds.,
           Pp. 205-213. Boulder, CO: Westview Press.
Franzel, S.
   1984   Comparing Results of an Informal Survey: A Case
           Study of Farming Systems Research and Extension
           (FSR/E) in Middle Kirinyaga, Kenya.
Jamaica Department of Statistics
   1979   Agricultural Census. Kingston, Jamaica:
           Government Printing Office.
Jamaica Ministry of Agriculture
   1982   An Agro-socioeconomic Survey of Banana Farmers.
           Data Bank and Evaluation Division. Kingston:
           Government Printing Office.
Louden, J.
   1985   Women in Production and Utilization of Plantain
           and Bananas. Paper presented at Third Conference of
           International Association for Research on Plantain
           and Bananas: Abidjan, Ivory Coast. May 28-June 1.
Mathurin, L.
   1975   The Rebel Woman in the British West Indies
           During Slavery. Kingston, Jamaica: African
           Caribbean Publications.
Powell, D.
   1984   Women in Agriculture in the Caribbean. Paper
           presented at conference on Women's Action for
           Progress-Caribbean, Central America. Miami, FL. May
           20-23.
Shaner, W.W. and P.F. Phillip
   1982   Scheme Farming Systems Research and Development:
           Guidelines for Developing Countries. Boulder CO:
           Westview Press.

Smikle C. and Taylor, H.
   1977   Data Bank and Evaluation Division. Kingston,
       Jamaica: Ministry of Agriculture.
Strueing E. L. and M. Guttentag.
   1975   Handbook of Evaluation Research. Vol. 1.
       Beverly Hills, CA: Sage Publications.
USAID
   1979   Planning a Women's Component – Integrated Rural
       Development Project Two Meetings and Pindars
       Watersheds. Jamaica: Office of Women in
       Development.

8

A Comparison of Rural Women's
Time Use and Nutritional Consequences
in Two Villages in Malawi

Lila E. Engberg, Jean H. Sabry, and Susan A. Beckerson

The primary purposes of this paper are to present a production activity model for categorizing and comparing time use in rural households and to illustrate the use of the model. A secondary purpose is to discuss the relationship of women's time use to the food supply and nutritional status of household members in two Malawi villages. The study of time use is not an end in itself, but is aimed at increasing our understanding of the relationship of time use to human well-being. According to Ettema and Msukwa (1985:49), "Little is yet known about the dynamics of cropping patterns of the smallest farmers, or for that matter, of related issues like the relationship between cash cropping, the cultivation of food crops destined for home consumption, and the nutritional status of the family."

First the paper explains the theoretical perspective that led to the use of the activity model, then it describes the research methodology and methodological problems related to collecting time use data and the results. The picture is incomplete for truly accurate time-use records as only two seasons were examined (pre-harvest and post-harvest) and only the work of a husband and wife in each of twenty-eight households were measured. The sample size was small because of the detailed measures taken: measures of time-use, food stored, food consumed, and nutritional status. Measurement of each category of variables is complex. Details will not be described, but some of the results will be presented with the hope that further interdisciplinary research of this nature will be encouraged.

THEORETICAL PERSPECTIVE

Study of the formal, monetized economic sectors rather than the informal economies of the household or family is conventional in the field of economics. Cash crop agriculture is visible and recognized but two sectors of economic activity (subsistence food production and household production) tend to be invisible and overlooked (Evenson 1981). Chipeta (1981:6-7) uses the term "indigenous economies" when referring to subsistence production. He argues that conventional economics does not help in understanding how the micro-economic unit of the family or household operates. A broader production model is needed.

Market production and home production are two main categories of production identified in the production model adapted from Beutler and Owen (1980:17). The production activity model attempts to integrate social and economic theory by taking into account the social relationships involved in production. (1) Market production involves exchange of money, goods, or services. Home production does not. The outputs of home production have "use" value rather than "exchange" value. Home production is divided into separable and inseparable components. Household and subsistence production are separable or market replaceable. Reid's (1934:11) classic definition of household production has been considered most appropriate:

> Household production consists of those unpaid activities which are carried on by and for the members, which activities might be replaced by market goods, or paid services, if circumstances such as income, market conditions, and personal inclinations permit the service being delegated to someone outside the household group.

Subsistence production has the same characteristics. The activities can be delegated to paid workers, the outputs have exchange value on the market and are conceptually different from outputs that have "use" value alone.

In addition to market production, household, and subsistence production, household members engage in another kind of production referred to in the model as social production. The latter type of production is called "inseparable" because of the social relationships involved (Beutler and Owen 1980:18). Activities identified as social production are carried out only by a member of the family and cannot be handed over to paid workers. Activities are organized by reciprocity and are called grants (Boulding

1970). Grants are one-way transfers of goods and services. Bivens (1976) gives examples and indicates their "use" value. The major function of grants is to integrate family members into the family system. Beutler and Owen (1981) subdivide the inseparable into three categories: intra-household grants, inter-household grants, and community service. Activities such as caring for, socializing and educating children, caring for the sick and elderly, and serving as mediators inside the household, would fit the first category. Helping kin or neighbors in the village or helping at weddings or funerals would fit the second category. Community self-help, voluntary work, and social participation as representatives of the household would fit the third. All are family obligations and essential to indigenous economies.

A production activity model such as the one just described could characterize the various types of production activities carried out at the level of the family. It could capture more fully the activities and responsibilities of women and asses in us recognizing and better understanding the allocation of their labor.

MEASURING HOME PRODUCTION

Measuring the various types of home production is very complex. The invisible nature of the activities, the seasonality of some activities, the definition of household, the number of people involved, and the problems of assigning a monetary value to non-market activities are among the problems. Reports by the International Center for Research on Women (ICRW 1980) and Minge-Klevana (1980) provide reviews of methodology related to measurement procedures. The study of time allocation is the method generally used for gaining information about all aspects of life. The time-diary method is common but is not useable in a semi-literate population. Only two approaches, the interview recorded recall and observation, are available. Day-long observations were chosen for the Malawi research in order to gain a more complete picture of the actual amount of time spent on each particular activity (Beckerson 1983).

Methodology

The study in Malawi was divided into two phases. The first phase, carried out in August 1981, was a general survey of rural household functioning in two nearby villages, Mkwinda and Patsankhondo. For this study, a household was defined as a unit that may include both family members and

persons other than kin, who occupy a housing unit as a
social unit in terms of division of labor, social interaction, and sharing of benefits (Janelid 1980:85). In the
case of Mkwinda village, thirty-two registered tobacco
growing households were included, and in Patsankhondo,
thirty-one subsistence farming households. Additional criteria for household selection were that the age of the
eldest child born to the resident wife be between twelve
and sixteen years of age and that the child be presently
living in the selected household. This was to obtain
comparability of households in terms of the age of the
female respondent and the developmental stage in the family
life cycle. As it turned out, all thirty-two tobacco producing households and twenty-two subsistence producing
households were headed by males at the time of the survey
even though 29 percent of Malawi's rural households are
female-headed (Beckerson 1983: 78).

Information collected in Phase 1 of the study was used
to develop the approaches to the second phase. Phase 2 of
the study was carried out in February and July, 1982,
during the height of the growing season and after the harvest. For this more detailed part of the study a smaller
sample of fourteen households was randomly selected from
each of the two original sets of households for a total of
twenty-eight households (Beckerson 1983:87).

Interviews were carried out in Chichewa by five Bunda
College of Agriculture students. Four of the students (two
men and two women) continued the survey work throughout all
phases of the study, working as male/female teams with the
researcher (Beckerson 1983:86).

Within each season, identical questionnaires were used
to interview the men and women about specific pre-cate-
gorized activities. The respondents were asked to recall
three weekdays of activities within each category and to
estimate an approximate block of time spent on the activity. The interviewer translated the time into hours and
minutes on the record sheets. Table 8.1 contains the list
of activity categories and the tasks for February. A different set of crop production activities was designed for
use in July. Since each list of activities was pre-
determined, based on information from Phase 1 of the study,
it was easier for the interviewers and respondents to
remember and record the tasks (Beckerson 1983:93).

Recall is not necessarily the best data collection procedure because of the difficulty of estimating the amount
of time spent on each activity. The information was supplemented by observation of seven women in each of the two
villages (Beckerson 1983:94). A random morning and random
afternoon were selected to observe each woman in each

TABLE 8.1

PRODUCTION ACTIVITIES

---

1. Market Production
    a. Tobacco production (weeding, fertilizer
    application, banking, removing flowers, sucker
    control, harvesting, transporting leaves, other
    tasks).
    b. Other Income-Earning Tasks (making goods, collecting
    produce for sale, work for wages).
    c. Marketing (selling or trading).
2. Subsistence and Semi-Subsistence Production
    a. Hybrid Maize Production (weeding, banking,
    fertilizer application, other tasks).
    b. Production of Local Maize (tasks as above).
    c. Production of Groundnuts, Beans, Peas, Vegetables.
    d. Sweet Potato Production (as above)
    e. Animal Care (grazing cattle or goats; feeding
    fowl, pigs, rabbits; preparing feed; milking,
    collecting eggs, slaughtering animals, cleaning animal
    shelters, other).
3. Household Production
    a. Food Related Activities (gathering fruits or wild
    vegetables, collecting insects, planting and caring
    for fruit trees, collecting vegetables from the
    garden, beermaking for home use, storage of food
    items, shelling and pounding maize, meal preparation,
    cooking and serving, washing-up and storing utensils,
    going to the maize mill).
    b. Shopping (for food and other household needs).
    c. Obtaining Fuel and Water (collecting firewood and
    the domestic water supply, planting trees and care of
    wood lot).
    d. Household and Farm Maintenance (cleaning the house
    and surroundings; building or repairing housing, food
    stores, animal shelters, fences, garden structures,
    garden tools, furnishings or equipment).
4. Inseparable Home Production
    a. Intra-household production
    Child care (feeding, bathing, punishing, teaching and
    playing with children).
    b. Health Care (tending to the sick at home; taking
    the sick to the health center, clinic, hospital, or
    healer, fetching medicines).

Note: Inter-household grants and community service not
recorded by Beckerson (pp. 222-223).

season. Both primary and secondary activities were
recorded at five minute intervals.

There were no significant differences in mean hours per
day obtained by the recall method as compared to the observation method except for the report of domestic activities
of tobacco growing in households in July, and of subsistence households in February. (2)

Food Supply, Food Consumption, and Nutritional Status

The major emphasis of this paper is the presentation of
the theoretical model for examining production activities,
therefore, no details will be provided regarding the
methodology for the study of food supply, food consumption,
and nutritional status of household members. It is important to note that the output of subsistence and household
production is food and that it has "use" value. There may
be no need to calculate or emphasize the monetary value of
labor time if it can be demonstrated that various labor
inputs result in decreased or increased outputs of food and
differences in nutritional well-being. The ultimate "use"
value is the nutritional status of family members.

RESULTS

Time Spent by Husbands and Wives

The mean-time spent by husbands and wives in the two
seasons in each farming system is presented in Table 8.2.
Home production is separated into three categories:
subsistence, household, and inseparable or social
production. The following comparisons are based on the
production activity model:

(1) Both husbands and wives were involved in market pro-
    duction as well as in home production but the busiest
    work period for the men was tobacco production in
    February.
(2) Household production took the largest proportion of
    women's time in each of the two villages in each
    season.
(3) The mean-time spent daily by women on all production
    activities added together did not vary much from season
    to season (about twelve hours per day). Meal prepara-
    tion tasks took the longest period of time (1.4 to 2.9
    hours per day).
(4) In comparison to the women, the men spent a total of
    four to six hours in production per day.

TABLE 8.2

MEAN TIME SPENT IN HOURS BY HUSBANDS AND WIVES IN TWO SEASONS IN EACH FARMING SYSTEM

Seasons	Farming System	Sex of Respondent	Market Production (Separable Hours)	Home Production			Total Hours Per Day
				Subsistence (Separable Hours)	Household	Social (Inseparable Hours)	
February	Tobacco	M	4.1	0.4	0.6	1.3	6.4
		F	3.8	2.5	4.1	1.8	12.2
	Semi-subsistence	M	0.3	2.4	0.6	1.3	4.6
		F	1.6	4.3	5.1	1.3	12.3
July	Tobacco	M	0.2	2.4	0.9	0.6	4.1
		F	0.2	5.5	6.1	1.2	13.0
	Semi-subsistence	M	0.1	5.5	0.2	0.4	6.2
		F	0.4	4.9	5.4	1.3	12.0

(5) The women shifted their work time away from home production to market production during the tobacco growing season in February.
(6) The men and women in the semi-subsistence village shared their subsistence production tasks more equally than was the case in the tobacco growing village.
(7) Only two activities were accounted for within the category "inseparable" or social production in this study: childcare and health care. Time spent helping kin or neighbors, in attendance at funerals, or in community self-help were not recorded, but these are important family obligations. Funerals, for example, took respondents away from home and all other work for one or two days at a time. Funerals were more frequent in February than in July and were a constraint to the demands of market work during the growing season.

It is clear that the separation of market production and home production and the categorization of men's and women's time use provided a more comprehensive picture of a household's economic endeavors. Findings indicate that the women in the sample had a heavier work load than their husbands. This could be a major constraint when attempting to increase local food production.

Outputs of Production

The amounts of traditional food crops stored by the sample households are shown in Table 8.3. The amounts of maize, groundnuts, beans, and peas were greater in the fourteen semi-subsistence households than among the tobacco producers. It might be assumed that tobacco producers would use cash to purchase food, but Table 8.4 shows that they used less maize flour for their staple food nsima and ate somewhat fewer meals per day.

Nutritional Status

Finally, the members of sample households from Patsankhondo, the semi-subsistence village, were somewhat better off in nutritional status than were those from Mkwinda (Beckerson 1983:184). Findings favoring Patsankhondo included:

(1) Significantly fewer children were malnourished in July.
(2) The female adults had significantly higher mean weights in February and in July.
(3) Male and female adults had significantly less weight change between February and July.

(4) Female adults were less lean in February, as indicated by a significantly higher mean body mass index.

(5) Significantly fewer female adults were underweight in February and in July.

(6) A larger proportion of the children in the semi-subsistence households maintained normal weight for height between February and July as compared to the children in tobacco producing households (Table 8.4).

Beckerson cautions readers about drawing conclusions regarding the reasons for the disadvantaged food position of the tobacco growing households. A number of possible variables, such as size of land holding, use of hired labor, storage problems, family structure related to polygnous marriage, access to technology, and amount of household capital, were not explored (Beckerson 1983:186). Nevertheless, the comparisons have shown that women are indeed contributing a large proportion of their time to household and subsistence production. During the tobacco growing season when labor demands are heavy, women tended to shift their time away from home production towards cash crop or market production. This shift in rural women's time use could contribute to food deficits in their house-holds and in the local community.

TABLE 8.3

MEAN AMOUNT OF TRADITIONAL FOODS STORED BY TYPE OF FARMING HOUSEHOLD IN TWO SEASONS

Season	Household Type	Maize (Cubic Meter)	Groundnuts (Cubic Meter)	Beans/ Peas (kg.)	Dried Leaves (cu.cm.)
February					
	Tobacco	0.71*	0.0	0.0	0.0
	Semi-subsistence	0.87**	0.12	0.80	0.0
July					
	Tobacco	10.62	1.73	16.65	356.4
	Semi-subsistence	14.64	2.41	22.62	312.2

Note: *Seven out of the fourteen households had no maize remaining in storage when data were collected and had emptied stores between five and nine weeks earlier.

** Four out of the fourteen households had no maize, and had emptied stores between two and four weeks earlier.

TABLE 8.4

MEALS AND MAIZE FLOUR CONSUMED AND PERCENT OF
CHILDREN WITH NORMAL WEIGHT FOR HEIGHT BY TYPE
OF FARMING HOUSEHOLD AND SEASON

	Tobacco-Growing		Semi-Subsistence	
	February	July	February	July
Number of Meals (Per Day)	2.1	2.4	2.4	2.7
Number of Cups of Flour (Per Day)	12.1	16.3	18.4	21.2
Percentage of Children With Normal Weight for Height (N = 73)	15.1	6.9	28.8	28.8

Source: Beckerson p. 136, 162.

CONCLUSIONS

The major purpose of this paper was to draw attention
to the fact that household production and subsistence production are important components of the family economy.
Women contribute a large proportion of their time to these
two sectors of production, but their work tends to be
unrecognized because it is a part of the informal or non-
monetized economy. A comparison in two Malawi villages of
work done by a sample of twenty-eight households suggests
that women are contributing a larger proportion of their
time to these sectors as compared to their husbands. Along
with their husbands, women are also contributing to cash
crop production through the work done on tobacco. During
the tobacco growing season in February the women tended to
shift their time away from subsistence and household production towards tobacco production. It is suggested that a
shift in rural women's time use towards cash crop production could lead to lower rather than higher standards of
nutritional well-being of all household members.

The use of a production activity framework for categorizing types of production has advantages for economies
such as those in rural Malawi. Market production and home
production are the two major types of production to be
recognized and valued, especially with respect to women's
role. Home production cannot be easily valued in monetary
terms, but could be valued in terms of outputs such as

109

health and well-being or standards of living of households.
Social production such as care of family, kin, and community is called inseparable home production because of
household members' obligations to such activities. It is
not market replaceable and requires more attention in
research because of the trade-offs in terms of time
utility.

There are only twenty-four hours in a day. More time
spent in one type of production means less time available
for another type. If rural women shift their time towards
cash crop production and reduce time spent in home production there may be drastic consequences in terms of food
shortages during the transition. There is a need for
research which examines more thoroughly the relationships
between various types of production, activity patterns of
men, women, and children in small-scale farming systems,
and the production outputs.

NOTES AND ACKNOWLEDGEMENTS

(1) This is not the entire production activity model
(Beutler and Owen 1980). Inputs in terms of human and
material resources and outputs of utility and levels of
living are included in the model.

(2) To check the validity of the three-day recall data,
Beckerson used two-tailed non-paired student's t-tests at
the one and five percent levels of significance. The mean
hours for one day as observed in each of three activity
areas (cash crop activities, food crop activities, and
household activities) for the women was compared to mean
hours per day of the recall data (Beckerson 1983:104).
Results indicated that the women respondents may have
over-estimated the time spent on household activities.

REFERENCES

Beckerson, S. A.
    1983   Seasonal Labour Allocation, Food Supply, and
        Semi-Subsistence Farming Households in Malawi,
        Africa. Unpublished Master's Thesis. Guelph,
        Ontario: University of Guelph.
Beutler, I. F. and A. J. Owen
    1980   A Home Production Activity Model. Home
        Economics Research Journal 9:1:16-26.

110

Bivens, G.
     1976   The Grants Economy and Study of the American
             Family: A Possible Framework for Trans-Disciplinary
             Approaches. Home Economics Research Journal 5:6:
             70-78.
Boulding, K. E.
     1970   Beyond Economics. Ann Arbor, MI: University of
             Michigan Press.
Chipeta, C. H. R.
     1981   Indigenous Economics -- A Cultural Approach.
             Smithtown, NY: Exposition Press.
Ettema, W. and L. Msukwa
     1985   Food Production and Malnutrition in Malawi.
             Zomba, Malawi: University of Malawi, Centre for
             Social Research.
Evenson, R. E.
     1981   Food Policy and the New Home Economics. Food
             Policy 6:3:180-193.
ICRW
     1980   The Productivity of Women in Developing
             Countries: Measurement Issues and Recommendations.
             Washington, D. C.: Office of Women in Development,
             USAID. Prepared by the International Center for
             Research on Women.
Janelid, I.
     1980   Rural Development and the Farm Household as a
             Unit of Observation and Action. In The Household,
             Women, and Agricultural Development. C. Presvelou
             and S. Spijkevs-Zwart, eds. Wageningen, The
             Netherlands: H. Veenman and B.V. Zonen.
Minge-Klevana, Wanda
     1980   Does Labour Time Decrease with Industriali-
             zation? A Survey of Time Allocation Studies.
             Current Anthropology 21:3:279-297.
Reid, M. G.
     1934   Economics of Household Production. New York,
             NY: Wiley Publishers.

9
Correcting the Underestimated
Frequency of the Head-of-Household
Experience for Women Farmers

Art Hansen

Women's contributions to agricultural production are
being increasingly recognized and documented (Boserup 1970;
Spring 1986). The extent to which women farmers make independent decisions about farming systems is still being dis-
puted, especially when the women are wives and their husbands are in residence. When women are heads of their own
households, it is generally accepted that women make independent decisions, and the frequency of women-headed households is a widely accepted indicator of the number of women
acting as decision-makers. This paper focuses on how the
frequency of women-headed households is determined and suggests that current methods underestimate the extent and
significance of the head-of-household experience for women
farmers. Corrected frequencies have important implications
for extension and credit policies and practices. The conceptual and methodological changes suggested in this paper
also illuminate some discrepancies between how farming
systems research and extension is described in theory and
applied in practice.

Before proceeding, two disclaimers are necessary.
First, focusing on the head-of-household experience does
not deny that women make many independent decisions when
married and living with their husbands, but the status of
head of household is an important and accepted key indicator. Information about headship status is usually collected during surveys, and the household head is often the
only or preferred individual to be interviewed. Second,
using the household as a unit of analysis does not imply
that the household is the best or only social and production unit that farming systems researchers should be
studying (Berry 1986:74-76; Cernea 1985; Moock 1986; Spring
1986:333-334). In reality people allocate land, labor, and
capital— the key factors of agricultural production —

within and among households and many decisions are made
because of intra-household and supra-household relationships (Behnke and Kerven 1983; Guyer 1981; Kerven 1979;
McMillan 1986). This paper has a restricted focus on the
household and the head of household because these are
presently the accepted standard units of data collection
and analysis (Low 1986; Moock 1986; Netting et al. 1984b;
Shaner et al. 1982; Wilk and Netting 1984; Yanagisako
1979).

The accepted way to estimate the frequency of women-
headed households underestimates the occurrence and
importance of this experience. Current estimates are
almost always based on surveys that note status at one
time; this is static information (McMillan 1984). Even
when historical changes or trends are desired, the same
static surveys are usually used to create a time series.
Frequencies or relationships are compared from different
surveys conducted at different times. This method may
easily misrepresent social dynamics since it only estimates
the frequency of women household heads at any given time,
but the status of household head need not be permanent.

Becoming a household head may be a phase that many
married women experience during their lifetimes as they
divorce or are widowed, or as their husbands emigrate for
shorter or longer periods of time (Carter 1984). Surveys
might show that one-fifth of all households are headed by
women every year. The easily drawn conclusions are that
only one-fifth of the women are household heads and experience independence and disadvantages, while four-fifths of
women are wives and do not experience being heads of household. The missing data, however, refer to the turnover of
status, that is, the extent to which women change from
being members of male-headed households to being heads of
their own households or vice versa. If the turnover rate
is rapid and continual, then many more than one-fifth of
the women will be heads of households over a period of
years.

The number of households headed by women at any given
time may not be the important statistic. What may be much
more important from a standpoint of policy, research, and
extension may be the percentage of adult women who are
heads of households over a ten year period or the likelihood that an adult women will be a household head during
her lifetime. These statistics recognize that more women
will be heads of households over their lifetimes than the
number existing at any given time. If women-headed
households share any characteristics affecting their
farming practices and standards of living, then recognizing
that more women cycle through this phase increases the

importance of working with women who are not now heads of households but might occupy that decision-making status in the future.

Changing the method of measuring the frequency of women-headed households has implications for the importance of ensuring that wives as well as husbands receive training and experience with agricultural innovations, and that wives as well as husbands establish credit records and contacts with change agents and institutions. Wives with more training and experience and better credit records and contacts may be able to maintain higher levels of productivity and living standards when they cycle through the disadvantaged head-of-household phase.

DISADVANTAGES OF WOMEN-HEADED HOUSEHOLDS

The farming systems research and social science literature increasingly documents the disadvantages experienced by women-headed households (Spring 1986; Spring and Hansen 1985). At minimum, household labor power is reduced since women-headed households rely in many cases on a single adult, the woman, while households headed by men usually have available the labor of at least two adults, the husband and wife. Labor shortages and capital shortages are intimately linked because labor is a commodity in all agricultural societies today. People are able to earn money through their labor, and money may be used to hire labor for agriculture. Capital is not as short as labor for all women-headed households. Often the availability of capital is associated with an absent husband who still sends remittances to his wife. Women-headed households with remittances from absent husbands may be more similar in capital availability and hiring labor to men-headed households than they are to women-headed households without remitting husbands.

Another disadvantage common to many women-headed households is reduction in the amount of land cultivated. This may be due to shortages of labor and capital rather than to land itself, that is, the land may still be controlled by the woman but she does not control the resources to cultivate it. The amount of land itself may actually be reduced for several reasons: (1) access to land may be through a larger supra-household kinship group that reallocates land away from a household that is reduced in labor and consumption units; (2) a household containing both a wife and husband may be able to acquire land through both families, whereas upon divorce or widowhood the woman may have access to land only through her own family; and (3) in many societies, the husband and wife cultivate land controlled only by the husband or his family, and upon divorce or widowhood the woman must return to her family which then has to reallocate its land to allow her access to land to cultivate. These reasons all apply in societies in which the married couple does not have joint freehold title to farmland.

Another disadvantage is the reduced access that many women household heads experience in their relationships, or lack of them, with governmental and other institutions. Women are often discriminated against by extension agents, predominantly men, and by credit programs. This discrimination is sometimes overtly sexual when male agents do not want to interact professionally with women farmers. Discrimination may be also a consequence of the other characteristics of women-headed households already mentioned: their shortages of labor, land, and capital.

Governmental policies could reduce or eliminate the disadvantages experienced by women-headed households. Making credit available to hire labor could transform capital and labor shortages. Issuing an explicit directive to male extension agents to work with women farmers could improve women's access to training, credit, and extension advice (Spring 1985).

Why should governments and their extension and credit agencies change their policies toward women? Although morality and equity have a place in this argument, economics and productivity are more effective reasons. The argument has to be made that women need extension and credit resources to make, or continue to make, important contributions to national agricultural production.

A ZAMBIAN CASE

Surveys usually note household composition and marital status only for the time of the survey. This synchronic or ahistoric data may easily mislead observers into under-estimating social dynamics. First practical awareness of the discrepancy between aggregate statistics and social dynamics occurred in the early 1970s during a study of socioeconomic change in smallholder communities in north-western Zambia (Hansen 1977, 1981). In order to learn rates of change in demographic factors, the same terri-torial population was surveyed twice. A 100 percent sample of 117 matrilineally based villages (called matri-villages here) was surveyed in 1971 and again, 12 months later, in 1972. These matri-villages comprised approximately half of one rural settlement stretching for more than a mile along a road.

Both surveys showed the same total population of 1,223
people, adults and children, in the same 117 matri-villages
(see Table 9.1). Apparently, this was a stable population
to judge by the aggregate statistics. The appearance was
illusory. Since the surveys were part of a larger ethnographic study, personal names and relationships of all the
people were collected and the named individuals were all
traced over the year to learn about social dynamics:
marriages, divorces, births, deaths, migrations, etc.
Although a total of 1,223 was counted both times, only
1,006 people were present for both surveys: 217 people had
emigrated in the intervening 12 months and an equal number
immigrated. Thus, only 82 percent of the 1971 population
remained in 1972: 1,440 people in all were residents
during one or both surveys — 434 were there for only one
survey. If individual names had not been collected and
compared, aggregate statistics would have camouflaged the
movement of people and projected a misleading image of
stability.

There were biological reasons for some instability: 70
of the 434 transitory people represented births into and
deaths out of the population. Social and economic reasons
were more important. Among these was the changing marital
status of women: 22 percent of the movement (97 cases) was
due to women entering and leaving the locality because of
marriage, divorce, or widowhood. This underestimated the
extent of marital change because the 97 did not include
women who remained within the 117 matrilineally-based villages while changing their marital status. Unfortunately,
the data did not note whether or not the 97 individual
women who changed marital status also changed the status of
their own households from men-headed to women-headed,
although this was undoubtedly the most common outcome.
This cannot be verified, however, because status of
household head in 1972 was not collected for those who left
the locality in the interim.

To place the 97 women in perspective, there were only
466 women of marriageable age in the locality in 1972
(marriageable age being defined as 15 years of age and
older). The high turnover of marital status is common in
this matrilineal area of southern Africa and does not infer
that high rates of marital instability are common around
the world. This case shows the extent to which aggregate
statistics and ahistoric data may mislead observers into
underestimating the frequency with which women in this area
experienced changes in marital status, an important causal
factor for changes in head-of-household status.

TABLE 9.1

MARITAL CHANGES AS EXPRESSED IN MIGRATION OF
WOMEN IN AND OUT OF RURAL AREAS OF
NORTHWESTERN ZAMBIA, 1971 TO 1972

	1971	1972
Population of 117 Matri-villages (Adults and Children)	1223	1223
People Present at Both Surveys	1006	
Emigrants	-217	
Immigrants	+217	
TOTAL PEOPLE SURVEYED	1440	
Reasons for Moving		
Women Marrying and Joining Husband	42	
Women Divorcing or Being Widowed	+55	
MARITAL SUBTOTAL	97	
Births and Deaths	70	
Children Moving With Relatives	131	
Other Reasons	+136	
	434	

Note: 1972 Population (N = 1223) contained 466 women of
marriageable age (15 years or older).

A MALAWIAN CASE

Data from a neighboring country (Malawi), collected as
part of a farming systems research program, also show high
social mobility for women. This research confirmed the
discrepancy between actual social dynamics and the estimates of dynamics that are based on aggregate statistics
from time series surveys. Before presenting the Malawi
data, some background information is necessary about survey
procedures in that country.
  Much of rural Malawi is incorporated into integrated
rural development projects that are routinely surveyed
annually for evaluation purposes. Each project has its own
evaluation unit that conducts these annual surveys of small

holder agriculture during the cropping season, essentially from October through May. Sampling is both stratified and random. Each project is divided into extension planning areas (EPAs), that were subdivided into enumeration areas (EAs) for the purpose of the 1977 national population census. Each year each project randomly selects a number of its EPAs and from these EPAs selects a number of EAs. Within each of the selected EAs an enumerator collects a list of all households, and 20-25 households (plus some alternates) are randomly selected for that year's survey. This procedure results in randomly selected clusters of households. Each year the EPAs, EAs, and households are newly randomly selected. This excessive reliance or trust in random sampling absorbs all survey resources, so no samples are followed longitudinally to examine actual social and agricultural dynamics.

The National Sample Survey of Agriculture (NSSA) of 1980-1981 changed this procedure. The NSSA is conducted at approximately ten year intervals over the entire country-side wherever there are smallholders and includes non-project as well as project areas. In the projects, the NSSA absorbs the evaluation survey staffs, while special staff are hired for the nonproject areas. The 1980-1981 NSSA was elaborate and required project evaluation staff to continue collecting NSSA data through December 1981. This meant that the evaluation staffs had no time in which to select new random samples for the 1981-1982 cropping season surveys of project areas. Thus the same EPAs and EAs were used again as well as half of the 20 NSSA households in each EA. The other ten households in each EA were newly randomly selected from the NSSA lists of households. This gives analysts the opportunity to examine for two consecutive years some large smallholder populations.

One large project covers much of the Lilongwe plain and extends south and west of the national capital. This Lilongwe Rural Development Project (LRDP) has been the object of several studies (Kydd 1982; Lele 1975; Spring n.d.). The NSSA surveyed 540 households in LRDP, or 27 clusters of 20. In 1981-1982 the LRDP evaluation staff resurveyed half of that sample, or ten households in each of 27 clusters, plus a newly selected ten households in each cluster. Thus, 270 households were surveyed for two years in a row in LRDP.

Two research units joined forces to conduct a follow-up study of these LRDP households. One unit was the national farming systems research unit, directed by the author -- the Adaptive Research Program of the Department of Agricultural Research, Ministry of Agriculture, and the other unit was the Women in Agricultural Development

Project directed by Dr. Anita Spring. These units wanted
to take advantage of this unprecedented opportunity to
utilize two years of survey data to demonstrate to other
Malawian researchers the advantages of longitudinal
analysis and the savings in governmental research resources
when agencies shared survey data and used each other's
samples (Hansen and Ndengu 1983; Spring n.d.).

By comparing the names of individuals and the coded
household numbers, 267 of the potential 270 households were
identified that had been surveyed for two years. In 1980-
1981, 22 percent of these households (58 of 267) were
headed by women (see Table 9.2). In 1981-1982, 23 percent
(60 of 266) were women-headed. The observer is led to
believe that the number of women-headed households is
fairly stable and consists of one in every four or five
households. A further conclusion might be that 22-23
percent of women experience being heads of their
households.

Comparing individual households from one year to the
next gives another impression. Instead of just two
households changing (58 to 60), 17 changed sex of head —
ten maleheaded households becoming female-headed, and seven
female headed households becoming male-headed. Thus,
one-sixth of the women who were household heads in
1981-1982 acquired that status only during the preceding
year (see Table 9.2). A total of 67 households (50, 10, 7)
were noted as being headed by women at some time in the
slightly more than one year of the two surveys, or 25
percent of the 267 households. Assuming that this change
from one year to the next is a normal occurrence, more
women are heads of household during their lifetime than are
indicated by the frequency of women-headed households at
any given time.

If this social dynamic is confirmed by testing against
other Malawian data, then women-headed households are more
common and more important than is currently recognized.
The LRDP data show approximately three percent of households shift each year in each direction (woman to man, and
man to woman). More testing and more longitudinal information are needed to distinguish the number of "repeaters,"
or households that shift sex of household head more than
once; the number of "stables," or households that remain
headed by the same sex for many years; and the extent and
variability in these percentage shifts. If the number of
repeaters is low, then the annual combined shift of six
percent in both directions means a rapid increase in the
cumulative number of women who experience being heads of
households.

TABLE 9.2

CHANGES IN GENDER OF HOUSEHOLD HEAD
IN LILONGWE RURAL DEVELOPMENT PROJECT, MALAWI
1980/81 TO 1982

	1980/81	1981/82	1982
Larger Subsample			
Women Household Heads	58	60	
Unchanged Over Year	—	50	
Changed From Men	—	10	
Men Household Heads	209	206	
Unchanged Over Year	—	199	
Changed From Women	—	7	
Total Households	267	266	
Intensively Surveyed Subsample			
Women Household Heads	21	20	16
Unchanged Over Year	—	18	12
Changed From Men	—	2	4
Men Household Heads	80	80	85
Unchanged Over Year	—	78	76
Changed From Women	—	2	9
Total Households	101	100	101

Note: Missing 1981/82 data on one household that was
headed by a woman in 1980/81 and 1982.
Interval between 1980/81 and 1981/82 was 12 months;
data collected during rainy agricultural season
(November through April). Interval between 1981/82
and 1982 was six months; data for 1982 collected
during dry agricultural season.

LRDP Subsample

A subsample of the 267 households in LRDP was chosen
for more intensive study during 1982. Seventeen of the 27
EAs were randomly selected, and six of the ten resurveyed
households in each of the 17 EAs; this created a subsample
of 102, of which 101 were successfully located and interviewed approximately six months after the 1981-1982 survey.

The subsample resembled very closely the larger sample from which the subsample was drawn. In the subsample, the number of women-headed households was fairly stable from one year to the next (21 and 20 percent) and appeared at approximately the same levels as in the larger sample. Again, as in the larger sample, this apparent stability concealed the actual dynamics of shifting household head-ship which changes at the annual rate of two percent each way (man to woman and vice versa) in the subsample.

The subsample revealed an unexpectedly large shift in sex of household headship between the 1981-1982 cropping season and the resurvey six months later in the dry non--agricultural season. Only 16 percent of households in the subsample were headed by women during the 1982 dry season survey (see Table 9.2). In the six months between the cropping season survey, 13 percent of households had changed the sex of the household head, most of them (69 percent) changing from women-headed to men-headed. These shifts were real, i.e., only one household was a repeater showing more than one change over the three surveys (see Table 9.3).

The move toward men-headed households probably reflected husbands returning who had been away from their wives during the agricultural season. These husbands probably migrate as seasonal laborers every year, which is why the number of women-headed households during the rainy season remains stable from one rainy season to the next but varies widely each year from the dry season frequency. This seasonal variability was not discovered in the usual annual surveys conducted during the cropping season. Not yet examined is the extent to which the presence of a resident husband during the dry season affects the agri-cultural decision-making process or the extent to which agricultural and farming systems decisions are made during the actual cropping season by the wife.

The additional shifts in sex of household head shown in the subsample (see Tables 9.2 and 9.3) mean that although no more than 21 percent of the households were headed by women in any given survey, 27 percent of the households in the subsample had a woman head over the 18 month period from the first survey in 1980-1981 to the third survey in 1982. The larger sample showed 25 percent of the households with women heads over the 12 month period from the first to second surveys. Both sample and subsample confirm the underestimates of the number of women-headed households given by surveys that only report the frequency at one time.

TABLE 9.3

PATTERNS OF CHANGES OF GENDER OF HOUSEHOLD HEADS
IN INTENSIVELY SURVEYED SUBSAMPLE

Household Head 1980–81	Household Head 1981–82	Household Head 1982		
Man	Man	Woman	N=4	(25%)
Man	Woman	Woman	N=1	
Man (2 Changes)	Woman	Man	N=1	
Woman	Woman	Man	N=8	
Woman	Man	Man	N=2	
			16	(100%)

Note: Although 20–21% of households are headed by a woman during any given rainy agricultural season, 27% of households had a woman head over 18 months.

IMPLICATIONS

The Zambian and Malawian cases document the discrepancy between the frequency of women–headed households at any given time and the cumulative number of women who are household heads over time. Shifts from one season to another and one year to another mean that many more women experience heading households during their lifetimes than are indicated by the current survey methods that only measure the frequency at one time.

Women–headed households are a category of disadvantaged households in terms of labor, land, capital, and governmental services. Correcting the underestimated frequency of women–headed households means recognition that more women experience the disadvantages for longer periods of time. Recognizing that many households may pass through a phase of being headed by a woman, while a smaller number are permanently women–headed, draws attention also to ways to ameliorate the disadvantages by preparing women before they become household heads.

Wives should be considered as potentially independent heads of households, rather than believing that they are dependent upon their husbands in men–headed households and do not need separate instruction or access to services. Women need to receive their own training, experience agricultural innovations, and establish their own credit records and contacts with extension and credit agents.

Gaining these skills and advantages while still in a household headed by someone else means that these women will be better equipped to cope and may be able to continue a more productive agriculture if they do become heads of households.

The current farming systems research and extension model utilizes a simple household unit that underestimates many social factors. Although in theory farming systems are biosocial systems that include social factors, in practice these factors are ignored or treated as parameters while research concentrates on biological and biotechnical relationships (Due this volume; Shaner et al. 1982). Although farming systems research and extension are supposed to be conducted by multidisciplinary teams, in practice the disciplines and interests represented are more technical than social.

Treating the head-of-household phenomenon as a phase rather than a permanent status for many women is a conceptual shift that requires the integration of: (1) farming systems research and extension with its technology focus and reliance on a simple, synchronically-surveyed household unit and (2) the rich social science literature on life and domestic cycles, marital forms and instability, and the variable forms of distribution of wealth, labor, and responsibility between husband and wife (Guyer 1981; Gray and Gulliver 1964; Moock 1986; Netting et al. 1984a; Norman et al. 1982). This integration is needed to rescue farming systems research and extension from collapsing into a locality specific Green Revolution activity (Fresco and Poats 1986:329; Hansen 1986; Jones and Wallace 1986).

Correcting the underestimated frequencies of women-headed households cannot be accomplished without assessing local variation in several social variables. These variables include: frequencies of women-headed households at any given time, especially during the cropping or other active seasons; frequencies of shifting from one household head to another, annually and seasonally; disadvantages facing women-headed households in labor, land, capital, and access to services; and the homogeneity of types of women-headed households. Matrilineal areas in southern Africa, such as the Zambian and Malawian cases referred to here, have relatively high rates of marital instability and male labor migration, that make the argument particularly relevant.

The argument in this paper focuses on women who are not currently heads of households. These women need to receive training and services as if they were already independent decision-makers. Obviously, women who are currently heads of households also need to receive training and resources. This focus on women who are or will become heads of households was chosen because household-head status is a widely accepted indicator of decision-making power and responsibility. It is not implied that women only make decisions when they are household heads. What is argued is that the extent to which women experience being heads of their households and thereby exercise the power and responsibility of independent decision-making is underestimated and that current one-time survey methods inherently underestimate social dynamics.

REFERENCES

Behnke, R. and C. Kerven
    1983   FSR and the Attempt to Understand the Goals and Motivations of Farmers. Culture and Agriculture 19: 9-16.

Berry, S.
    1986   Social Science Perspectives on Food in Africa. In Food in Sub-Saharan Africa. A. Hansen and D. E. McMillan, eds., pp. 64-81. Boulder, CO: Lynne Rienner Publishers.

Boserup, E.
    1970   Women's Role in Economic Development. New York: St. Martin's Press.

Carter, A. T.
    1984   Household Histories. In Households. R. McC. Netting, R.R. Wilk, and E.J. Arnould, eds., pp. 44-83. Berkeley, CA: University of California.

Cernea, M.
    1985   Alternative Units of Social Organization Sustaining Afforestation Strategies. In Putting People First: Sociological Variables in Rural Development. M. M. Cernea, ed., pp. 267-293. New York, NY: Oxford University Press.

Due, J. M.
    1987   Intra-Household Gender Issues in Farming Systems in Tanzania, Zambia, and Malawi. In Gender Issues in Farming Systems Research and Extension. S.Poats, M.Schmink, A.Spring, eds., Boulder, CO: Westview Press.

Fresco, L. O. and S. V. Poats
    1986   Farming Systems Research and Extension: An Approach to Solving Food Problems in Africa. In Food in Sub-Saharan Africa. A. Hansen and D. E. McMillan, eds., pp. 305-331. Boulder, CO: Lynne Rienner Publishers, Inc.

Gray, R. and P. Gulliver
  1964   The Family Estate in Africa: Studies in the Role
          of Property in Family Structure and Lineage
          Continuity. London: Routledge and Kegan Paul.
Guyer, J. I.
  1981   Household and Community in African Studies.
          African Studies Review 24:2/3:87–137.
Hansen, A.
  1977   Once the Running Stops: The Socioeconomic
          Resettlement of Angolan refugees (1966 to 1972) in
          Zambian Border Villages. Ithaca, NY, Ph.D.
          Dissertation, Cornell University, Department of
          Anthropology.
  1981   Refugee Dynamics: Angolans in Zambia 1966 to
          1972. International Migration Review 15:1–2:
          175–194.
  1986   Farming Systems Research in Phalombe, Malawi:
          The Limited Utility of High Yielding Varieties. In
          Social Sciences and Farming Systems Research. J. R.
          Jones and B. J. Wallace, eds., pp. 145–169.
          Boulder, CO: Westview Press.
Hansen, A. and J.D. Ndengu
  1983   Lilongwe RDP: Cropping Patterns Information
          from the National Sample Survey of Agriculture.
          Presented to National Meeting of Evaluation Officers
          and Planning Officers. Lilongwe, Malawi: Ministry
          of Agriculture.
Jones, J. R. and B. J. Wallace, eds.
  1986   Social Sciences and Farming Systems Research:
          Methodological Perspectives on Agricultural
          Development. Boulder, CO: Westview Press.
Kerven, C.
  1979   Urban and Rural Female-Headed Households'
          Dependence on Agriculture. Gabarone, Botswana:
          Rural Sociology Unit and Central Statistics Office.
Kydd, J.
  1982   Measuring Peasant Differentiation for Policy
          Purposes: A Report on a Cluster Analysis
          Classification of the Population of the Lilongwe
          Land Development Programme, Malawi, for 1970 and
          1979. Zomba: Government Printing Office.
Lele, U.
  1975   The Design of Rural Development: Lessons from
          Africa. Baltimore, MD: Johns Hopkins University
          Press.

Low, A.
    1986   Agricultural Development in Southern Africa:
    Farm Household-Economics and the Food Crisis.
    London: James Curry.
McMillan, D. E.
    1984   Monitoring the Evolution of Household Economic
    Systems Over Time in Farming Systems Research.
    Presented at the Workshop on Conceptualizing the
    Household: Issues of Theory, Method, and
    Application. Cambridge, MA: Harvard Institute for
    International Development.
    1986   Distribution of Resources and Products in Mossi
    Households. In Food in Sub-Saharan Africa. A.
    Hansen and D. E.McMillan, eds., pp. 260-273.
    Boulder, CO: Lynne Rienner Publishers, Inc.
Moock, J. L., ed.
    1986   Understanding Africa's Rural Households and
    Farming Systems. Boulder, CO: Westview Press.
Netting, R. McC., R. R. Wilk, and E. J. Arnould, eds.,
    1984a Households: Comparative and Historical Studies
    of the Domestic Group. Berkeley, CA: University of
    California Press.
    1984b Introduction. In Households: Comparative and
    Historical Studies of the Domestic Group. R. Mc. C.
    Netting, R. R. Wilk, and E. J. Arnould, eds., pp.
    xiii-xxxviii. Berkeley, CA: University of
    California Press.
Norman, D. W., E. B. Simmons, and H. M. Hays
    1982   Farming Systems in the Nigerian Savanna:
    Research and Strategies for Development. Boulder,
    CO: Westview Press.
Shaner, W.W., P.F. Philipp, and W.R. Schmehl
    1982   Farming Systems Research and Development:
    Guidelines for Developing Countries. Boulder, CO:
    Westview Press.
Spring, A.
    1985   The Women in Agricultural Development Project in
    Malawi: Making Gender Free Development Work. In
    Women Creating Wealth: Transforming Economic
    Development. Rita Gallin and Anita Spring, eds.,
    pp. 71-75. Washington, D.C.: Association for Women
    in Development.
    1986   Women Farmers and Food in Africa: Some
    Considerations and Suggested Solutions. In Food in
    Sub-Saharan Africa. A. Hansen and D. E. McMillan,
    eds., pp. 332-348. Boulder, CO: Lynne Reinner
    Publishers, Inc.

126

Spring, A.
    n.d.   Agricultural Development in Malawi: A Project for Women in Development. Boulder, CO: Westview Press, forthcoming.
Spring, A. and A. Hansen
    1985   The Underside of Development: Agricultural Development and Women in Zambia. Agriculture and Human Values 2:1:60-67.
Wilk, R. R. and R. McC. Netting
    1984   Households: Changing Forms and Functions. In Households. R.McC. Netting, R.R. Wilk, and E.J. Arnould, eds., pp. 1-28. Berkeley, CA: University of California Press.
Yanagisako, S. J.
    1979   Family and Household: The Analysis of Domestic Groups. Annual Review of Anthropology 8:161-205.

10

An Evaluation of Methodologies
Used in Time Allocation Research

Eva Wollenberg

Time allocation studies are frequently used to describe gender and age based labor patterns. In farming systems research (FSR), labor patterns have been analyzed to support a wide range of findings including the determination of peak labor periods (Maxwell 1984; Price and Barker 1978), income opportunities for female farmers (Burfisher and Horenstein 1985), the contribution of children to farm production (Navera 1978), crop labor investments (Barlett 1980), seasonal fluctuations in agricultural and non-agricultural activities (Norman et al. 1981), and inter-household differences in the family cycle (Cadelina 1985).

The strengths, weaknesses, and variations of time allocation methodologies are rarely discussed. This paper evaluates commonly used methods of data collection in time allocation studies. The first section of the paper is a review of general characteristics of time allocation studies found in the literature. Issues requiring closer scrutiny are identified. The second section describes four methodologies that were used in a 1984–85 study of shifting cultivators in Negros Oriental, the Philippines. Each of the methodologies is evaluated according to how well it contributed to the understanding of upland farming patterns and the organization of household labor. Finally, some implications for the use of time allocation data in farming systems research projects are discussed.

Why should researchers do time allocation studies? Time is a resource. Individuals must make decisions about how to manage their time. These decisions reflect the constraints and opportunities surrounding human goals. Farmers lead an economic existence in which they must allocate their time among agricultural activities. Poor farmers generally have better access to labor than to land

or capital. Information about time allocation can illuminate facets of farm household behavior that are not obvious
from other types of data. Most small farmers in developing
countries operate at the household level as a unit of both
production and consumption. As a result, the activities of
all household members are tied to the farm enterprise.
Nevertheless, the roles of women, children and the aged are
often overlooked or discounted because the analysis focused
on income generation or male "heads of households" (Hayami
et al. 1978; Maxwell 1984).

Time allocation studies can demonstrate the type and
quantity of labor that different members of the household
contribute. Most subsistence level farm households engage
in more than one means of support. Their livelihood might
be based on crops, fishing, livestock, off-farm wage labor,
cottage industry, and forest products. Time use data can
show how a household distributes its labor resources among
such activities and how subsistence activities compete with
time needed for childcare, cooking, cleaning, sleeping,
leisure, or education (Mueller 1979).

GENERAL CHARACTERISTICS OF THE METHODOLOGY

Methods of time allocation research applicable to
gender and farming systems can be described according to
the information objective of the research, the way in which
that information is collected, and the way in which the
data are interpreted.

Information Objective

The purpose of the research determines the information
objective. This purpose will affect how the information is
structured, and whether or not the focus is on certain
tasks, certain people, certain locations, or all of these
factors at once. A research design might include all of
these variables, but is likely to focus on only those
variables of principal interest. For example, a corn
production study might use crop activities as an "organizing variable." Data on location or people would be
collected only as they relate to crop activities. Such a
focus minimizes the amount of nonrelevant information, but
maximizes the probability that some unanticipated but
significant factors will be overlooked. The comprehensive
method allows for the unanticipated by requiring the
researcher to examine all aspects of farm time allocation.
Costs are directly proportional to the amount of information collected; therefore, the comprehensive approach might
be best applied for a limited time as an initial baseline

study to aid in the identification of variables demanding
more intensive study.

Research designs for time use studies must consider
which levels of disaggregation are appropriate to the
population and the research objectives at hand. Information can be disaggregated by social units (community,
household, individual), demographic criteria (sex, age,
position in household), socioeconomic criteria (land
tenure, income, assets), biophysical characteristics
(climate, soil, vegetation) and farming system criteria
(market crops, subsistence crops, livestock). The temporal
units and boundaries of the research need to be chosen in a
way that best reflects the many rhythms of a farmer's
experience: the fluctuations of the year, season, day;
labor supply and demand; a given crop schedule; the family
development cycle; or the local school calender.

Information Collection

Time allocation information has been collected with the
use of: (1) respondent recall (Mueller 1979; Moji 1985);
(2) direct observation by the investigator (Johnson 1975);
and (3) farm household record keeping (Price and Barker
1978). Recall and record keeping are convenient for large
samples since relatively little time is required per case.
Direct observation tends to be more accurate, but also more
time consuming both in collection and analysis. (See Anker
1980, The Asia Society 1978, and Birdsall 1980 cited in
Acharya and Bennett 1982: 64, for further discussion of the
differences between recall and observation.)

Each method of gathering information is subject to the
biases introduced by either the enumerator or the farmer.
Recalled information passes through the filter of the
farmer's memory as well as the filter of the enumerator's
ears and interests; it is therefore subject to the potentially greatest bias. Direct observation data have the
potential of less "filter" bias, but are influenced by the
presence of the investigator. The extent of this influence
depends on the nature of the relationship of the enumerator
with the community as well as the manner in which local
households treat guests. Long periods of acquaintance are
likely to minimize inadvertent influence on respondent
behavior.

It is difficult to follow all members of a household to
the locations of their activities. Researchers typically
stay in the home, observe comings and goings, and ask questions about activities that they were not able to witness
firsthand. In time allocation studies for FSR, an alternative approach might be adopted, since many agricultural

tasks may take place out of view from the home. One option
might be to have more than one enumerator per household so
that enumerators could accompany household members to their
plots.

Sampling

Sampling strategies can help to minimize bias in time
allocation research. By using instantaneous observations,
Erasmus (1955) avoided collecting data that reflected the
enumerators' presence. Johnson (1975) and Acharya and
Bennett (1982) randomized observations to make statistical
estimates of the distribution of time among activities.
The longer the period of an activity, the greater the
probability that it was observed in random spot checks.
Time allocation with this technique is determined by the
frequency of observing a given activity. Costs associated
with lengthy periods of observation are minimized. Whether
or not sampling is random or systematic, instantaneous or
continuous, the investigator still needs to determine how
often he or she will take samples -- hourly, daily, weekly,
or seasonally. Trade-offs between costs and detail in
information must be considered in every aspect of time
allocation research.

The use of random observations lends itself to a representative description of farmers' time allocation. Most
research is designed to describe "representative" or
"average" conditions. Outlying cases are often removed
from analysis in order to achieve a better fit between
observations and a calculated mean or regression line.
Yet infrequent or irregular activities may be significant
to an understanding of farm behavior. How does a household
recoup losses after a once-in-a-lifetime typhoon has
destroyed an entire standing crop of corn? How do religious holidays affect the balance of leisure and agricultural production? Do the activities surrounding childbirth
disrupt a household's agricultural work schedule?

Sample size is another critical element of the research
design. Case studies are not necessarily representative of
the larger population, but may offer in-depth information
that is missed in large surveys (Maxwell 1984). Both
surveys and case studies have been used successfully
although often with different purposes (see Lewis 1959 and
Erasmus 1955 for examples of case studies; Acharya and
Bennett 1982; Hayami et al. 1978; and Hart 1980 for
examples of surveys). Most reports on farm time allocation
mention the total number of individuals or observations as
evidence of the study's statistical weight (Mueller 1979;
Johnson 1975). While these figures give the reader a

better picture of the sample, the number of households seems to be the critical variable in populations where farms are household enterprises.

Interpretations of Time Use Data

Time allocation data are assumed to offer useful information to the FSR practitioner. The data are sometimes used as an indication of labor or energy "expenditure." Input-output labor budgets have been calculated to determine the productive efficiency of different means of subsistence (Sahlins 1972; Rappaport 1968). In other cases, wages have been imputed to estimate a farm's economic efficiency (Price and Barker 1978). Care should be taken not to equate time and labor. Time is only a convenient indicator of labor inputs. It indicates the allocation and duration of labor, not the intensity of effort or the economic value of that effort.

The amount of time spent in any one activity may not be an adequate indication of the importance of that activity to the farm economy. For example, market transactions take relatively little time, yet the household might depend on the cash income for tools, medicine, or clothes. Furthermore, activities often occur in sequences; the events of one day may affect the activities of the following week. The farmer who was not able to sleep one night might sleep twelve hours the following day. A week of rain and inactivity may be offset by long hours spent weeding on the first sunny day. The methods commonly used to study time allocation do not easily account for such activity sequences.

Time use studies have typically treated an individual's daily routine as a series of single tasks carried out by one person (Johnson 1975). Joint activities such as child care and weeding, or fuelwood gathering and walking home from the market are not treated adequately. At what point does an activity become a joint use of time if the individual alternates repeatedly between tasks? Activities involving more than one person, such as cooperative harvesting groups, are also not described well in most time allocation studies. Even when the occurrence of joint activities is recognized, it is still often analyzed as if the tasks were performed separately (Acharya and Bennett 1982).

Farm behavior is not discretely organized into pre-coded categories. Categorization is a device used by the investigator to reduce the details of everyday farm life into manageable units. The arbitrary definition of these categories and the sometimes biased process of classifying activities can lead to an equally arbitrary or biased interpretation of the data. Johnson (1975) discusses how the relative amount of time spent by men and women in "productive" labor varies with the definition of production.

To compensate for these shortcomings, various authors have stressed the need to consider time allocation in the context of other information (Chibnik 1980; Stone and Campbell 1984; Acharya and Bennett 1982). Time use data can be used as the quantitative counterpart of a broader examination of household decision-making. Ethnographic descriptions derived from participant observation and in-depth interviews can serve as a complement to and crosscheck on data limitations. Detailed time allocation studies are expensive; the same conclusions can sometimes be drawn from other, cheaper sources of information. Using participant observation to note which activities tend to be done by men or women may be sufficient for most purposes (Wiley 1985: 184) To minimize the expense associated with the collection of time use information and to avoid the accumulation of detailed data that "never gets processed anyway," it is suggested that researchers streamline time allocation studies according to the purpose at hand. The systems perspective may be maintained by backing up time data with information from cheaper sources. With the exception of baseline studies, the more focused the study and the fewer the variables, the more useful information the researcher is likely to derive from the data.

Summary

This first section of the paper has presented a conceptual framework for the analysis of time allocation methodologies. The framework is derived from the wealth of experience that other researchers have had and therefore includes ideas that are probably familiar to those who have worked already with time allocation methods. The framework is most useful to those designing new studies or analyzing previous studies. A series of methodological questions has been posed: What is the information structure of the time study? What are the variables of greatest interest? What is the information source — recall, direct observation, or household records? What is the sampling strategy? What are the biases, costs, and benefits of the method? How are the data to be interpreted? Is time a surrogate measure of another variable of interest? How does the method account for spurious events, joint activities, sequences, or different scales of time? Are there other sources of

information to serve as cross-checks? The second half of
the paper gives the reader an opportunity to apply the
conceptual framework to an FSR project in which four time
allocation methodologies were used.

TIME ALLOCATION AT LAKE BALINSASAYAO

This section of the paper discusses the four approaches
that were used in the collection of time use data in
Balinsasayao in the province of Negros Oriental. Data on
the household distribution of time, broken down by task and
gender, are presented. The four approaches are compared
according to how well they contributed to the cost-
effective understanding of farm behavior.

Site and Project Background

The Balinsasayao Rainforest is at 9 degrees 21' N
latitude, 123 degrees 10' E longitude. Lowland farmers
began migrating to the forested mountains of Negros
Oriental province in significant numbers in the 1950s.
Several scattered settlements exist in the Balinsasayao
Rainforest. Land ownership is not possible since the area
has been designated as a critical watershed by the national
government.

The Balinsasayao community and project activities are
concentrated around a lake approximately ten hectares in
size. Shifting agriculture, or kaingin, is practiced.
Corn, root crops, and chayote are the predominant crops.
Average landholdings are five hectares with approximately
twenty percent under intensive cultivation at any one time.
Most households control three to five different plots.
Plots are managed differently according to the distance of
the field from the household and the plot's productive
capacity. Fallow-time ranges between one and twenty-six
years.

The nearest markets are in the lowland coastal towns,
San Jose and Dumaguete, about 12 kilometers from Balinsasayao. Farmers walk this distance since there is no public
transportation. The average household sells produce in the
market once a week. Chayote (Sedium edule), squash, root
crops, and abaca (Musa textilis, Manila hemp) are commonly
sold in the lowlands. All of these items have relatively
low market values and the amount of produce that can be
sold in any one trip is limited by the number of pounds the
farmer can carry on his or her back.

Time Allocation Methods

Four time allocation studies were administered in the Balinsasayao community between April 1984 and July 1985. Each of the studies was designed to answer specific questions about the organization of household activities. The first method, participant observation, was used to determine norms regarding the division of labor within households. In-depth, open-ended interviews and residence in the community provided data over a 16-month period. Participant observation was also used as a cross-check for subsequent studies.

The second method was a comprehensive case study of four households. Sample households represented the different stages of the family cycle. The investigator lived with three of the households for five consecutive days, and one household for six days, resulting in 21 days of observation. The time of day, duration, person, task, and location of task were observed. When the investigator could not observe all of the household members at once, he or she had to rely on the farmers' descriptions of their activities. The objective of the case studies was to collect a comprehensive set of data as a baseline for subsequent research.

The case study data showed that men's and women's time use differed by field type. At Balinsasayao, fields may be classified according to the intensity of cultivation. Intensively cultivated fields include crops such as corn, beans, and peppers. These fields tend to be close to the home and have soils with a high productive capacity, high levels of solar insolation, and low risk of loss due to pests or theft. Non-intensively cultivated fields include root crops, chayote, squash, abaca, and other low input crops. The case study profiles showed that women spent more of their time on intensively cultivated fields while men spent more of their time on the less intensively cultivated fields (see Results section).

Differential allocation of time by field type was examined in more detail in the third and fourth studies. These studies were used to test hypotheses about gender and age-based patterns in agricultural activities, particularly differences between use of intensively and non-intensively cultivated fields. The third study was a 24 hour recall of household time use. Respondents were asked to recall the time of day and the duration of activities of each member of the household. The fourth study was a one week recall of field activities. In-depth information about the field location, activity (planting, weeding, harvesting, etc.),

participating household members, crops planted or harvested, and the date of the field visit were noted. This
was the only method that did not record the duration of the
activity.

The data for the third and fourth studies were gathered
during a seven-month survey of ten households. The sample
represented 71 percent of all households within one kilometer of the lake. Households were visited approximately
every four weeks. A total of 53 days was recorded for all
households. Female heads of households were interviewed
because they were the ones usually at home. In two instances an adolescent and in 16 cases male heads of households
were interviewed. Since households were unaware of the
interview schedule, there were several occasions when no
one was home. One household shifted its residence to a
distant field shortly after the survey began and was
therefore only interviewed twice.

Comparison of Methods: Quality of Data

The survey and field data gave the best representation
of the community since they were based on the largest
sample size. Most of the respondents had no difficulty
remembering the previous day's activities, but everyone had
difficulty recalling one week of field visits. Field
recall seemed easier when questions were structured by crop
related tasks, rather than by field visits, that is, asking
"what did you plant yesterday," rather than "what did you
do when you went to your corn field?" The field results
were important in the way they corroborated and expanded
upon the survey data. Respondents answered in more detail
about what they were doing, who was doing it, and which
crops were involved. The field questions also served as a
baseline for a more in-depth longitudinal study of labor
inputs on different field types.

Survey responses were short. The previous day was
described according to what the respondent considered to be
the relevant activities of the day. This was useful
information in itself. Respondents recalled primarily
agricultural tasks, watching children, marketing, and food
preparation. Non-work activities and tasks of short
duration were conspicuously absent.

The comprehensive case study data show all activities
for all members over three years of age of a limited number
of households. Information was recorded for such non-work
activities as sleeping, eating, personal hygiene, socializing, resting. Short tasks such as fetching water,
chopping wood, feeding livestock, sweeping the floor, and

business transactions were also observed. Any misinformation in the case study data is due to the investigators' interpretations of behavior or the inability to view all acts.

The comprehensive case study method provided the most complete description of multi-person and joint activities. In the survey, respondents tended to underrepresent the number of people and the number of tasks per activity. Children were underrepresented as well. There are a number of explanations for this. First, since informants might consider children as helpers "in training," informants would be more likely to list activities according to the person in charge, usually an adult. This distinction is useful to an understanding of household decision-making. Informants generally neglected to mention the participation of nonhousehold members, such as visiting relatives, a neighbor who helped cook dinner, or a group of friends who helped cut timber. Direct observation and participation observation showed that many activities were indeed joint or involved more than one person (see Results section).

Information derived from participant observation helped to compensate for methodological shortcomings in the other approaches, namely, the limited number of households in the case study and survey and the limited duration of all the studies. Through participant observation, inferences could be made about time use during the entire agricultural year, even though only seven months were formally sampled. Likewise, familiarity with the community and the idiosyncrasies of households and individuals facilitated extrapolation about time use for the entire Balinsasayao community. As has been noted elsewhere (Colfer 1985), participant observation provides the fundamental background information necessary for a systems view of human behavior.

Bias

Living with a household while trying to observe its behavior introduces the bias of the investigator's presence. This bias was minimal due to the close rapport that both investigators had with the community as well as the fact that having visitors is a common, culturally encouraged event in the Balinsasayao community. One of the investigators (the author) had lived in the community conducting field surveys and interviews for nine months previous to the time allocation study. She had lived with one of the case study households for six of those months. The other researcher had conducted interviews and lived in the area during the previous four months. During the time allocation studies, acts of hospitality, such as offering morning snacks and afternoon coffee, were more frequent than they probably would have been without visitors, and consequently were overrepresented in the case studies. Five days was a relatively long time to observe the activities of every individual. It was exhausting for the investigators as well as for the household. After two days, most of the households seemed anxious to regain their privacy.

Costs and Benefits

There are certain costs associated with the collection of data for each of the methods. These are the costs of the investigators' time, the costs of the farmers' time, and the costs of analyzing the data. These costs must be weighed against the benefits of information according to its depth, accuracy, precision, and quantity. On a per day basis, the comprehensive case study approach was the most expensive method. The time required to collect and analyze the quantity of information was not offset by the amount of information gathered that was relevant to understanding household land use and farming practices. However, the nature of participation in some activities would not have been quantified as well through any other method. For example, the joint activity of childcare and food preparation by women was not recalled by respondents in the survey (see Results section).

Participant observation was the least expensive method on a per day basis, but the level of information was too general to be of use in a sophisticated analysis. Quantitative data are useful for comparative work. The survey and field methods provided the greatest amount of relevant information at the lowest cost. The four methods were complementary and served their particular purposes well.

There are benefits associated with time studies that are uniquely valuable to FSR since the process itself generates an on-going link of communication between the researchers and the farmers. That link could be used for informal exchange of extension information, expression of farmers' changing attitudes, or simply to show that the researchers are interested and concerned about how the farmers live their lives.

ANALYSIS AND RESULTS

Since the emphasis of this paper is on time allocation methods, the results presented are a brief summary of some

of the findings. This section will focus on a comparison
of the comprehensive case studies and the survey.

Data were analyzed on a computer with Statistical Package for the Social Sciences (SPSSX) programs. Activities
for each study were grouped into thirteen major categories:
land preparation, weeding, planting and harvesting, livestock care, fishing, gathering and forest use, processing/
manufacture, marketing, income generation and exchange
labor, domestic chores, daily needs, leisure, and visiting.
The time recorded for each activity included associated
travel time. Time values were rounded off to the nearest
hour in the survey, but were left in fractions of hours for
the case study in order to avoid eliminating activities of
short duration.

Activities Breakdown

Table 10.1 shows the time distribution among activities
for the comprehensive case studies and the survey.
According to the case studies, households spend most of
their time in daily needs, marketing, and domestic chores,
respectively. The survey shows that the most time is spent
in weeding, domestic needs, and land preparation. There
are several explanations that might account for the discrepancies between the two sets of results. The timing of the
case study and survey reflect differences in the agricultural calendar. The survey occurred during the dry season
when land preparation, weeding, and planting were common.
The case studies occurred one month earlier, when the rainy
season was still in progress (see Fortmann 1985; Chambers
et al. 1981 for further discussion on the importance of
seasonality). Daily needs, such as eating and sleeping,
were generally not mentioned by survey respondents, and
therefore were underrepresented in the survey results.

Table 10.1 also shows the distribution of time by
activity broken down by gender. Since the data were not
weighted for the relative frequencies of males and females,
the figures must be interpreted carefully. For example,
one may correctly say "of all the time spent in weeding, 64
percent was by males and 36 percent was by females," but
not "that weeding is more of a male role than a female
role." The results of the case studies and survey both
show substantial differences between male and female
participation in land preparation, harvesting/planting,
fishing, visiting, and income. Statistical significance
exceeds the ten percent confidence level for harvesting/
planting in both data sets.

TABLE 10.1

DISTRIBUTION OF TIME BY ACTIVITY AND GENDER

Activity	Case Study			Survey		
	Percent of Time[a]	Breakdown by Sex[b]		Percent of Time[a]	Breakdown by Sex[b]	
		Male	Female		Male	Female
Livestock	1.3	55	43	1.7	31	69
Land preparation	2.8	100*	0*	12.7	93	7
Weeding	1.4	64	36	26.0	60*	41*
Harvesting/planting	1.9	89*	11*	8.3	60**	41**
Fishing	1.2	96	4	3.8	70	30
Gathering/forest use	0.3	93*	7*	1.5	48	52
Processing/manufacture	2.6	98	2	2.7	93*	7*
Income/exchange labor	3.4	100**	0**	7.6	87	13
Marketing	14.2	94**	2**	7.2	46	54
Domestic chores	8.9	44	50	19.2	42	58
Daily needs	50.3	60	34	5.6	43	57
Leisure	2.3	51	47	1.3	49	51
Visiting	5.3	33*	66*	.7	69	31

Note: aPercent was calculated as time spent by all individuals in each activity divided by time spent by
all individuals in all activities.
bPercent was calculated as amount of time spent in activity by gender group divided by amount of
time spent in activity by all individuals.

* Indicates t-test significant at .1 level.
**Indicates t-test significant at .01 level.

The differences between the case studies and survey findings may be partially explained by several intervening events. One of the households rebuilt its roof during the survey; consequently, women spent large amounts of time gathering cogon grass (Imperata cylindrica), while the men thatched the roof. Another household had two women who were sick for several days during the survey. This probably skewed the daily needs category in the survey towards females.

Two of the results in Table 10.1 can be explained by information collected through other sources. First, the case study data in Table 10.1 shows that women are found in daily needs activities more often than men. Experience from participant observation suggests that males and females spend nearly equal amounts of time in sleeping, dressing, and personal hygiene, but women and children tend to spend more time eating. Of the 60 percent of time spent by males in daily needs, 83 percent is by males under 16, that is, children. Second, visiting is probably higher for men in the surveys because informants tended to mention major trips out of the community (a pre-dominantly male activity), but not casual visits within the community (a pre-dominantly female activity). Casual visits by women within the community were observed directly in the case studies and in participant observation. The discrepancies in the analysis support the contention that time allocation data should not be interpreted in isolation. Multiple sources of information can compensate for the inadequacy of time use data in explaining activity sequences, sporadic events, or special subgroups of the population.

Joint and Individual Activities

The case and survey data show that certain activities tended to be performed in combination with other people or other tasks. Joint activities included weeding and land preparation, weeding and planting, fetching fuelwood and water, weeding and off-farm labor, land preparation and off-farm labor, food preparation and house watching, and child care and house watching. Other joint activities that were noted with the comprehensive case study method, but not in the survey, were washing clothes with bathing, childcare with food preparation, and any field activity that required some travel with fuelwood collection.

Activities that typically involved more than one person were weeding, harvesting, marketing, relaxing or socializing, eating, and childbirth. There was no significant difference between men's and women's participation in activities involving more than one person. Some activities were carried out only by individuals: care of livestock, personal hygiene, and making tools or baskets. Other activities involved single individuals at least 80 percent of the time: land preparation, planting, childcare, food preparation, and laundry.

Location of Activities

Information on the location of activities was analyzed for the survey data. The findings are summarized in Table 10.2. Women contributed sixty-four percent of the time spent on activities in the home. Men performed the activities located at some distance from the home 71 percent of the time. At Balinsasayao, this division of labor is probably due to the childcare and domestic responsibilities that women assume. There is very little time available for these women to participate in the "outside world."

TABLE 10.2

PERCENT OF TIME SPENT BY GENDER, LOCATION, AND FIELD TYPES

Location[1]	Sex	
	Male	Female
Home	7.9	14.2
Fields	40.2	18.3
Intensively Cultivated Fields	27.5	14.8
Non-intensively Cultivated Fields	12.7	3.5
Lake	3.3	1.3
Forest	1.8	0.0
Out of Community	5.3	4.7
Other	1.1	1.8
Field Type[2]	Sex	
	Male	Female
Intensively Cultivated	68.5	80.8
	56.3	43.7
	41.3	32.1
Non-Intensively Cultivated	31.5	19.2
	71.4	28.6
	19.0	7.6

[1]Percent is calculated as amount of time spent by gender classification divided by time spent by both sexes in all locations.

[2]Percent time arranged as row percent, column percent and total percent, respectively.

According to the survey data, women spent 19 percent of their field time on non-intensively cultivated fields and 81 percent of their time on the intensively cultivated fields. Men spent 69 percent of their field time on the non-intensively cultivated plots and 32 percent of the intensively cultivated plots. Women are spending significant (chi-square test significance of .09) amounts of time on intensively cultivated fields, probably because these fields usually are located close to the home. The large amount of time probably also reflects the amount of weeding required in these plots, and the tendency for women to spend most of their field time weeding (Table 10.1).

Despite the greater time spent, women do not necessarily have more control than men over crop production on these fields. Table 10.1 shows that men do most of the planting and harvesting — activities that affect production at least as much as weeding. Furthermore, Table 10.2 shows that of all the time spent on intensively cultivated fields, 56 percent was by men and 44 percent was by women. On the non-intensively cultivated fields, 71 percent of the time spent was by men and 29 percent was by women. A clear dichotomy exists, but the conclusions must be stated with care. Men appear to have more control of the less intensively cultivated fields, but women probably do not enjoy a parallel control over the intensively cultivated plots for the reasons cited above.

Substitutability of Household Members

The participant observation information shows what respondents considered to be the typical division of labor, and what the researcher observed to be the usual flexibility among these divisions. A comparison of the case study, survey, and participant observation data shows that certain tasks are more substitutable than others. For example, fetching water, weeding, and gathering fuelwood were jobs that anyone in the household could and would do. Activities that were more restricted to a particular position in the household, or gender/age group, included washing clothes (women), home repairs (men), and fishing (children). Information about substitutability is relevant in the determination of labor constraints.

IMPLICATIONS FOR FSR

Time allocation studies are useful to the analysis of farming system behavior, yet the methodology used to gather these data is not always made explicit (International Labor

Organization 1981: 40; Smith and Gascon 1979). The methods
used must be described to facilitate interpretation of the
findings and to guide researchers in the future design of
other time studies.

The Balinsasayao research shows that time allocation
studies provide useful information about gender to the FSR
and extension practitioner. The inside/outside the household locus of women's and men's activities suggests that
gender must be considered in the design, introduction, and
evaluation of new technologies. Technological innovations
that affect non-intensively cultivated fields will involve
men more directly than women. Likewise, the predominance
of one sex in certain activity categories, such as men in
land preparation and women in food preparation, indicates
the need to introduce technologies suited to the activity's
gender classification.

Communication links are vital to the process of
research and extension that is at the core of FSR. Since
men do more visiting and are involved in out of home activities more often than women, it is more likely that they
have better access to information. Since settlement at
Balinsasayao is scattered, women who stay at home must rely
on visitors and family members to bring them information.

The experience of the Balinsasayao research suggests
that 24 hour long comprehensive case studies can be used to
gather reliable baseline information. Short duration (four
to five day) case studies might be conducted as part of a
rapid rural appraisal to provide data for the diagnostic
stage of FSR. The all-encompassing nature of a comprehensive, full day case study insures that the non-obvious
details of household behavior will not be overlooked. The
comprehensive case studies may be followed up with more
task focused or detailed studies during the latter stages
of the project.

Time allocation methods are well suited to the interdisciplinary nature of FSR. The data can provide information about planting and harvesting schedules for the agronomist, details on labor supply and demand for the agricultural economist, patterns of food preparation for the
nutritionist, or insights about household communication
networks for the anthropologist and community organizer.
The analysis of time use is a way of charting farmer
behavior in all of its dimensions. The flexibility of time
allocation research permits investigators to pursue
questions of a specific or a general nature.

The conceptual framework presented in the first part of
this paper can be used to develop methods of time allocation research that create the closest fit between the
purpose of the research and the data generated as well as

to evaluate whether or not the methodology achieves this fit with a minimum of cost and bias. The framework helps to structure the comparison of the four methods used at Balinsasayao. The Balinsasayao project illustrates that a range of methods can produce a corresponding range of results. Methods do not have to be used in isolation. Complementary approaches can be used together. While there were strengths and weaknesses associated with each approach to the study of time use, the combination of all methods provided a comprehensive understanding of the organization of time in the farm household.

NOTES AND ACKNOWLEDGMENTS

Jeff Romm and Louise Fortmann provided useful comments on an earlier draft of this paper. Ramon Somido helped to collect the data. Thanks are due to the eleven households that generously gave us their time and allowed us to invade the privacy of their daily lives. The author is also grateful for the assistance given by the University Research Center, Silliman University in Dumaguete City, Philippines. The research was funded by a Fulbright-Hays grant.

REFERENCES

Anker, R.
  1980   Research in Women's Roles and Demographic Change: Survey Questionnaires for Households, Women, Men and Communities with Background Explanation. Geneva: International Labor Organization.
Acharya, M. and L. Bennett
  1982   Women and the Subsistence Sector: Economic Participation and Household Decision making in Nepal. World Bank Staff Working Paper No. 526. Washington D.C.: The World Bank.
The Asia Society
  1978   Time Use Data: Policy Uses and Methods of Collection. Report of an Asian Development Seminar sponsored by the Asia Society, September 1978, NY: The Asia Society.
Barlett, P. F.
  1980   Cost-Benefit Analysis: A Test of Alternative Methodologies. In P.F. Barlett, ed., Agricultural Decision-Making, pp. 137-160. San Franciso, CA: Academic Press.

Birdsall, N.
   1980   Measuring Time Use and Non-Market Exchange. In
      P.W. McGreevey, ed., Third World Poverty. Lexington,
      KY: D.C. Heath and Company, Lexington Books.
Burfisher, M. E. and N. R. Horenstein
   1985   Sex Roles in the Nigerian Tiv Farm Household.
      West Hartford, CN: Kumarian Press.
Cadelina, R.
   1985   Production Patterns, Household Developmental
      Cycle Stages and Participation of Household Members:
      The Case of the Lake Balinsasayao Lowland Migrant
      Upland Farmer. Working Paper, University Research
      Center. Dumaguete City, Philippines: Silliman
      University.
Chambers, R., R. Longhurst, and A. Pacey, eds.
   1981   Seasonal Dimensions to Rural Poverty. London:
      Frances Pinter Limited.
Chibnik, M.
   1980   The Statistical Behavior Approach. In P.F.
      Barlett, ed., Agricultural Decision Making, pp.
      87–114. San Francisco, CA: Academic Press.
Colfer, C. J. P.
   1985   People Factors in Farming Systems: An
      Anthropologist's View. Paper presented at the
      Workshop on FSR/E and Agroecosystem Research, 12–15
      August, Honolulu, HI: East–West Center.
Erasmus, C.
   1955   Work Patterns in A Maya Village. American
      Anthropologist 57:2:322–333.
Fortmann, L.
   1985   Seasonal Dimensions of Rural Social Organization.
      Journal of Development Studies 21:3:377–389.
Fresco, L.
   1984   Comparing Anglophone and Francophone Approaches
      to FSR and Extension. Paper presented at the Fourth
      Annual Conference on Farming System Research. Kansas
      State University, Manhattan, Kansas, October.
Hart, G.
   1980   Patterns of Household Labor Allocation in a
      Javanese Village. In H.P. Binswanger, R.E. Evenson,
      C.A. Florencino, and B.N.S. White, eds., Rural
      Households Studies in Asia, pp. 188–217. Singapore:
      Singapore University Press.
Hayami, Y., P. F. Moya and L. M. Bambo
   1978   Labor Utilization in a Laguna Rice Village. The
      Philippine Economic Journal 17:1–2:244–259.

International Labor Organization
    1981   Women, Technology and the Development Process.
        In R. Dauber and M.L. Cain, eds., Women and
        Technological Change in Developing Countries,
        pp. 33–47. AAAS Selected Symposium No. 53. Boulder,
        CO: Westview Press.
Johnson, A.
    1975   Time Allocation in a Machiguenga Community.
        Ethnology 14:3:301–310.
Lee, R. B.
    1968   What Hunters do for a Living, or How to Make Out
        on Scarce Resources. In R.B. Lee and I. DeVore,
        eds., Man the Hunter, pp. 30–48. Chicago, IL:
        Aldine.
Lewis, O.
    1951   Life in a Mexican Village: Tepoztlan Restudied.
        Urbana, IL: University of Illinois.
    1959   Five Families. NY: Basic Books Inc.
Maxwell, S.
    1984   The Role of Case Studies in Farming Systems
        Research. IDS Discussion Paper No. 198.
Moji, K.
    1985   Labor Allocation of Sundanese Peasants, West
        Java. In S. Suzuki, O. Soemarwoto and T. Igarishi,
        eds., Human Ecological Survey in Rural West Java in
        1978 to 1982. Tokyo: Nissan Science Foundation.
Mueller, E.
    1979   Time Use in Rural Botswana. Presented at a
        Seminar on the Rural Income Distribution Survey in
        Gabarone, Botswana, 26–28 June.
Navera, E. R.
    1978   The Allocation of Household Time Associated with
        Children in Rural Households in Laguna, Philippines.
        The Philippine Economic Journal 17:1–2:203–223.
Norman, D.W., Newman, M.D. and Oedrago, I.
    1981   Farm and Village Production Systems in the
        Semi–Arid Tropics of West Africa: An Interpretive
        Review of Research. Research Bulletin 1:4
        Pantancheru, A. P. India. ICRISAT.
Price, E.C. and R. Barker
    1978   The Time Distribution of Crop Labor in Rice–Based
        Cropping Patterns. The Philippine Economic Journal
        7:1–2:224–243.
Quizon–King, E.
    1978   Time Allocation and Home Production in Rural
        Philippine Households. The Philippine Economic
        Journal 17:1–2:185–202.

Rappaport, R.
   1968   Pigs for the Ancestors. New Haven, CT: Yale
     University Press.
Rosaldo, M.Z. and L. Lamphere, eds.
   1974   Women, Culture and Society. Stanford, CA:
     Stanford University Press.
Sahlins, M.
   1972   Stone Age Economics. Chicago, IL: University of
     Chicago Press.
Smith, J. and F. Gascon
   1979   The Effect of the New Rice Technology on Family
     Labor Utilization in Laguna. IRRI Research Paper
     Series No. 42. Los Banos: IRRI.
Stone, L. and J. G. Campbell
   1984   The Use and Misuse of Surveys in International
     Development: An Experiment from Nepal. Human
     Organization 43:1:27–37.
Wiley, L.
   1985   Tanzania: The Arusha Planning and Village
     Development Project. In C. Overholt, M.B. Anderson,
     K. McCloud and J.E. Austin, eds., Gender Roles in
     Development Projects: A Case Book, pp. 163–184.
     West Hartford, CT: Kumarian Press.

11
Gender, Resource Management and the Rural Landscape: Implications for Agroforestry and Farming Systems Research

Dianne E. Rocheleau M.

Agroforestry is a form of land use and management familiar to millions of farmers and forest-dwellers throughout the world. Formally, agroforestry is any system of land use in which woody plants are deliberately combined, in space or over time, on the same land management unit as herbaceous crops and animals (Lundgren 1982). This definition applies to a variety of land use systems ranging from very intensive farming to extensive pastoral systems, including: bush fallow farming; management of fodder trees in private or communal grazing lands; planting of trees and shrubs as live fences on farm boundaries for fuelwood, small timber, and other useful products; intercropping of tree cash crops with food, timber, fodder and soil improving crops; intercropping of hedges with grain crops for leaf mulch; home gardens of all types where trees and annual crops are mixed; and many other systems where farmers and herders combine trees with field crops or animals (Rocheleau 1986). In many of these systems women are primarily responsible for planting, tending, gathering, harvesting, processing, and using woody plants, in addition to performing their roles in crop and animal production and consumption within the larger agroforestry system.

Agroforestry systems reflect the prevailing sexual division of labor, skill, responsibility, and control within the larger society. In cases where new systems are introduced, precedents may be set for the sexual division of costs and benefits from new classes of plants or types of work not previously known in the same way. The success or failure of future research efforts to improve existing agroforestry systems or to develop new ones will depend largely on the ability of researchers to serve the social objectives of diverse groups of rural producers and to reconcile or accommodate the conflicts between men and women and between classes of rural clients.

While an overall farming systems approach is an appropriate starting point, an effective, equitable approach to agroforestry requires something more. Among those aspects that demand a broader approach are the system-wide scope of the topic, the variable scale of the land units involved (plot to watershed or community), the variety of clients and land managers, the diversity of activities involved, the combination of production and environmental objectives, the time factor required for testing and growing trees, and the relative ignorance of researchers about the past and current use of woody plants by farmers and herders (Rocheleau 1986). These characteristics overlap to some extent with gender and class issues. All of these factors combined require a more comprehensive and complex approach than might be needed to deal with gender issues in crop-based farming systems research and extension (FSR/E).

The problems and opportunities inherent in the gender division of access to land, labor, cultivated and wild plants, and products present a special challenge to agroforesters. They require specific consideration and programs not yet part of the mainstream approach to agroforestry research and development projects. The implications of these differences extend to the content of technology designs and social contracts for management as well as to the way that research and development is conducted with women clients. Gender based differences in legal status, use of and access to space, type of activities, and control over labor and resources, all have a direct bearing on what type of plants can be planted, managed, used and harvested, in terms of place, person, purpose, and benefit (Rocheleau 1987b).

Whether or not women are considered apart or as a distinct client sub-group within the larger population, the terms of their participation will usually be distinct from that of men. This is especially true with regard to the quantity, quality, and terms of access to land. Women's access to other productive resources (water, draft power, agrochemicals, labor, information) also differs from men's. Moreover, women's control over the components (animals, crops, trees, shrubs, pasture) and the products (food, fodder, fuel, timber, cash, fiber, medicine) of agroforestry systems is often subject to rules distinct from those governing men's actions. All of these differences are expressed in the existence of men's and women's separate places and activities, in nested complementary roles in the same places and activities, or in sharing of interchangeable roles.

While these differences may limit the scope and nature of agroforestry technology and project design, there are also distinct advantages and opportunities for agroforestry within women's separate domains of space, time, activities, interests, and skills. Women may also have special knowledge, rights, and obligations relating to certain categories of artifacts (tools), natural objects, and phenomena (water, fire, plants, animals).

Agroforestry may impose new demands on women clients such as the need to negotiate new arrangements for use and management of shared lands, labor, or capital inputs, to learn new skills, and to observe more careful management of soil, water, plants, and animals in existing woodland, cropland, pasture, or boundary lands. Agroforestry may also validate women's land and tree use rights or ownership, increase production and decrease gathering time, and reconcile conflicting objectives for shared household or community plots.

A few project histories (Hoskins 1979; Scott 1980; Wiff 1984; Jain 1984; Fortmann and Rocheleau 1984) and a wealth of experience in traditional and evolving agroforestry systems suggest that rural women and agroforestry programs have much to gain from a well-informed and well-defined association. Among the explicit gender issues of relevance to women's participation in agroforestry projects are women's legal status and access to productive resources, and the division of space, time, knowledge, and decision-making.

This heightened awareness of gender issues has surfaced at a time when agroforesters and social foresters are still learning to involve the population at large (FAO 1985) and to think in terms of "clients" rather than "targets." Much of the action research and organizational experiments required to find viable rules of tree and land ownership, as well as access to and management by women, can be nested within broader programs based on a land user perspective.

A general land user perspective for agroforestry research and development programs should consider multiple uses, multiple users, landscape as a major focus in a larger context of sliding scale analysis and design, and consideration of indigenous knowledge as science. In addition to these four conditions for understanding and serving users' interests, the terms of client participation must be considered. Treating land users as clients is a critical ingredient in the successful integration of users' concerns into analysis, design, and action. "Clients" may be seen as passive recipients of services or active participants. Incorporating clients as active participants produces the best results for development projects and builds local capabilities for continued agroforestry analysis, design, and management efforts (Rocheleau 1987b).

An overall land user perspective constitutes a necessary but not quite sufficient condition for serving women's
interests in agroforestry. The explicit acceptance of
women as valid clients in their own right would permit the
broadest participation of women whether or not they are in
households headed by men or heads of their own households,
and are artisans, processors, merchants, smallholders, or
landless laborers.

This is not to say that women's issues should be
absorbed into a single homogenized agenda. There is still
a need to disaggregate information, decisions, and action
to assure reasonable and equitable distribution of land,
trees and their products, and program costs and benefits to
all clients. A land user perspective with equity must deal
with women's relationship to the larger community as well
as with the very real differences between groups of women,
based on class, age, ethnicity, and sources of livelihood.

A brief outline of a land user perspective for agroforestry research and development illustrates how it can
serve various women's interests.

A LAND USER PERSPECTIVE FOR EQUITY IN AGROFORESTRY

Multiple Uses

Agroforestry is a land use system, not a commodity. It
can address a wide range of rural people's priorities for
fulfillment of basic needs (Raintree 1983). Agroforestry
practices may apply to cash crops, subsistence crops, animal production, and gathered products, as well as to farm
infrastructure and to the soil, water, and natural vegetation on the site. Among the major needs that are affected
by agroforestry are: food, water, fuel, cash income, shelter and infrastructure, savings/investment, and resources
to meet social obligations.

A land user approach to the development and implementation of agroforestry technologies requires design and
evaluation according to a complex set of criteria that goes
far beyond simple economic cost and benefit.
Considerations of need, preference, and multiple use must
be balanced against available resources, required inputs,
risk, and expected yields.

In rural areas all over the world, women are providers
of a wide range of subsistence goods, including water,
fuel, food, fodder for confined animals, fiber for handicrafts and other "minor" products of range, forest and
fallow lands. As such, they have much to gain from
development approaches like agroforestry that incorporate a

wide range of products, services, and concerns beyond cash crops, livestock, and staple grains.

Beyond their concerns in crop and livestock production, women's responsibility for household water and energy supplies gives them a special interest in the long term maintenance of the natural resource base (soil, water, vegetation). Researchers must pay specific attention to the history of resource use and condition, and to potential improvements in soil conservation, watershed management, and management of range and forest lands (Rocheleau and Hoek 1984).

Multiple Users

Any program that purports to serve the majority of rural people is by definition dealing with a diverse array of land users, many of whom are women. Even projects specifically geared to "target groups", such as farmers, will find that their target group may include non-farming land users. Farmers also depend on a number of items they do not produce, such as gathered products. Farmers' livelihoods may be inextricably tied to those of gatherers, processors, merchants, artisans, farmworkers, herders, or forest dwellers.

Within a target group loosely defined as farmers there may also be several types of actors, often women, who neither own nor manage the farm. Women and children in farm households perform such essential operations as paid and unpaid farm labor, child care, home management, domestic and commercial processing, gathering of goods for farm household use or sale, and management of livestock and household gardens. If all women land users are to be fully served as clients, then agroforestry research and action must also address gathering, processing, trade and consumption as well as production processes in the farm household and community system.

For example, agroforestry technology design should consider landless men and women, who tend to depend more heavily on gathering than the population-at-large in many farming communities. Whereas wealthier women may gather more on their own land or have easy access to other lands, landless people and women small-holders share problems of insecure access to shared collection areas. A change in the cropping system, a new chemical herbicide, or a change in tree species in bush-fallows may have important "side effects" on gatherers. Agroforestry technologies can be specifically tailored to maintain or increase the flow of "by-products" to particular groups, including women.

154

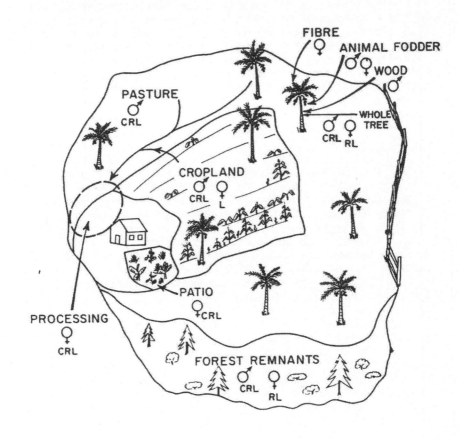

FIGURE 11.1

PANANAO SIERRA, DOMINICAN REPUBLIC

This figure demonstrates the multi-purpose use of land and trees in Pananao assuming that both men and women are present in the household. R = responsibility to provide a product thereof to household; l = labor input for establishment, maintenance or harvest; c = control of resource or process.

Source: Rocheleau, D. (1987a)

A multiple user approach also allows for separation between spheres of activity and control between men and women, between age groups and between classes of households. "Management" of specific plants or places may be subdivided between these same groups. Examples abound of the need for agroforesters to deal with multiple users as clients even with respect to single tree species. The case of Pananao in the Central Mountains of the Dominican Republic (see Figure 11.1) illustrates the multiple and sometimes conflicting uses of individual palm trees by men and women. The same tree or parts thereof can be used for fiber by women, cheap construction wood by men, and animal feed by men and women.

In Pananao, the distinct division of control and responsibility over resources and labor extends to spaces and activities as well as to plants or specific products. Women's processing activities require products from men's fields, herds, and woodlands. While women control the processing enterprise, they do not manage source areas of raw materials. In this community, some cassava bread enterprises have been severely curtailed by fuelwood shortages resulting from rapid conversion of woodlands to cropland and pasture by men (Rocheleau 1984). Women's handicraft enterprises also suffered from raw material shortages when swine fever reduced demand for palm fruit for hog feed and men felled the palms for cheap building materials or cash (E. Georges 1983, pers. comm.).

Agroforestry design in such situations clearly requires consultations with both men and women to design agroforestry practices that address the needs of each group, whether separately or jointly. Agroforestry technologies for multiple users can accommodate separate, fully shared, or interlocking (partially shared) use depending on the compatibility of both the uses and the users.

THE RURAL LANDSCAPE AS CONTEXT AND FOCUS

The landscape embodies spatially and over time rural people's ideas of their relation to each other and to the natural environment. Visible landscape patterns and features provide an excellent point of departure for determining the spatial distribution of men's and women's domains of activity, responsibility, control, and knowledge. During the past few years, many societies have experienced dramatic changes in the division of space and activity due to the introduction of cash cropping, commercial logging, and other enterprises.

The process of "landscape domestication" in rural areas presents a challenge for agroforestry design and practice.

While this aspect of rural development has been left in the gap between natural resource management, farming systems research, and rural women's programs, it is precisely at this level that many rural people integrate trees, crops, and livestock with personal and community needs and objectives. It is also the site of many gender-based land use conflicts. In many areas women are moving rapidly into activities and spaces formerly occupied by men, though often with less security of access to productive resources.

The rural landscape is the drawing board for integrated agroforestry diagnosis and design beyond the single farm or the individual plot (Rocheleau and Hoek 1984). Since women are responsible for collecting water, fuelwood, and other "off-farm" resources, they have a vested interest in the planning of the larger landscape. Women's access to off-farm lands, woodland and water resources, and gathered products can be better addressed when landscape is fully integrated into agroforestry analysis and design.

Tenure is inextricably tied to the evolution and design of landscape, and to the place of women's resources and interests in the landscape. Land and tree tenure are particularly important for tree planters and managers, compared to annual cropping that is more ephemeral or animal husbandry which is a more mobile enterprise. Where agroforestry designs apply to several categories of land, land use, and plants in a complex landscape, then tenure assumes even greater importance.

Community development cycles (settlement, expansion, diversification, land use intensification) will determine in large part the future availability of landscape niches for women's agroforestry activities at the community level. Oral history and discussions of possible future scenarios with women and the community at large may provide some insights into current trends. The choice of agroforestry practices and landscape designs appropriate for rural women requires their involvement from the beginning in whole commuity applications as well as in individual farm planning.

Within the context of landscape planning and design, a diverse array of agroforestry technologies can address a wide range of land use and production units. These units may range from small plots to farms, watersheds, communal holdings, and public lands. The managers may be men or women, acting as individuals/households or as whole ethnic groups, cooperatives, communities, or larger political units. Land use planning at multiple scales requires an integrated social and ecological approach to agroforestry that deals with the division of labor, responsibility, expertise, control, and interests at the intra-household, inter-household, and community level. To deal with this complexity, a user approach in agroforestry research and development must stratify clients by class, sex, household composition, and social organization, as it affects access to resources and spatial patterns of activity and resource use.

An example from Bhaintan watershed in the Lower Himalayas, Uttar Pradesh State, India (Raintree et al. 1985) illustrates the role of gender in the interplay between multiple users and landscape units in analysis of agroforestry potentials. The landscape sketch (Figure 11.2) shows the distinct division in land use and cover which is closely related to tenure. There is a pronounced division of use control, and access to specific landscape features, based on sex.

The relative share of production (and land use pressure) from a given area also varies by user group (Figure 11.3). In turn, the relative importance of particular areas to each user group also varies. In this case, the forest reserve is most heavily used by men, yet it is most important to poor women in terms of its relative contribution to their livelihood. While women's harvest from the forest may be "minor" compared to men's timber offtake, the forest products are major components of women's total income. Moreover, poor women's interest in renewable use and sustained yield may be more compatible with national and village level objectives for the commons and forest reserves.

The potential for commercializing minor forest products versus timber resources in the Himalayan foothills is a good example of this (Surin and Bhaduri 1980). Women are already interested and involved in cash enterprises based on gathering, processing, and retailing of many forest products, and might be best served by projects to improve and sustain that activity rather than by planting new stands of trees that will not yield products for processing by women. Since women's enterprises depend largely on renewable products, this presents an opportunity for an agroforestry system to serve women-as-gatherers, while ensuring sustainable, renewable resources.

THE ROLE OF INDIGENOUS KNOWLEDGE

Agroforestry as a "formal" science is in a unique position to learn from and to improve upon traditional knowledge and practice and to combine forces with indigenous experimental initiatives (Rocheleau and Raintree 1986). The relative ignorance of the research community about woody plants used by rural people implies a special need

158

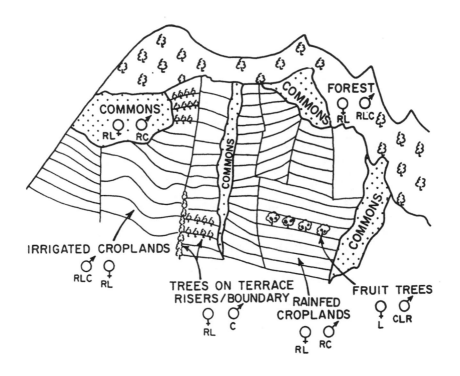

FIGURE 11.2

FAKOT VILLAGE, BHAINTAIN WATERSHED, UTTAR PRADESH, INDIA

R, L, and C have the same meaning as in Figure 11.1.

Source: Rocheleau, D. (1987a)

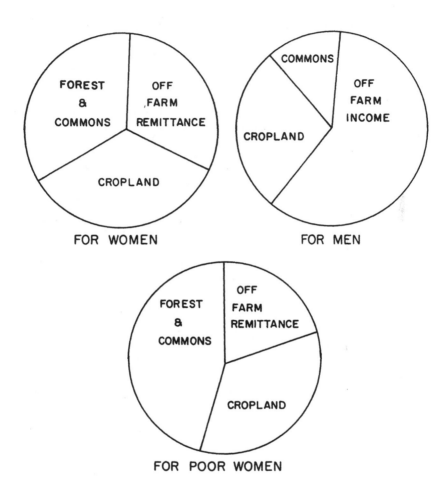

FIGURE 11.3

SOURCES OF LIVELIHOOD (CASH & KIND) IN FAKOT
(by relative importance to land user).

Source: Rocheleau, D. (1987a)

for ethnobotanical research to identify promising species
(woody and herbaceous) for agroforestry systems and to
understand what is already known about these plants' interaction with soil, animals, other crops, and their uses,
ownership, and management. Within the context of such
initiatives, women's knowledge, skills and interests can
change the content and approach of future agroforestry
research and action programs to serve women and the rural
population at large better.

There is a tremendous depth of indigenous knowledge
about particular traditional agroforestry systems under
very site-secific circumstances (von Maydell 1979;
Fernandes and Nair 1986; Flores Paitan 1985; Brokensha et
al. 1983; Budowski 1983; Okafor 1981; Weber and Hoskins
1983; Clay 1983). Existing knowledge can span the full
range of design and management considerations: site
selection, preparation and management; plant selection
and/or breeding; plant propagation, establishment and
management; plant combinations and spatial arrangements;
plant-soil-water interactions; pest management; technologies for processing and use of products; and market
conditions at local and regional levels. Men's and women's
knowledge of these various aspects of traditional agroforestry systems is often quite distinct and may require
separate documentation and discussion (Hoskins 1979).

Existing and potential agroforestry systems include a
particularly diverse array of species, both woody and
herbaceous, many of which are wild or only semi-
domesticated thus far. In cases where agroforestry is not
well developed as such, the local people may still have a
wealth of knowledge about useful plant species, including
source areas of superior parent material, the ecology of
the plant habitat, compatibility with other plants, interaction with animals and insects, growth rate, method of
regeneration, and response to variation in site conditions
and management practice. Women and men will often have
distinct skills and knowledge for use of natural vegetation
in forests and rangelands. They may each have different
knowledge about the same plants and places, or their experience may be divided by species or by ecosystem.

Rural people can also play a key role as consumers in
deciding the criteria for the selection and improvement of
agroforestry germplasm and in judging the likelihood of
domesticating particular species. Women's knowledge as
consumers and processors of many tree products should
figure strongly in any user-focused program of germplasm
selection and improvement (Hoskins 1983).

The incorporation of both men's and women's knowledge,
experience, and experimental initiatives into agroforestry

research plans is illustrated by research on the chitemene system of shifting cultivation in the miombo woodland of north east Zambia. The area is rich in examples of indigenous knowledge as well as its dynamic application to technology innovation.

The classic chitemene system involves the felling and harvest of woody vegetation on a 1-5 hectare plot, followed by piling and burning the collected wood from the entire area on a sub-plot approximately one-fifth of the total area. The combination of high heat and woody biomass results in higher soil pH and fertility on the burned plot (Mansfield 1975). The crop rotation follows a six year cycle beginning with finger millet, maize, cassava, and perennial sorghum intercropped with yams, gourds, pumpkin and cowpea on the periphery or on termite mounds. Groundnuts are planted next, followed by cassava maturation and harvest, and two to four years of bean cultivation (Figure 11.4) after which the plot is left in woody fallow for several years. Most households maintain at least four fields in different stages of the cycle so as to produce the full range of major crops (millet, groundnut, cassava, and beans) in any one year (Stollen 1983; Vedeld 1981; Haug 1981). The long term effects of this system on soil fertility vary with the length of the fallow with a trend toward shorter fallows and sharp declines in site productivity (Mansfield 1975).

An informal survey of the land users in the vicinity of the Misamfu Research Station revealed a wealth of information and opportunities for collaborative experiments on farmer-initiated innovations and farmer-defined lines of research. The survey incorporated a user perspective, which included consultation with both men and women land users as clients, consideration of multiple uses, multiple users, and a sliding scale of analysis from region to plot with emphasis on landscape features and land use at the farm and community level. The method and content of these consultations encouraged people to draw upon and explain specific items from traditional bodies of knowledge, as well as their methods and rationale for developing or adapting new technologies (Huxley et al. 1985; Mattson, in press).

Several points of information proved to be critical for the design of new agroforestry technologies for testing on-farm or on-station and for agroforestry research planning at the Misamfu station.

First, many farmers are actively engaged in experimentation with mounding as a way of incorporating plant biomass (usually grass, with some tree and shrub parts) into the soil. The mounding of loose topsoil over plant biomass

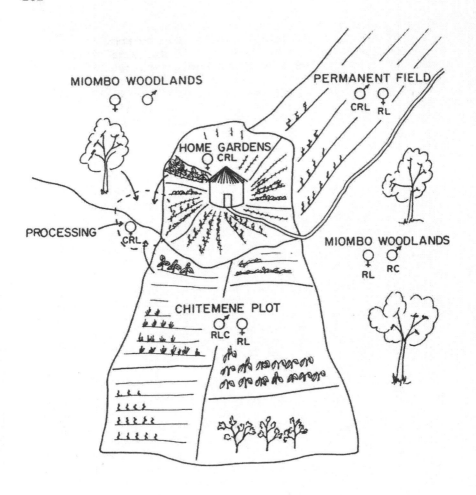

FIGURE 11.4

MISAMFU, N.E. ZAMBIA

This figure shows the Chitemene system in northeast Zambia,
including new practices observed near Misamfu. Note that
women control the millet crop (one of several in the
intercrop rotation.) R, L, and C have the same meaning as
in Figure 11.1.

Source: Rocheleau, D. (1987a).

has been adapted from the grass mounding technology of a neighboring savannah group. It is being tried in permanent or long term plots planted to beans, or beans and cassava, in women's home gardens planted to beans, cassava, fruits, vegetables, and specialty crops, and it is used in the latter part of the cycle on chitemene plots to prolong the useful life of the plot for bean and/or cassava production (Huxley et al. 1985; Haug 1981). Second, women's home gardens are becoming increasingly important for food production and cash income and are being diversified to include fruit trees. Some women are experimenting for the first time with tree planting in such gardens. Mounding, raised beds, and clean tilled plots are all being tried, with a tendency toward mounding in the larger gardens. Women heads of household rely heavily on cassava home gardens for food production to supplement what they can buy with wages. For women without male household labor for chitemene clearing, the home garden, beer making, and cassava processing are important alternatives to earn cash to buy food. Most garden experiments reflect a desire to intensify land use on small plots and to diversify processing enterprises (Stollen 1983; Huxley et al. 1985).

Two women farmers' experiments were especially noteworthy. One woman conducted a trial with low level fertilizer application on a clean tilled plot with millet and cassava, with a partial control (clean tilled, no fertilizer). This trial combined the site preparation technique for monocropped maize with lower fertilizer levels and traditional chitemene crops. The result was increased millet yield, with lower cost and less risk than maize. The woman who conducted the experiment wanted millet for home consumption and beer brewing. Another woman planted soybeans on clean-tilled plots for soya milk, prompted by a concern for nutrition and by free seed provided through her daughter's participation in an urban women's program in the mining district (Huxley et al. 1985).

Both men and women indicated several important roles of woodland and fallowland products in the household diet and economy (both commercial and subsistence). Woodland and fallow areas are major sources of wild leafy vegetables, and exclusive sources of mushrooms and caterpillars, that occur mostly on one tree species, Julbernardia paniculata (S. Holden, pers. comm.). Caterpillars and wild leafy vegetables are major sources of protein and both mushrooms and caterpillars are important sources of cash income for most households. All three products fall within women's domain of responsibility as providers and may be processed or sold by them. Timber (men's responsibility), fuelwood (women's responsibility), and wild fruits were also cited as important woodland products, with supply problems occurring mainly near towns and old villages (Huxley et al. 1985; Mattson, in press).

Trees play an important part in the land use system, including those planted or "kept" in home compounds, fallows, and cropland as well as those found in woodland. A considerable body of knowledge and experience exists with respect to both indigenous and exotic, wild and domesticated species. Some men had extensive knowledge of exotic fruit tree horticulture, including layering and grafting techniques. Both men and women readily identified their respective favorite non-domesticated tree species by use, those species in short supply, and those that they would consider planting now, or in the future in the event of limited supplies or access (Huxley et al. 1985). Both men and women also provided information on site requirements, potential for management (tolerance to coppicing, pollarding), relative growth rates, and relative leafy biomass production for several species that occur in miombo woodland succession (Mattson 1985; Huxley et al. 1985). While both men and women knew the miombo woodland ecosystem well, their experience tended to be divided by species.

In spite of the extent of the surrounding woodlands, many farmers surveyed were often conscious of relative land limits based on proximity to markets, rivers, and roads. Most people were concerned about defining and securing their land rights prior to the imminent return of the mining population to their home area. People's decisions to intensify cropping in place or to expand their cropland varied mainly with household composition, village development cycles, and the quality of the village site and services. In many cases people were unwilling to move out and away into outlying woodlands, and they chose instead to intensify production.

Many women heads of households and sub-households cited woodland gathering and home garden intensification as their best strategies to supplement household food supply and cash income. As a group they were less able to move into new woodland areas and they expressed a greater interest in more intensive use of both woodlands and farmlands.

If national programs are prepared to follow the lead of rural land users, knowledge of indigenous science and users' initiatives may alter national agricultural and rural development policy. Useful information and techniques can best flow from the scientific community to the rural land users once its known what they already know, and what else might be most useful to add to their store of knowledge and tools. A well-informed basis for agroforestry research and action programs must incorporate and address women's and men's distinct domains of both knowledge and concern.

CONCLUSION

Women's interests in agroforestry research and development will not be the same everywhere. They will usually be nested within a larger tangle of conflicting and complementary relationships between and within rural households. Whether or not ownership is legally demarcated, most rural people operate in overlapping domains of access and control on a variety of resources involving a complex array of activities and purposes. Technological changes in domains controlled by men may drastically alter the terms of access, control, production, and ecological stability on shared lands and resources, or in women's separate places and activities. Aside from the differential effects of technology and land use change on men and women, the interests of different groups of women may diverge significantly. Among the factors that may divide women's interests are age, class, household composition, ethnic group, location, and sources of livelihood.

The proposed land user perspective can incorporate women as one of a number of valid client groups and active participants in agroforestry research and action programs. This approach can address women's distinct needs, constraints, opportunities, and interests in agroforestry technology and land use innovations. Since it is based on a premise of dealing with multiple users and multiple interests in any given place, the land user perspective c'an also accommodate both women's relationship to the larger community and the differences between groups of women within a given community. This approach combines an explicit concern for women's interests with a commitment to address those interests within the larger web of social and ecological relationships in which they live.

NOTES AND ACKNOWLEDGEMENTS

A similar version of this paper was published in a book edited by H. Gholz, 1987. Agroforestry: Realities, Possibilities and Potential. Boston, MA: Martinis Nijhoff, in the chapter entitled: The User Perspective and the Agro-forestry Research and Action Agenda.

REFERENCES

Budowski, G.
 1983   An Attempt to Quantify Some Current Agroforestry
     Practices in Costa Rica. In Plant Research and
     Agroforestry. P. Huxley, ed., pp. 43-62. Nairobi:
     ICRAF.
Brokensha, D., B. W. Riley and A.P. Castro
 1983   Fuelwood Use in Rural Kenya: Impacts of
     Deforestation. Washington, D.C.: USAID.
Clay, J.
 1983   A Bibliography of Indigenous Agroforestry
     Systems. Draft, unpublished manuscript.
FAO
 1985   Tree Growing by Rural People. Review Draft.
     Forestry Policy and Planning Series. Rome: FAO.
Fernandes, E.M.C. and P.K.R. Nair
 1986   An Evaluation of the Structure and Function of
     Tropical Home Gardens. ICRAF Working Paper 38.
     Nairobi: ICRAF.
Flores Paitan, S.
 1985   Informe Sobre el Papel de Umari en los Sistemas
     de Produccion Agroforestal en Fincas de la Poblacion
     Indigena y los Mestizos en la Zona de Iquitos.
     Manuscript. Peru: University of Iquitos.
Fortmann, L. and D. Rocheleau
 1984   Why Agroforestry Needs Women: Four Myths and a
     Case Study. Unasylva 36:146.
Haug, R.
 1981   Agricultural Crops and Cultivation Methods in
     the Northern Province of Zambia. Occasional Paper
     1, Dept. of Agricultural Economics: Agricultural
     University of Norway.
Hoskins, M.
 1979   Women in Forestry for Local Community
     Development: A Programming Guide. Washington,
     D.C.: USAID Office of Women in Development.
 1983   Rural Women, Forest Outputs and Forestry
     Projects. Rome: FAO.
Huxley, P.A., D. E. Rocheleau and P.J. Wood
 1985   Farming Systems and Agroforestry Research in
     Northern Zambia. Phase I Report: Diagnosis of Land
     Use Problems and Research Indications. Nairobi:
     ICRAF.
Jain, S.
 1984   Standing Up for the Trees: Women's Role in the
     Chipko Movement. Unasylva 36:146.

Lundgren, B.
    1982   Introduction. Agroforestry Systems 1:3-6.
Mansfield, J. E.
    1975   Summary of Research Findings in Northern
            Province, Zambia. Supplementary Report 7. Land
            Resources Division, Ministry of Overseas
            Development, United Kingdom.
Mattson, L.
    1985   Summary Report on Survey of Farm Level Problems,
            Needs, Existing Strategies and Knowledge of Miombo
            Woodland Species. Misamfu, Zambia: Zambia Ministry
            of Agriculture and Water Development SPRP.
    n.d.   Landscape Analysis of Agroforestry Systems in
            Northeast Province. Agricultural University of
            Norway, As. (In press).
Maydell, H. von
    1979   Agroforestry to Combat Desertification: A Case
            Study of the Sahel. In Agroforestry: Proceedings
            of the 50th Symposium on Tropical Agriculture, pp.
            11-24. Bulletin 303. Amsterdam, The Netherlands:
            Department of Agricultural Research, Konink Lijk
            Institute voor de Tropen.
Okafor, J. C.
    1981   Woody Plants of Nutritional Importance in
            Traditional Farming Systems of the Nigerian Humid
            Tropics. Ph.D. Dissertation, University of Ibadan,
            Nigeria.
Raintree, J. B.
    1983   A Diagnostic Approach to Agroforestry Design.
            Proceedings of the International Symposium on
            Strategies and Designs for Afforestation,
            Reforestation and Tree Planting, Hinkeloord,
            Wageningen, The Netherlands, Sept. 19-23.
Raintree, J.B., D. Rocheleau, P. Huxley, P. Wood, and F.
    Torres
    1985   Draft Report on the Joint ICAR/ICRAF Diagnostic
            and Design Exercise at the Bhaintan Watershed in the
            Outer Himalayan of Uttar Pradesh. Nairobi: ICRAF.
Rocheleau, D.
    1984   An Ecological Analysis of Soil and Water
            Conservation in Hillslope Farming Systems: Plan
            Sierra, Dominican Republic. Ph.D. Dissertation,
            Department of Geography, University of Florida,
            Gainesville, FL.

1986   Criteria for Re-appraisal and Re-design: Intra-household and Between Household Aspects of FSR/E in Three Kenyan Agroforestry Projects. In Selected Proceedings of the Annual Symposium on Farming Systems Research and Extension, Oct. 7-14 1984. C.B. Flora and M. Tomacek, eds., pp. 456-502. Manhattan, KS: Kansas State University.

1987a   The User Perspective and the Agroforestry Research and Action Agenda. In Agroforestry: Realities, Possibilities and Potentials. H. L. Gholz, ed., pp. 59-87. Dordrecht, The Netherlands: Martinus Nijhoff/D. R. Junk Publishers.

1987b   Women, Trees and Tenure: Implications for Agroforestry Research and Development. In Land, Trees and Tenure: Proceedings of an International Workshop on Tenure Issues in Agroforestry. Madison/Nairobi: LTC/ICRAF.

Rocheleau D. and A. van den Hoek
1984   The Application of Ecosystems and Landscape Analysis in Agroforestry Diagnosis and Design: A Case Study from Kathama Sublocation, Machakos District, Kenya. Working Paper No. 11. Nairobi: ICRAF.

Rocheleau, D. and J.B. Raintree
1986   Agroforestry and the Future of Food Production in Developing Countries. Impact of Science on Society 142:127-141.

Scott, G.
1980   Forestry Projects and Women. Washington, D.C.: The World Bank.

Stollen, K.A.
1983   Peasants and Agricultural Change in Northern Zambia. Occasional Paper 4, Department of Agricultural Economics: Agricultural University of Norway.

Surin, V. and T. Bhaduri
1980   Forest Produce and Forest Dwellers. Proceedings of the Seminar on the Role of Women in Community Forestry, Dec. 4-9. Dehra Dun, India: Forest Research Institute and Colleges.

Vedeld, T.
1981   Social-Economic and Ecological Constraints on Increased Productivity among Large Circle Chitemene Cultivation in Zambia. Occasional Paper 2, As, Norway: Department of Agricultural Economics, Agricultural University of Norway.

Weber, F. and M. Hoskins
    1983   Agroforestry in the Sahel. Blacksburg, VA:
        Virginia Polytechnic Institute and State University,
        Department of Sociology.
Wiff, M.
    1984   Honduras: Women Make a Start in Agroforestry.
        Unasylva 36:146.

12
Farming Systems Research in the Eastern Caribbean: An Attempt at Analyzing Intra-Household Dynamics

Vasantha Chase

The objective of this paper is to describe and discuss the attempts of the Caribbean Agricultural Research and Development Institute (CARDI) at incorporating intra-household dynamics in farming systems research (FSR) in the Eastern Caribbean.

CARDI was established in 1975 to serve the agricultural research and development needs of the twelve member countries of the English-speaking Caribbean community: Guyana, Trinidad and Tobago, Barbados, Grenada, St.Lucia, St. Vincent and the Grenadines, Dominica, Montserrat, Antigua and Barbuda, St. Kitts-Nevis, Jamaica, and Belize. In 1983, CARDI initiated a Farming Systems Research and Development Project (FSR/D) with USAID assistance.

CARDI's FSR/D project is designed to facilitate the growth of a more diversified agriculture consistent with the changing political and economic requirements of the Eastern Caribbean states. The project is particularly concerned with developing technologies appropriate to the circumstances of target groups of farmers (CIMMYT 1985). The direct beneficiaries of the project are typically the small farm household and the farm family is the unit of analysis. However, until very recently, CARDI has been emphasizing understanding farmer circumstances at the expense of understanding the nature of the dynamic interactions that exist within the farm family.

Several studies carried out in the Caribbean (Buvinic et al. 1978; Ellis 1982; Knudson and Yates 1981) have shown that the different roles of household members in food production for income generation and household requirements directly affect the farming systems in which they participate. Women in the Caribbean confront special problems in becoming more efficient food producers because of their multiple roles in agriculture, child care, and home maintenance (Narendran 1982). Although a substantial number of women are engaged in farm work, women farm operators on the average receive less income than men and many of the women classify themselves as "housewives" rather than as "farmers." Women also receive less attention from the extension service than do male farmers.

Furthermore, there are distinct areas of farm decision-making in which women play an important role. In many Eastern Caribbean farming systems, adult males have primary responsibility for deciding about the time and nature of operations related to the cultivation of bananas in the Windward Islands (Grenada, St. Lucia, St. Vincent, Domini-'ca), of sugar cane and cotton in the Leeward Islands (Antigua, St. Kitts-Nevis, Montserrat), of other cash crops, and of the maintenance and sale of livestock. While women contribute much of the labor in such activities as planting, weeding, harvesting, and marketing for domestic and regional markets, the allocation of labor is determined by the husband or a male relative. The wife generally has virtual autonomy in decisions about the cultivation of legumes, vegetables, and root crops for home consumption and for sale in domestic markets. Moreover, decisions about family food consumption and nutrition lie solely within the domain of women.

However, agricultural development projects and researchers largely work with male farmers and field pro-duction. Traditionally female dominated aspects of farming systems have gone without much attention. Given the exten-sive participation of women in small farm agriculture in the Eastern Caribbean and its own experiences conducting research with small farmers, CARDI has now embarked on refining its methodology to include an understanding of intra-household dynamics. It is felt that assuming the male farmer makes all the decisions or accurately repre-sents the interests and intentions of other members of the farm household may yield the wrong answers to research questions.

AREA FOCUSED STUDY: MABOUYA VALLEY

As its first attempt at including intra-household vari-ables in the design of appropriate technologies, CARDI con-ducted an Area Focused Study (AFS) in the Mabouya Valley in St. Lucia to gather original micro-level data in several closely situated communities. The goal of the study was to provide information for designing, testing, documenting, and evaluating improved production systems. Close atten-tion was placed on analyzing the farm household and its patterns of labor allocation, decision-making, and resource flows.

Following the recommendations of McKee (1984), two
issues were highlighted in the analysis:
(1) The household was defined in terms of its composition and its production and consumption functions. This
definition was useful for the formulation of household ty-
pologies which were used as one of the criteria in selecting the sample for on-farm trials.
(2) The farming system was defined in terms of the
overall household production/consumption system. This
helped identify the appropriate decision marker(s), the
different sources of labor available to the farm family,
the allocation of resources, and consequently, the major
areas for intervention.

THE MABOUYA VALLEY

The Mabouya Valley was selected as the research site
because the valley represents a wide range of farming systems found in St. Lucia; it encompasses a number of agroecosystems; and its agricultural and political history has
been influenced both by a large plantation and by numerous
small holders with varying forms of tenure. Traditionally,
this valley has been an area that has experienced male
migration which in turn has resulted in matrifocal residence. Women in the Mabouya Valley have strong incentives
-- the need to work coupled with low levels of education --
to take low paying jobs in the marginal sectors. Finally,
in 1984 the Government established the Dennery Basin
Advisory Committee to advise on an integrated development
proposal for the valley. One of the many proposals
includes the subdivision of approximately 100 hectares of
hillside lands into small farms to be leased to local
farmers.
The Mabouya Valley is located in the center eastern
part of the island and it contains the largest expanse of
flat fertile land on the windward side of St. Lucia,
comprising 4,043 hectares. Dennery Farm Company (FARMCO),
a Government owned estate, constitutes approximately 26
percent of the valley. The present land use in the most
easterly and wind-driven parts of the catchment area is the
intensive cultivation of slopes with banana and root crops.
In many cases this had led to the complete removal of the
topsoil.
A rapid appraisal of the Mabouya Valley was carried out
by CARDI and personnel from the Ministry of Agriculture.
Linkages were set up with the Central Planning Unit and the
Organization of American States, that provided the relevant
background information on settlement patterns in the Valley
and with recently drawn land-use maps.

Discussions were also held with officials of the Dennery FARMCO, who also provided information on the agricultural history of the valley. This information was confirmed in discussions and interviews with a number of prominent farmers in the area. The rapid appraisal identified three agroecosystems whose characteristics are profiled in Figure 12.1. There are twelve communities in the valley but only eight were included in the Area Focused Study. Tables 12.1, 12.2, and 12.3 describe the agroecological and socioeconomic characteristics of each of the communities.

Out of the population of 810 households, a stratified sample of 155 households (19 percent) were interviewed. Seven of the 155 questionnaires had to be discarded because of inconsistent or incomplete data. Of the sample interviewed, 92 (62.2 percent) were male farmers while the remaining 56 (37.8 percent) were female farmers.

Basic demographic and farm data revealed no significant gender differences. Even when labor utilization by agricultural activity was compared by gender, the variation was significant only for specific cultural practices. The findings of the Area Focused Study are very detailed and cannot be presented completely here due to a lack of space, but it should be mentioned that while a formal survey revealed no gender differences, in-depth informal discussions and case studies revealed otherwise.

For example, patterns of the utilization of hired versus family labor did reveal gender differences. The percentages of respondents who utilized family labor for planting was 89 percent for females and 43 percent for males. In-depth discussions to ascertain the reasons for this variation revealed that capital resources available to the female farmer are limited, thus every attempt is made to use family labor. In many instances the entire land available to the female farmer is not always planted with crops. At least 75 percent of the female respondents claimed that they cultivated only the amounts of land that family labor could maintain.

Although survey results show no difference in types of crops cultivated, the female farmers claimed that they tend to emphasize short term crops such as roots and vegetables. It is their contention that these crops can be easily maintained by family labor only. Forty percent of the female respondents said that they had fairly regular access to a pool of family labor. This labor pool was made up of an assortment of relatives, not all of whom necessarily lived in the same household.

It is interesting to note the variation in labor utilization by gender observed in the lower valley where nearly all the holdings (including FARMCO) are more than 25

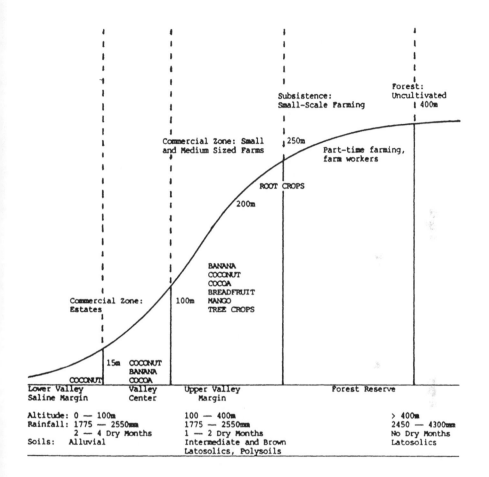

Forest:
Uncultivated
400m

Subsistence:
Small-Scale Farming

Commercial Zone: Small   250m
and Medium Sized Farms

Part-time farming,
farm workers

ROOT CROPS

200m

BANANA
COCONUT
COCOA
BREADFRUIT
MANGO
TREE CROPS

100m

Commercial Zone:
Estates

15m   COCONUT
       BANANA
COCONUT   COCOA

Lower Valley Saline Margin	Valley Center	Upper Valley Margin	Forest Reserve
Altitude: 0 — 100m		100 — 400m	> 400m
Rainfall: 1775 — 2550mm		1775 — 2550mm	2450 — 4300mm
2 — 4 Dry Months		1 — 2 Dry Months	No Dry Months
Soils: Alluvial		Intermediate and Brown Latosolics, Polysoils	Latosolics

FIGURE 12.1

AGRO-ECOLOGICAL TRANSECT OF MABOUYA VALLEY OF ST. LUCIA

TABLE 12.1

COMMUNITIES SAMPLED BY AREA FOCUSED STUDY IN MABOUYA VALLEY

Community	Agroecological Niche	Approx. Area (Acres)	Total Population	Total Households	Sample Households Male	Female
Gadette	Lower Valley	64	500	109	13	8
Richefond	Lower Valley	32	528	115	5	2
Grande Riviere	Lower Valley	50	714	155	45	28
La Resource	Lower Valley	23	405	88	1	3
Derniere Rivere/ Belmont	Central Valley	168	1,785	388	45	28
Grande Ravine	Central Valley	42	381	83	6	2
Aux Lyon/ Despinoze	Upper Valley	109	1,333	290	12	9
La Perl/Lumiere	Upper Valley	57	138	30	—	—

TABLE 12.2

AGROECOLOGICAL CHARACTERISTICS OF THE MABOUYA VALLEY

Characteristic	Lower Saline/ Central Valley	Upper Valley	Forest Reserve
Ecological Zone	lowland humid tropics	mid-elevation humid tropics	high elevation humid tropics
Farming System	irrigated commercial	small/medium scale commercial	small scale subsistence; shifting cultivation
Principal Crops	banana, coconut	banana, coconut, cocoa, yam, breadfruit, mango, plantain, pigeon peas	root crops, plantain
Land Preparation	plough cultivation	plough cultivation on medium-sized farms; manual	clearing by fire; manual; no tillage
Chemical Inputs	intensive	limited	extremely rare
Technology Level	modernized	semi-modern; traditional	traditional
Cropping Pattern	monocropping	multiple cropping	multiple cropping
Agricultural Calendar	fixed dates; regulated by commodity association	fixed dates; individual decision	dates highly variable; individual decision

TABLE 12.3

SOCIOECONOMIC CHARACTERISTICS OF THE MABOUYA VALLEY

Characteristic	Lower Saline/ Central Valley	Upper Valley	Forest Reserve
Settlement Type	linear	linear/nucleated	scattered
Social Organization	altruistic; marked economic stratification which affects political affiliation	divisive; economic and political stratification	cohesive
Economic Organization	primarily agricultural, limited commercial activities	agricultural, commercial, and service oriented	only agricultural
Labor Organization	hired labor (full-time)	self-employed, hired labor (part-time)	self-employed
Kinship Organization	nuclear, stem families, female-headed households	extended families, female-headed households	single member, nuclear
Political Organization	leadership influenced by economic status, power and influence; political affiliation influenced by economic stratification	leadership influenced by economic status, power and influence; political affiliation influenced by economic stratification	leadership influenced by knowledge of farming; apathy towards political affiliation
Land Tenure	own, family land	own, family land, rent, squat	squat
Type/Condition of Housing	wooden houses in very poor condition	concrete houses, well built and in good condition; wooden houses in good condition	wooden shacks in poor condition

hectares. All of these holdings are owned by men. The seven female respondents interviewed in this zone are all estate laborers who cultivate small parcels of land either within the boundaries of the estate in which they work or higher up in the valley. Their production is primarily for farm family consumption and for sale within the community. In the lower valley, 58 percent of the females used family labor while only 44 percent of the male farmers used family labor. Some of the male farmers claimed that their utilization of hired labor would have been higher if most of their holdings had not already been established in bananas and other tree crops. This meant that labor (both hired and family) was only being utilized for the maintenance and harvesting of crops. In contrast, female farmers claimed that their utilization of hired labor was high because family labor was either unavailable at all or not available on time for land clearing and preparation activities.

Other variables which showed gender differences included those related to farm management practices. Specifically, female farmers felt that although they grew the same crops as their male counterparts, outputs were markedly lower. They attributed this to their low level of literacy, their difficulties in obtaining capital for farm improvement, poor extension information, and their inability to carry out profitable market transactions, primarily because of a lack of market intelligence.

IMPLICATIONS OF THE AREA FOCUSED STUDY

There were many implications of the area focused study to CARDI's agricultural and livestock specialists for technology generation. One of the recommendations that is aimed at women farmers is the Integrated Backyard System. The major function of the Integrated Backyard System will be to produce food and essential nutrients on a day-to-day basis for immediate family consumption. CARDI's decision to include nutrition in one of its recommendations for the Valley stemmed from the realization that a number of its programs to improve the productivity of small farmers have had little impact on the nutritional status of their families. There may be nutritional imbalances in the valley because of the preponderance of root crops grown in the area, the limited numbers of livestock, the very low production of vegetables and legumes, and the absence of cereal production.

CARDI believes that these nutritional imbalances could be corrected by introducing backyard gardens that would encompass both plant and animal husbandry in relatively confined areas close to family dwellings. The backyard systems have been designed to be small-scale, clearly defined, and controlled by women. They should be unaffected by broader market trends which renders them an area for development that offers the highest pay-offs for limited inputs.

CARDI decided to work with the women in the farm household because the data revealed that women make decisions pertaining to family consumption and food preparation. They have also expressed concerns over family nutrition. If additional labor is needed to maintain the backyard gardens, the women themselves assured CARDI researchers that the work will be considered as labor used in normal household chores if the backyard system provides for the kitchen.

The technology that has been designed for the Integrated Backyard System includes: (1) a fenced-in vegetable plot (the size depends on land availability and farm size); (2) preparation of a compost heap as a source of organic material; (3) preparation of pens for small livestock designed to facilitate easy feeding, watering, and collection of pen manure and other livestock products such as eggs; (4) design of a simple irrigation system to collect rain water from the roofs of livestock pens using bamboo guttering and oil drums; and, (5) planting of small fodder plots as feed banks for livestock, particularly rabbits.

Once the system has been introduced to a sample of approximately 15 households, each household will be closely monitored to identify changes in consumption patterns and household expenditures, and the division of labor by gender that emerges in the activities needed to set up the system and maintain the different crops and livestock components. Experience in Dominica has shown that men do the heavier tasks of land preparation and pen construction while the women do the remaining tasks.

As part of the technology transfer phase, CARDI will prepare leaflets on how to: establish an Integrated Backyard System; maintain the continuity of the system; process and preserve surpluses; prepare food from the products of the system; and on the nutritive value of the crops and livestock. An important component of this phase will be the development of simple record-keeping methods.

Finally, the evaluation will monitor the farm household before and after the introduction of the technology to measure changes in consumption patterns, labor utilization patterns of different household members, and farm family savings through the reduction of food purchases and the sale of surpluses from the system.

Farm households in the Eastern Caribbean in general and St. Lucia in particular have food production systems that make use of a variety of staple and non-staple food crops. Unfortunately, as more and more of these households become integrated into national, regional, and international markets, their consumption preferences and food patterns are becoming altered. Cash cropping has resulted in a decline in production of food for home consumption and a consequent decline in nutrition. Thus CARDI has recognized the urgency for incorporating a food consumption perspective into its farming systems methodology. It is CARDI's contention that while the primary goal of FSR is to increase the overall productivity of farming systems to enhance the welfare of the farm household (Norman 1982:2), improved farm technologies are not always compatible with the goals of the farm family. Therefore, CARDI's decision to include nutrition in one of its recommendations for the Mabouya Valley stems from the realization that a number of its programs to improve small farmers productivity had not positively impacted on the nutritional status of their families. In addition, traditionally female-controlled aspects of farming systems, especially the cultivation of subsistence oriented secondary crops that provide for the nutrition of the farm family, have not received sufficient attention within agricultural development projects. The proposed Integrated Backyard System is an attempt at incorporating both gender and nutrition issues into CARDI's programs.

REFERENCES

Buvinic, M., N. Youssef and B. Von Elm
    1978   Women-Headed Households: The Ignored Factor in
        Development Planning. Washington, D.C.: ICRW.
CIMMYT Eastern African Economics Programme
    1985   Teaching Notes on the Diagnostic Phase of
        OFR/FSP Concepts, Principles, and Procedures.
        Nairobi, Kenya: International Maize and Wheat
        Improvement Centre (CIMMYT).
Ellis, P.
    1982   Agricultural Extension Services and the Role of
        Women in Agricultural Development in the Eastern
        Caribbean. St. Vincent: CARDI.
Knudson, B. and B. Yates
    1981   The Economic Role of Women in Small Scale Agri-
        culture in the Eastern Caribbean -- St. Lucia.
        Barbados: WAND.

182

McKee, K.
    1984   Methodological Challenges in Analyzing the
       Household in Farming Systems Research: Intra-House-
       hold Resource Allocation. In Animals in Farming
       Systems, Proceedings of the FSR Symposium, Kansas
       State University. Cornelia Butler Flora, ed., pp.
       593–603, Manhattan, KS: Kansas State University.
Narendran, V.
    1982   Agricultural Decision Making in the Eastern
       Caribbean. St. Lucia: CARDI.
Norman, D.
    1982   The Farming Systems Approach to Research. FSR
       Background Papers. Manhattan, KS: Kansas State FSR
       Paper Series.

13

Economic and Normative Restraints on Subsistence Farming in Honduras

Eunice R. McCulloch and Mary Futrell

In the last few years, the farming systems approach, the study of how farmers in particular environments manage pertinent resources such as soils, water, seeds, livestock, labor, etc., has become the predominant way in the agricultural disciplines to evaluate farming methods. These methods of analysis are not as useful when dealing with the economics of subsistence farming. Farmers who grow crops primarily for food operate according to a different rationale than do those who expect a monetary profit, however small.

Along the Pacific coast in Central America, in southern Guatemala, El Salvador, Honduras, and Nicaragua, there are areas where the principal crops are maize, sorghum, and beans, usually intercropped (Hawkins 1984). Many of the farming families in these areas attempt to subsist on very small farms and are very poor. Family members, particularly small children, are often malnourished.

In assessing such systems, that operate largely outside the market, farm outputs must be gauged in other than monetary terms; social and cultural factors assume a larger role than might otherwise be the case.

The approach discussed here was developed for research on the Sorghum Millet Collaborative Research Support Project (INTSORMIL) in Southern Honduras. The area of concern is how cultural and economic factors contribute to the farming system that Ruthenberg (1976) would describe as a "low level steady state", a condition to be avoided. The analysis here strengthens this negative view. Such a subsistence system appears to be in equilibrium since the food supply is barely adequate to sustain the population but subsistence occurs at inadequate levels of nutrition. Although actual starvation is not a problem, food shortages and particularly shortages of certain essential nutrients have a notable impact on the health of family members with a resultant high mortality among infants and young children. In addition to limited food production, food habits and intra-household distribution patterns appear to be significant factors affecting population numbers.

THE AREA AND PEOPLE

In the southern district of Choluteca, the Choluteca River cuts through a fertile alluvial valley bordered by mountains as high as 3,000 feet. Vegetation consists of deciduous tropical forest and savannah. Annual rainfall is between 60-80 inches with a marked dry season between the months of December to April. Mean annual temperatures are between 79 to 82 degrees Fahrenheit, but high temperatures in March and April often exceed 100 F.

In this southern province, farms range from large commercial sugar cane plantations or cattle ranches made up of thousands of acres to the smallest minifundia, where farmers strive to simply provide enough food for their families. Official records on production are incomplete on these small farms since most of the produce does not enter the market.

The field research under discussion centered around three village locations. Two of these, El Corpus and Guajiniquil, were in mountainous areas. The other, Pavana, was on the lowland plains.

The life style of these small farmers is characterized by a few material possessions. Houses are built of adobe or wattle and daub. Roofs are made usually of baked clay tiles or, less commonly, of thatch. The houses have earth floors and consist of only one room with one or two small square windows. The main room is often divided into separate areas by a partition made of slats and cardboard that is often "papered" with pages of old magazines, newspapers, or even grain sacks. Furniture is scarce, consisting of one or two wooden folding chairs, a crudely built wooden table and a bench, and a wooden bed with a twine platform on which straw mats are laid. Every house has one or two hammocks that are used for sleeping, resting during the day, or lulling a fretful baby. Kitchen areas are found on one wall of the house or in a separate lean-to structure. Food is prepared at a roughly constructed work counter. A large tin or clay olla (pot) of water is at hand. Cooking utensils are few, consisting of one or two clay or aluminum pots, several plastic containers and baskets, a grinding stone and a clay or iron comal (a slightly rounded pan) for

cooking tortillas on the raised hearth stove which is
constructed of dried mud.

Unlike neighboring Guatemala where Indian groups have
retained local customs, dress and language, the people of
Honduras have become creolized, exhibiting both Spanish and
Indian traits, and using only the Spanish language.
Western clothes are worn; in the countryside, women wear
simply cut dresses, usually of polyester, often with an
apron. Men wear cotton shirts, work trousers, and straw
hats. Men, women, and children go barefoot much of the
time but women also wear plastic scuffs and men work shoes.
The work shoes may be leather or plastic molded with mock
laces and tongues to resemble leather shoes. Men's hair
styles are short and similar to those in the United States.
Unmarried girls have loose shoulder length hair, while
married women wear their hair long and pinned up.

The predominant household grouping is the nuclear family, although some young single women with babies continue
to live at home with their parents. Extended families are
also quite common, as are households in which some other
relative resides with a family. Marital alliances are usually extra-legal; that is, few couples have had religious
or civil ceremonies. Nevertheless, such alliances are
fairly stable, the cooperation of both man and woman being
essential for survival. For this society, the basic unit
of production and consumption is the household.

FARMING TECHNOLOGY

Two different technological systems were encountered
during the study. The farming techniques used were governed by the terrain, but both systems were rudimentary.

In the mountains, farmers use only a planting stick and
a machete as they have done for generations. The planting
stick, or chuzo, is a long pole with a blade set into the
end. Cultivable land available to the small farmers is
scarce and very little land or crop rotation is practiced.
Land is prepared by cutting brush and weeds that had grown
since the previous season and sometimes burned when dry.
Burning was being discontinued on the advice of government
agricultural experts in an effort to halt serious soil
erosion.

Several methods of seeding are practiced. Maize is
planted first in a diagonal pattern by dropping seeds in a
hole made with the planting stick. Sorghum is usually
planted between the maize plants when the maize is a few
inches high or it is planted at the same time in the same
hole or hillock. Occasionally, sorghum seeds are simply
broadcast. Beans are planted among the maize and sorghum a

few days after the initial planting. Cultivation and cutting weeds is done with a machete. Herbicides may be used
to control weeds if the farmer is working a considerable
amount of land and can afford it. Fertilizer and insecticides are too expensive for these farmers to use. In El
Corpus, but not Guajiniquil, some of the farmers are
organized into a liga, a cooperative league.

In the lowlands around Pavana, the level terrain permits the use of oxen and plow. Wooden steel-tipped plows,
carved from large pieces of timber, are used for planting.
Only a few farmers own teams and hire them out to others
for planting. Sorghum and corn are interplanted without
beans because both insects and weeds are worse here. Since
fertilizer is not used and legumes are not grown, soil fertility is low. There are two ligas in Pavana. One of
these, the Esperanza de Pavana, was experiencing almost
total crop failure and did not have the resources to
replant the urgently needed sorghum crop. This cooperative
had received assistance from government sources but was not
being aided at this time. Members were dispirited and work
was ineffective. Both groups were cropping land that had
been in pasture until a few years previously. The
Esperanza land had been cultivated for two years and still
retained a measure of fertility, but the Central farm land
had been continuously cropped for five years and soil fertility had become greatly diminished.

Farm sizes for these subsistance farmers are small,
ranging from 0.5 to 5 manzanas (one manzana = 1.7 acres or
0.7 hectare; 0.5 manzana to 5 manzanas = 0.9 to 8.5 acres).
The cooperative leagues assigned plots of from one to two
manzanas to each household depending on household size and
available labor. One manzana is considered the amount of
land one adult male could work reasonably. Labor appeared
to be the determining factor in the cooperatives since the
harvests were divided communally.

THE SUBSISTENCE CONDITION

Even though data collected indicate that similar arguments could be made for both lowland and upland farming
systems, the remainder of the discussion will focus on the
upland system of intercropping maize, sorghum, and beans.

Production Capability

It is fairly obvious that such small farms with primitive technology would not produce much in the way of cash
income even if the primary goal were other than food
production. In order to evaluate this type of system, an

approach that looks at production in terms of providing adequate nutrition for family members appears reasonable. Existing cultural factors such as dietary preferences and food habits need to be included. Surpluses to be sold for cash income can exist only after basic food needs are met.

A schematic diagram of what such a farming subsistence system might look like for a population such as has been described is shown in Figure 13.1.

Household size ranged from 3 to 13 persons with an average of 7 persons. Labor averaged approximately 1.2 labor units per household. Land in crops averaged 1.1 manzanas per household. Average yields are shown in Table 13.1.

Before considering other farming enterprises and outside resources, the nutritional value of the crops produced will be discussed. It is argued that livestock raising, vegetable and fruit production, and outside economic opportunities are not sufficient to have an appreciable impact on overall nutritional status.

Diets

Maize, sorghum, and beans are staple items in the diets of these farmers. Sorghum and maize are consumed almost entirely in the form of tortillas. Although preferences may be expressed for a number of unaffordable food items, many members of this population must rely solely on tortillas and beans for sustenance for much of the year. During one phase of the field work research, that was during the planting season, it was documented that 8 percent of the families had only tortillas and salt and 51 percent had only tortillas and beans.

TABLE 13.1

ANNUAL AVERAGE YIELDS FOR UPLAND INTERCROPPING OF MAIZE, SORGHUM AND BEANS

Crop	Lbs./mz.	Cost/lb. if purchased	Value $US	Value (1000C.)*
Maize	1300	.12	156	2101
Sorghum	1800	.08	144	2802
Beans	240	.50	120	372
Total for 1.1 manzanas			462	5803

Note: *FAO/WHO 1954.

FIGURE 13.1

CONCEPTUAL MODEL OF SUBSISTENCE FARMING-NUTRITION SYSTEM

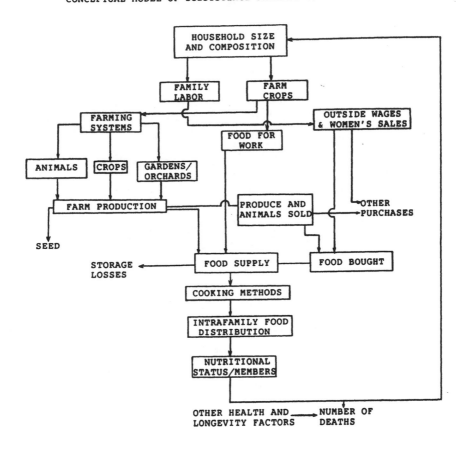

Nutritional Requirements

Household composition and yearly nutritional requirements were computed in terms of the age-sex categories
given in FAO tables (FAO/WHO, 1973). The requirements for
the average family of seven are given in Table 13.2.

Nutritional Output of Cropping Activities

A linear programming model (Beneke and Winterboer 1973)
was used to assess the cropping activities in terms of the
nutritional constraints previously computed and other constraints. Normative constraints included all of the maize
grown (1430 lbs.) since maize is preferred to sorghum for
tortillas; all of the beans (264 lbs., less than one lb.
per day); and fifty lbs. of purchased oil for refrying
beans.

Several runs were necessary and the constraints were
relaxed on scarce nutrients before a reasonable solution
could be obtained. Table 13.3 summarizes how well the subsistence activities are fulfilling nutrition requirements.
The only surplus the farmer produces above household needs
is 594 lbs. of sorghum (worth only $47.52). This means
that daily grain consumption of 7.7 lbs. of grain (informants gave eight lbs. as the daily consumption of grain)
and .7 lb. of beans would theoretically provide the
calories and proteins needed. However, aside from the
deficiencies of Vitamin A, riboflavin, ascorbic acid, and
calcium, calories are the limiting factor and low crop
yields or failures mean that although some of the surplus
sorghum might be used to offset a calorie shortage, serious
food shortages would most likely prevail.

So far it has been shown that the cropping activities
are capable of providing basic calorie and protein requirements. What of the other influential factors such as livestock and vegetable production, outside income, other
necessary cash purchases, food furnished by food for work
programs, food preferences and beliefs, and intra-family
food distribution patterns? These will be briefly discussed in turn.

Livestock

Most of the households kept a few chickens and a pig or
two. The major contribution to the regular diet was eggs
but not enough were produced to constitute a major item.
Chickens were eaten only when there were important guests
or occasionally by someone who was ill. Pigs were only

TABLE 13.2

RECOMMENDED DAILY AND YEARLY INTAKES OF NUTRIENTS FOR AVERAGE FAMILY[a]

Age Group (in years)	Number	Energy[b] (kilocal.)	Protein (g.)	Vitamin A (micrograms)	Thiamin (mg.)	Riboflavin (mg.)	Niacin (mg.)	Ascorbic Acid (mg.)	Calcium (g.)	Iron[c] (mg.)
< 1	.4	436	5.6	120	.12	.20	2.16	8	.20	4.0
1-3	.8	1088	12.8	200	.40	.64	7.20	16	.32	8.0
4-6	.8	1464	16.0	240	.56	.88	9.68	16	.32	8.0
7-9	.7	1533	17.5	280	.63	.91	10.15	14	.28	7.0
10-12 Males	.3	780	9.0	173	.30	.48	5.16	6	.18	3.0
13-15 Males	.3	711	11.1	218	.36	.51	5.73	9	.18	5.4
16-19 Males	.3	747	11.4	225	.36	.54	6.09	9	.15	2.7
> 20 Males	.4	976	14.8	300	.48	.72	7.92	12	.16	3.6
10-12 Females	.2	470	5.8	115	.18	.28	3.10	4	.12	2.0
13-15 Females	.2	416	6.2	145	.20	.30	3.28	6	.12	4.8
16-19 Females	.3	580	9.0	225	.27	.42	4.56	9	.15	8.4
> 20 Females	.3	552	8.7	225	.27	.39	4.35	9	.12	8.4
43.2 Male	1.0	2318	37.0	750	1.20	1.80	19.80	30	.40	9.0
37.2 Female	1.0	1840	29.0	750	1.30	1.30	14.50	30	.40	28.0
Total Persons 7.0										
Total Daily		13,911	193.9	3965	9.37	9.37	103.68	178	3.10	102.3
Total Yearly		5,077,515	70,774.0	1,447,225	2274.0	3420.0	37,843.0	64,970	1132.0	37339.0

Note: [a]Computed from Tables, FAO/WHO report, Energy and Protein Requirements, 1973. Rome.
[b]Ibid. Energy are requirements those for people from developing countries (average weight, adult male = 53 kg., adult female = 46 kg.)
[c]Requirements are those given for diets based on non-animal foods.

TABLE 13.3
SUBSISTENCE ACTIVITIES AND NUTRITIONAL ADEQUACY

Activities	Pounds Produced or Bought	Land Used (mz.)	Monetary Value (U.S. $)
Intercropping of:			
Sorghum	1,386	.76	110.84
Maize	1,430	1.09	171.60
Beans	264	1.10	132.00
Oil (purchased)	50	0.0	31.50
TOTAL		1.10[a]	445.94

Nutrient	Percentage of Needs Met	
Calories	100	(Limiting)
Protein	214	
Vitamin A	31	(Deficient)
Thiamin	270	
Riboflavin	54	(Deficient)
Niacin	107	
Ascorbic Acid	5	(Deficient)
Calcium	33	(Deficient)
Iron	132	
Histidine	403	
Isoleucine	318	
Leucine	627	
Lycine	158	
Methionine plus cystine	226	
Phenylalanine plus tyrosine	569	
Threonine	276	
Tryptophan	118	
Valine	363	

Note: [a]Intercropping total.

raised to be sold when money was needed for emergencies.
The farmers could not afford to keep beef cattle or milk
cows.

Gardens and Orchards

A few of the families in the population studied had
fruit orchards. These most often consisted of a few mango

or banana trees. Occasionally production was sufficient that fruit was sold but this was not a regular or dependable source of income. A few families had very small plots of field peas near their homes and the cooperative had a small patch of yucca.

It was apparent that the small amounts of such produce, both individually and cooperatively grown, were insufficient to supplement the diets in an appreciable way. One exception, perhaps, were the mango trees which were plentiful in all areas, growing near houses or along the side of the village streets. These were not considered as important dietary items, but children habitually picked up and ate this fruit which is a valuable source of Vitamin A and ascorbic acid. It was not possible to measure the contribution of this type of intake to the diets at the time of the study, but unless such items are regularly used they are of limited value in overcoming deficiencies.

Food for Work

Some of the families were receiving commodities from cooperative sponsors. Beans, maize, and rice were being furnished. These supplies did not enable the families to eat more as they were provided during a time of general food shortages. They were distributed twice a month but each time supplies lasted only a week. It was also noted during the research that some of the women were using the maize to make tortillas for sale instead of using it to feed the family. A comparison of the diets of those receiving these subsidies with those not receiving them showed that those who received the food rations did not have substantially better quality diets.

Food Practices

Food preferences and beliefs and food distribution patterns within the family were closely related. Adults most often mentioned meat as the food which is best for adults and the one which they like best. Beans and rice were also frequently mentioned. Tortillas were not mentioned but no meal was eaten without them. Presumably they are well liked but taken for granted. Eggs, milk, and cheese were considered good for adults, but there was a notable lack of appreciation for vegetables.

Data collected revealed a bias against high protein foods and vegetables in children's diets, particularly avocados and other foods green in color. Milk, eggs, and cheese were considered good for children and children were often given snacks of white cheese on tortillas.

A Complete Low Cost Diet

As part of the evaluation of the farming-nutrition system, a model incorporating the most common food items was tested to determine the combination of local foods that could furnish a balanced diet at the lowest possible cost. The on-farm production of eight lbs. of grain (maize and sorghum) per day and the yearly production of beans were entered. The optimum low-cost model that resulted included in addition tomatoes, cabbage, plantain, banana, powdered milk, and mango. The low cost diet is presented in Table 13.4. The cost of this nutritionally complete diet was $1,165.54 or $703.54 more than the value of the crops produced. The cost of coffee and sugar would need to be added to this amount since all families drank coffee each day, usually sweetened. If these items cannot be produced, this amount would have to be earned for household food intakes to be raised to the recommended levels of nutrition.

Money Income

So far discussion has been limited to food requirements with no consideration of other expenses such as clothing, transportation, schooling, etc. It has been shown that additional income is needed just for complete diets. What are the opportunities for these farmers who are largely untrained and not educated?

Small cash incomes are obtained by petty commercial activity and seasonal wage labor. Women make and sell food items and to a lesser extent, hand crafted goods such as mats woven from reeds. Men work on house construction, large commercial farms, or as occasional laborers in the cities when they do not need to work in the fields.

Wages are set by the government and are low. At the time of the study, agricultural workers were paid from $2.30 to $2.50 per day. The highest wage paid for day labor was $3.55 per day.

At $2.50 per day, a farmer could earn about $255 during the dry season. In order to furnish the amount needed to supplement the diet ($703), a man would have to work 281 days, leaving him insufficient time to produce the staples for the family diet if no other cash outlays were made. Wage work does not appear to solve the problem of inadequate nutrition.

TABLE 13.4

FOODS AND THEIR CONTRIBUTION TO THE LOW COST DIET

Food Item	Amount (lbs.)	Percentage of Nutrient Supplied									Cost ($)
		Cals.	Pro-tein	Vitamin A	Thiamin	Ribo-flavin	Niacin	Ascorbic Acid	Calcium	Iron	
Sorghum	1490	45.7	96.6	28.1	121.8	29.6	71.5	0.0	15.8	77.2	119.20
Maize	1430	45.5	87.1	0.0	128.3	20.5	34.4	0.0	4.2	40.8	171.60
Beans	264	8.0	37.4	.8	28.5	6.3	6.7	5.6	14.7	22.5	132.00
Oil	50	3.9	0.0	0.0	0.0	0.0	0.0	0.0	0.0	0.0	31.50
Rice	100	3.2	4.3	0.0	1.6	.4	1.9	0.0	.4	1.3	39.00
Eggs	147	2.1	11.6	16.4	3.0	5.8	.2	0.0	3.1	.4	121.99
Cheese	26	.7	3.0	2.5	.3	1.6	.1	0.0	.3	.3	39.00
Tomatoes	156	.3	1.1	9.9	1.9	.8	1.0	25.0	.7	1.3	42.12
Cabbage	63	.1	.4	.4	.5	.3	.2	15.5	1.0	.3	12.55
Plaintains	150	1.0	.8	3.2	1.2	.6	.7	11.6	.3	1.0	22.50
Bananas	150	.9	.9	2.2	.9	.8	.9	5.4	.4	.8	10.50
Powdered Milk	187	8.2	31.2	19.5	9.0	32.5	1.7	5.3	67.7	1.7	400.98
Mangoes	150	.5	.4	16.8	.9	.8	.7	31.5	.4	.4	22.50
Total Percentage of Amount Recommended		120.1	274.8	100[a]	297.6	100[a]	120.0	100[a]	109.0	148.0	$1165.44

Note: [a] Limiting Nutrient.

Family Size

How does such a subsistence system impact on family size? Nutrition can be expected to have an effect on child mortality and prenatal losses if dietary intakes are inadequate either in quantity or quality. In-depth studies of the health and nutritional condition of young children of this area were conducted over a period of three years.

Wilson et al. (1975) cites studies showing that the incidence of miscarriages, stillbirths, and deaths of newborns is higher for poorly nourished mothers. In this study, 61 percent of the women interviewed had had children die who were under the age of 5 years. Prenatal losses were also high. Twenty-eight percent of the women had lost babies before birth. The number of deaths averaged 1.3 children per women (with five losses for one woman). Most deaths occured before the child was one year old with 40 percent of the children having died in the first ten days. Other factors such as sanitation, disease, and the quality of care can contribute to such conditions.

The studies of the children (Jones, 1983; Simpson, 1982) included those who attended a feeding center in El Corpus. Anthropometric measures of height, weight, upper arm circumference, and triceps skinfold were used to assess degrees of malnutrition, blood and hair analysis, parasitic examinations, and analysis of diets.

It was found in 1981 that around 68 percent of the children (including those attending the feeding center) were malnourished. Severe malnutrition was at 57 percent for stunting and 48 percent for wasting. Children generally had inadequate intakes of calories, protein, vitamin A, and vitamin C. All children examined in 1983 for parasites were infected, many with more than one type.

Based on these results, malnutrition among young children in this population in Southern Honduras appears to be widespread. A greater incidence of malnutrition among children occurred during years of poor harvest. Although actual starvation does not seem to occur (only two cases of kwashiorkor were observed), malnourished children are more likely to succumb to infections, disease, and death than are well-nourished ones.

SUMMARY

The argument presented in this paper attempts to show how very small subsistence farm operations reach a low level steady state in which certain levels of malnutrition

196

become ingrained, at least for some members of the family.
The levels of malnutrition affect health and family size in
a substantial way through infant and child mortality and
prenatal losses. This paper has also tried to show how
such a system can be evaluated in a relevant way.

The cropping systems are viable in the sense that production can furnish the basic staples for the traditional
diet in sufficient amounts under normal conditions. The
typical small farm with primitive technology produces just
enough to supply the average size household with recommended allowances of calories, protein, and some of the
essential nutrients. However, because some nutrients are
not present in the staple foods in sufficient amounts,
additional foods are needed.

The absence of any significant surplus of the crops
produced makes the purchase of additional food difficult.
Other food-producing activities, such as growing small
plots of vegetables or fruits or raising chickens and pigs,
are engaged in on such a small scale that they do not
overcome the deficiencies. In addition, other household
needs (such as clothing or transportation) must be met by
selling part of the crop, leading to a situation where even
calorie requirements might not be met. Small cash enterprises such as the making and selling of food by the women
bring small amounts of cash into the household. Low wages
for men and the amount of time that can be spared from
cropping activities limit outside earnings.

Economic, technological, and nutritional factors interact to produce a self-sustaining adaptation at minimal
levels of nutrition, one in which a chronic state of under-
nutrition seems inevitable.

REFERENCES

Beneke, R. R. and R. Winterboer
    1973   Linear Programming Applications to Agriculture.
            Ames, IO: Iowa State.
Food and Agricultural Organization (FAO)
    1954   FAO Nutritional Studies No. 11. Food Composition
            Tables: Minerals and Vitamins for International Use.
            Rome: FAO.
    1973   FAO Nutrition Meetings Report Series No. 52,
            Energy and Protein Requirements. Rome: FAO.
Hadley, G.
    1962   Linear Programming. Reading, United Kingdom:
            Addison-Wesley.

Hawkins, R.
   1984   Intercropping Maize with Sorghum in Central
            America: A Cropping System Case Study.
            Agricultural Systems 15:79-99.
Jones, R. E.
   1983   Nutritional Status of Preschool Children in
            Honduras Where Sorghum is Consumed. Ph.D.
            Dissertation, Mississippi State University.
Ruthenberg, H.
   1976   Farming Systems in the Tropics. Oxford:
            Clarendon.
Simpson, C. K.
   1982   Nutritional Status of Honduran Children in Two
            Nutrition Intervention Programs. Unpublished MS
            Thesis, Mississippi State University.
Wilson, E. D., K. H. Fisher and M. E. Fuqua
   1975   Principles of Nutrition. New York, NY: John
            Wiley and Sons.
World Health Organization (WHO)
   1974   WHO Monograph Series No. 61. Handbook of Human
            Nutritional Requirements. Geneva: World Health
            Organization.

14
Phases of Farming Systems Research: The Relevance of Gender in Ecuadorian Sites

Patricia Garrett and Patricio Espinosa

In recent years Ecuador, like many Third World countries, has shifted its priorities for agricultural research and development, reassessing the traditional orientation towards export commodities and assigning increasing importance to improving agricultural production by smallholders. The Instituto Nacional de Investigaciones Agropecuarias (INIAP), a semi-autonomous institution within the Ministry of Agriculture that is responsible for approximately 85 percent of the agricultural research conducted in Ecuador, began working seriously on behalf of smallholders in 1976. The institution's general objective is to develop technological alternatives that will increase agricultural production and productivity, to the benefit of both producers and consumers.

With substantial support from the Agency for International Development (AID) and the Centro Internacional de Mejoramiento de Maíz y Trigo (CIMMYT), INIAP established the smallholders program within the Department of Agricultural Economics as the Programa de Investigación en Producción (PIP).

Since 1982, INIAP and scientists from Cornell have been collaborating on integrated, multi-disciplinary farming systems research in two zones of Ecuador with contrasting crops, ecology, and social organization. This work focused on improving the design and conduct of farming systems research. Through the Bean/Cowpea Collaborative Research Support Project (CRSP), diagnostic research identified problems for experimentation on station and on smallholders' fields; technological alternatives are currently being evaluated for their biological and economic characteristics. CRSP research, plus the experience of INIAP in other zones, suggests how gender can be incorporated into the generation and evaluation of agricultural technologies.

INIAP conducts three kinds of experimentation. On-station trials are conducted at the seven experiment stations and eight farms. Controlled conditions and proximity to logistical and support personnel permit complex experimental designs, especially concerning genetic improvement, cultural practices, and pest control. Ecuador is a country of dramatic ecological contrasts. Consequently, regional trials permit scientists to evaluate the appropriateness of management recommendations and the behavior of new varieties under a range of ecological conditions. These trials are normally conducted on medium to large scale farms because the variables of interest are responses to variations in climate and soil. Research in production or PIP trials are specifically designed to generate technological alternatives for smallholders. Experimental designs are more simple to permit their placement on small farms, and technicians attempt to elicit the active involvement of farmer collaborators.

Research in production or PIP activities usually begins at the initiative of a commodity program wishing to evaluate promising technological alternatives under smallholder conditions. An alternative approach is to organize research and development activities to meet the needs of the smallholders in a zone. Taking the latter approach, this paper will follow a chronological model of farming systems research activities (Figure 14.1) to illustrate how sensitivity to gender consistently helps programs benefit smallholders.

PROBLEM IDENTIFICATION

The CRSP project found it useful to assign one or two people to analyze information available about a region prior to initiating field research. This work includes the review of agronomic data (rainfall, soil types, etc.), socioeconomic data (agricultural censuses), and previous studies of the zone (government reports). A written report serves to brief the interview team and to tailor questions to a region. The document by Uquillas et al. (1985a) illustrates this approach.

The analysis of agricultural census data permits the identification of zones in which small scale, marginal farms predominate. In these regions, the role of women in agricultural production is likely to be important. Farm families who have limited resource bases frequently find that they do not have enough land to absorb family labor productively and to support the family adequately. They

PHASE I
PROBLEM IDENTIFICATION

PHASE II
IDENTIFICATION OF POTENTIAL
RECOMMENDATION DOMAINS

PHASE III
DEFINITION OF A STRATEGY
FOR EXPERIMENTAL RESEARCH

A
For problems which cannot be identified by farmers
but which scientists suspect exist.

B
For problems which have been identified
by farmers and/or scientists.

PHASE IV
EVALUATION OF TECHNOLOGY

PHASE V
REVISION OF RESEARCH STRATEGY

PHASE VI
TRANSFER OF TECHNOLOGY

FIGURE 14.1   PHASES OF FARMING SYSTEMS RESEARCH:
A SYNTHESIS OF INIAP AND ICTA PROCEDURES (1)

come to rely heavily on income from wage labor, especially wages earned by men during temporary or seasonal migration. Under these circumstances, both labor and land are constraints to agricultural production and to technological innovation (Garrett 1986a).

Zones in which semiproletarian households predominate can be identified during preliminary analyses of secondary data (Garrett et al. 1986a), and the composition of research teams can be modified to facilitate dialogue with both women and men. Planting dates in highland Ecuador, for example, frequently coincide with major religious holidays, and men normally return to plant with their families. Men who work in the cities may be absent during much of the agricultural cycle. During interviews, they "forget" the answers to questions about agricultural practices because their wives are more intimately involved in these tasks during their absence. Good information about agricultural practices, therefore requires the participation of both men and women.

Realistically, this is difficult to achieve. Agronomic scientists are paid to work during normal business hours, but agriculturalists need to work, not chat, in order to take advantage of daylight. Visits during weekends often are with men who are otherwise working in cities, and their presence may discourage their wives from contributing to the discussion. Finally, efforts to interview women, especially by groups of men, frequently fail because women are reluctant to cooperate. In Indian communities, women may be unable to communicate effectively in Spanish. Nevertheless, there are topics on which women's input is essential.

INIAP has found that research on seed production and on farm storage, both traditionally female activities, frequently provide promising approaches to improving the productivity of basic grains. Historically, INIAP'S concern had been limited to the cycle of production. Research on smallholders' fields identified critical post-harvest issues such as the use of by-products. These topics rank with other major issues covered in the interview guide developed during CRSP work and they are preferentially discussed with women (Garrett et al. 1986b; Uquillas et al. 1986) (2).

Interviewers are the principal tool for eliciting information during diagnostic research and their characteristics (including gender, age, ethnicity, and regional background) limit the kinds of information to which they have ready access. In Latin America, it is not deemed appropriate for men to take an interest in certain female matters such as seed selection. It is also difficult for men to enter female spheres such as kitchens and to observe relevant matters like food preparation for different age groups. Trained female interviewers can engage women in dialogue, thereby eliciting information about production, storage, and consumption practices. In this way, multi-disciplinary teams of men and women can achieve comprehensive coverage of matters of potential interest to farming systems research.

IDENTIFICATION OF POTENTIAL RECOMMENDATION DOMAINS

Preliminary research, both in the library and in the field, allows the team to identify major sub-regions and the principal resident social groups. These geographic zones potentially are recommendation domains. They are defined principally by their ecological characteristics, because physical properties limit land use patterns. The productive capacity of the land also limits production and the kind of livelihood that families can derive from agriculture. The ecological and socioeconomic characteristics of a region are often closely related and research is continuing to explore this relationship in all its complexity.

One theme of Cornell/INIAP collaboration has been the analysis of social stratification and its implications for farming systems research. One paper (Garrett 1986b) explains why different policy objectives can be achieved by working preferentially with identifiable strata of small holders. Another paper (Garrett 1986a) argues that the socioeconomic characteristics that make technologies appropriate vary by social strata and enterprise within the farm. This analysis provides a conceptual framework for selecting collaborating farmers and for evaluating agricultural technologies.

As the organization of agricultural production varies across social strata, so the role that women play also varies dramatically. Reports on field research in Imbabura (INIAP/Cornell Team 1982; Uquillas et al. 1985b) illustrate variation in just a few regions. Among petty commodity producers, women's work in the field is frequently supplemented or replaced by hired labor. Their labor may be important, especially in post-harvest food processing and storage, but their activities typically center on child care and homemaking activities. In peasant households, the labor of men and women is usually complementary, as are their spheres of financial responsibility. This may be further defined epecially in Indian communities, by lands and animals recognized to be "his," "hers," or "ours." In semiproletarian households, a different kind of complementarity exists. While men and women may share the responsibility for certain agricultural tasks, notably planting and harvesting, women may be left with the principal responsibility for the farm during the absences of their husbands (Garrett 1986a; 1986b). A general principle is that the feminization of agriculture is a consequence of poverty; consequently agricultural scientists have a greater need to deal constructively with women as they begin to serve increasingly marginal populations.

DEFINITION OF A STRATEGY FOR EXPERIMENTAL RESEARCH

Farming systems researchers must coordinate activities on-station with those off-station; this task is basically gender neutral. The problem is to determine how to take best advantage of the relatively controlled conditions of the experiment station while exploring the ecological and socioeconomic variety that exists on smallholders' fields. Each institution and program will develop its own strategy.

Gender is relevant in the selection of collaborating farmers for on-farm trials. In semiproletarian households, men who are engaged in seasonal wage labor are likely to be absent from the farmstead during the week. Technicians who are aware of this pattern can visit over the weekend to negotiate the placement of a trial with the male head of household. Technicians can reach an amicable agreement with husbands, only to encounter wives who were never consulted. Some women simply retreat into the house and never work with either the technician or the experiment. Others unceremoniously throw the technician off the farm.

In the Ecuadorian context, semiproletarian households present tractable problems. Technicians can include both male and female heads of households in initial explanations of research objectives, even if the request for permission to plant a trial is directed at the husband. Experience has demonstrated to INIAP that these early discussions are critical (Londoño et al. 1984; Moscardi et al. 1983). Many collaborators, both male and female, have not understood the experiment that was being conducted. Adequate understanding encourages intellectual engagement in the research enterprise, and this is essential to obtaining the active cooperation of farmers in on-farm trials. It is now INIAP policy to emphasize the importance of explanation and to negotiate with both members of a couple whenever appropriate.

This is particularly important in Indian communities where complex sets of property rights and obligations may be found within a single family. The traditional inheritance system is parallel -- daughters inherit from their

mother and sons inherit from their father. Consequently,
both men and women traditionally own land and animals.
Women ultimately control what they own, and their husbands
cannot alienate their property.

An unfortunate example illustrates this principle. In
one highland community, technicians negotiated one weekend
with men to build an erosion control system. When they
arrived during the week with the land-moving equipment,
they were met by a demonstration of women who refused to
permit their fields to be divided by the trench. The women
had the right to protest and they prevailed. If erosion
control was to be effected in that community, technicians
had to learn to behave in culturally appropriate ways and
to incorporate women into project design and decision-
making.

EVALUATION OF TECHNOLOGY

Scientists typically have expectations about how a particular innovation will fit in an existing farming system.
These expectations frequently delimit how the impact of the
new technology will be evaluated. New technologies, however, may be incorporated in unexpected ways for reasons
not anticipated by their designers.

The case of INIAP 101 maize illustrates this principle.
INIAP 101 is an early maturing maize with a white grain, a
flimsy architecture, and a maturation pattern that was
problematic in certain zones (Staver 1982). Experimentation demonstrated that its yield under farmers' conditions
was equivalent to local varieties. Because it matured more
rapidly and produced a delicious cob, the variety was
adopted for consumption and sale while still green. Men
initially reported that merchants paid poorly because they
confused 101 with an inferior variety from the coast. This
problem corrected itself, and 101 commanded a premium price
as merchants and their customers came to recognize its
superior qualities.

Input from women modified the scientists' expectations
for INIAP 101 in several ways. Plant breeders had been
concerned about tillering. By their criteria, multiple
shoots were negative characteristics of plant architecture.
Women tending animals regarded the shoots positively
because they provided additional stover. Breeders abandoned tillering as a problem. However, women pointed out
another characteristic of INIAP 101 that did pose a problem
as far as they were concerned. When kernels were removed
from the cob, a bit of the cob remained attached to the
bottom of the grain. This was problematic for the preparation of mote, a staple made from whole grains. When this

problem was brought to the attention of plant breeders, they were subsequently able to correct it.

Input from farm families allowed researchers to estimate adoption. INIAP 101 excelled as green cob (chochlo), and, after genetic improvement it made excellent mote. White was the wrong color for good tostado and the variety made poor flour. These characteristics facilitated adoption for certain purposes and limited adoption for other uses. Subsequent research on consumption patterns allowed scientists to anticipate that in specific zones 101 could displace 40 percent of the area planted in native varieties (30 percent for consumption and 10 percent for seed). This provided a criterion against which to measure actual adoption rates.

Mestizo professionals working in Indian communities came to realize that native speakers of Quechua were necessary if INIAP was to work effectively with all the people, men and women, who farm and cook in their native language. Bilingual employees have special social access. Consequently, INIAP has hired a few male technicians with high school educations and paired them with a college-educated agronomist. These teams have been quite effective, but they are few in number.

The CRSP project has also worked with one female Quechua speaker. Because she was hired as a translator, she was not incorporated fully into the research team. Nevertheless, the experience was positive. Because she maintained traditional dress, she had access to women who would have shunned other interviewers. She was also recognized in the zone to be from a privileged family, which had been able to provide her with a college education. This particular interaction of ethnicity and social class membership had both positive and negative consequences for rapport with informants.

This example illustrates a problem in hiring people who are likely to be effective with poor farmers. In Ecuador, the majority of poor farmers are Indians. The men tend to be bilingual in Spanish and Quechua, while the women are usually monolingual Quechua speakers. Most children born to poor households fail to achieve the level of formal education (available only in Spanish) that would make them eligible for employment by national agricultural research centers. Institutions desiring to serve disadvantaged populations should establish paraprofessional positions and actively recruit members of different ethnic groups. The recruitment of both males and females in paraprofessional positions is desirable because they have differential access to informants and information.

REVISION OF RESEARCH STRATEGY

The task of revising a research strategy is basically gender neutral. Some considerations may be related to gender, while others are not. One of the clearest examples of reorientating research at INIAP does not concern gender, but it does illustrate how on-farm research can dramatically influence priorities for commodity programs.

The evaluation of INIAP 101 under smallholder conditions is the best documented case of technological innovation in Ecuador. As Tripp (1985: 119-120) suggests, the Legume Program was concerned that a maize variety was being released without a compatible pole bean. Consequently, the perceived need for a compatible bean focused legume research on the evaluation of germplasm for precocious, non-aggressive cultivars.

Continual experimental research on INIAP 101 eventually suggested that its inability to support aggressive beans in a polyculture was actually not a problem. Many local varieties of maize were grown in monoculture, so it was not essential that all maizes support intercrops. Moreover, because 101 had a faster maturation cycle, it permitted intensified production allowing the benefits of poly-cropping through crop rotation. Under rainfed conditions, two crops a year were possible; with irrigation and increased inputs, three crops could be produced. Consequently, INIAP 101's inability to support aggressive local bean cultivars became relatively unimportant. Priorities for legume research were redefined.

TRANSFER OF TECHNOLOGY

A major purpose of farming systems research is to specify with some precision the agronomic and socioeconomic conditions under which an agricultural innovation will offer comparative advantages over existing technologies. Given a realistic evaluation of where and for whom the technology will work, extensionists can work more efficiently. Active collaboration between research and extension is often impeded by administrative problems. In Ecuador, as in many other countries, research institutions recruit the more talented scientists. Moreover, even well-trained extensionists are limited by inadequate resources. Incentives to serve small holders are few, and the logistical difficulties of reaching such farmers efficiently are enormous. These issues create objective problems for technology transfer.

The realities of smallholder work have compelled INIAP to exceed its mandate. As a research institution, INIAP is administratively enjoined from engaging in extension. Nevertheless, the evaluation of technology under small-holder conditions has effectively required that the PIP program initiate technology transfer.

Traditionally INIAP mobilized its technical resources when it called a field day. Representatives from all relevant departments made detailed presentations on the science behind each recommendation and the entire technological package was presented to assembled growers. This format proved overwhelming to small holders. INIAP consequently modified its program. Field days focused not on a package of recommendations but on a single message. The theme might be fertilization levels, planting distances, or weed control, but the field day focused exclusively on one topic.

INIAP used to invite only men to attend field days and to participate in tours to observe trials. As the PIP program worked with smallholders, technicians realized that it would be desirable to incorporate women. Initial experiences with mixed groups suggested that women separated themselves from men and failed to participate in discussions. Language was identified as one problem. INIAP changed the format of the field days and incorporated a Quechua/Spanish speaker with a megaphone into the presentation. This technician served as translator and broker between the community and the researchers, facilitating both feedback and technology transfer.

INIAP directs the evaluation and transfer of technology to specific recommendation domains. As Tripp (1985) emphasizes, on-farm research permits the multi-disciplinary team to specify the agronomically and socioeconomically relevant aspects of a particular production technology. Researchers' definitions of "recommendation domains," (groups of farmers in delimited areas) are technology specific.

Gender should be considered in the definition of recommendation domains because certain technologies are more likely to be used by men or by women. A central question is whether or not the innovation has been developed for traditionally male or traditionally female enterprises. The distinction between male and female enterprises, with their associated budgets, is currently made in the African literature (Garfield 1979). Field research demonstrates that it is also a relevant distinction for small holders in Ecuador.

The kitchen garden and the barnyard animals are usually female enterprises in Ecuador (and elsewhere), and many development projects focus on them. If the female head of the household can mobilize her labor and that of her children, and if she can control the proceeds of these enterprises, planners can anticipate a direct relationship between increased production and improved nutritional status.

Certain Integrated Rural Development Projects (IRDPs) in highland Ecuador have directed agricultural programs at women by focusing on improved production and consumption of guinea pigs. Several IRDPs built guinea pig improvement programs into the activities of mothers' centers that are multipurpose organizations that address women's concerns as homemakers and also serve as a vehicle for non-formal education. The existence of these women's centers facilitated efforts to improve production and consumption for poor rural families.

Other agricultural enterprises and tasks are definitely male and innovations affecting them should be brought to the attention of men. In Ecuador, men are principally responsible for the design and management of small-scale irrigation systems and they usually irrigate family fields. Just as men are relatively ignorant about guinea pigs, women are relatively uninformed about irrigation systems. Moreover, in some Andean communities there are complex and highly ritualized practices that regulate the contact women may have with irrigation systems. Such factors suggest that ordinarily men would be the desired target for innovations in irrigation.

These examples illustrate a general principle: the transfer process is technology specific. Efforts must be directed towards the individual who will actually use and adopt the innovation. Gender is one of several relevant variables.

CONCLUSION

Many people wear very strong blinders when it comes to gender and it is difficult to penetrate these ideological barriers. Agriculture may be a female activity in certain parts of Africa, but it is believed to be a male activity in Latin America. Colleagues often say that women do only light work, but often in field preparation the women performed labor as heavy, if not heavier, than the men. We repeat what we hear, despite what we see or do. Women also declared that they did not engage in field work, but just did a bit of planting, weeding, harvesting, etc.

It is a tribute to human imagination to persist in declarations and interpretations that are patently false. This may be a wonderful premise for science fiction, but it is an inadequate basis for research. If agricultural programs that actually serve people's needs are to be developed, research must proceed realistically. "Hay que partir de la realidad" is the way that Spanish speakers express the principle. Reality dictates that gender be included among the key variables to be considered in the development and dissemination of agricultural innovations.

NOTES AND ACKNOWLEDGEMENTS

The research on which this paper is based was conducted principally by the Instituto Nacional de Investigaciones Agropecuarias (INIAP) in Ecuador, primarily through the smallholder Programa de Investigación en Producción (PIP) program. Some work was supported under a Collaborative Research Support Program, the Bean/Cowpea CRSP (AID/DSAN/ XII G-0261).

(1) This scheme was developed during a project seminar held during August, 1985 in Quito, Ecuador. Participants in that seminar, who were effectively authors of this model, included: Instituto de Ciéncias y Tecnologias Agrícolas (ICTA)/ Guatemala: Selvin Arriaga, José Manuel Díaz, and Porfirio Masaya; Instituto Nacional de Investigaciones Agropecuarias (INIAP)/ Ecuador: José Acuña, Diana Barba, Victor Hugo Cardozo, Romulo Carillo, Edmundo Cevallos, Napoleón Chávez, Francisco Muñoz, Arturo Villafuerte, Cristóbal Villasís, and Ely Zambrano; Cornell University: Patricia Garrett, Judith Hall, Wesley Kline, and Jorge Uquillas.

(2) One innovative characteristic of the INIAP/Cornell interview guide is that it specifies before each set of questions what kind of person is the appropriate respondent. Introductory comments to each subsection explain why men/women or land owners/wage laborers or mestizos/indios might have different perspectives on the issue under consideration. These commentaries outline basic principles of social organization so that interviewers trained in agronomy can be more effective in taking interviews. The guide also suggests which member of multidisciplinary team should take leadership in questioning, so that the interview can really benefit from disciplinary expertise. English and Spanish versions are available (Garrett et al. 1986b; Uquillas et al. 1986).

REFERENCES

Garfield, E. B.
   1979   The Impact of Technical Change on the Rural
           Kenyan Household: Evidence from the Integrated
           Agricultural Development Program: A Research
           Proposal and Literature Review. Institute of
           Development Studies. Working Paper No. 358.
           Nairobi, Kenya: University of Nairobi.
Garrett, P.
   1986a   Social Stratification and Multiple Enterprises:
           Some Implications for Farming Systems Research.
           Journal of Rural Studies 2:3:209-220.
   1986b   Viable Objectives for Smallholder Programs:
           Variation by Social Strata. Agricultural Admini-
           stration 22:1:39-55.
Garrett, P., D. Golden, and J. D. Francis
   1986a   The Measurement and Analysis of Inequality Using
           Microcomputers. Social Science Microcomputer Review
           4:2:194-206.
Garrett, P., J. Uquillas, and C. Campbell
   1986b   Interview Guide for the Regional Analysis of
           Farming Systems. Working Paper 86.2E. Ithaca, NY:
           INIAP/Cornell Project.
INIAP/Cornell Team
   1982   Características de los Pequeños Productores en
           Zonas de Imbabura: Informe Preliminar.
           (Characteristics of Small Scale Producers in Zones
           of Imbabura: Preliminary Report.) CRSP Working
           Paper 82.2S. Quito, Ecuador: Proyecto
           INIAP/Cornell.
Londoño, D., P. Espinosa, R. Vinda, T. Bustamante, G.
           Diener y L. de la Torre
   1984   Evaluación Intermedia Parcial de los Proyectos
           de Desarrollo Rural Integral Salcedo y Quimiag
           Penipe. Report commissioned by the Agency for
           International Development/Ecuador (AID/E) and the
           Secretaría de Desarrollo Rural Integral.
Moscardi, E., V. H. Cardoso, P. Espinosa, R. Solís, and E.
           Zambrano
   1983   Creating an On-Farm Research Program in Ecuador.
           Mexico City: CIMMYT.
Staver, C.
   1982   Un Comentario Sobre la Tecnología Mejorada del
           PIP de la Provincia de Imbabura. Quito, Ecuador:
           Unpublished Manuscript.
Tripp, R.
   1985   Anthropology and On-Farm Research. Human
           Organization 44:2:114-124.

212

Uquillas, J., V. Arévalo, N. Chávez, y J. Arroyave
   1985a   Diagnóstico Agro–Socioeconómico de la Provincia
       de Manabí. (Agro–Socioeconomic Diagnosis of the
       Province of Manabí.) Documento de Trabajo ASE.6.
       Quito, Ecuador: Proyecto INIAP/Cornell.
Uquillas, J., D. Barba, P. Garrett y E. Zambrano
   1985b   Estratégias de Reproducción de la Economia
       Campesina en Imbabura. (Strategies for the Repro-
       duction of Peasant Economies in Imbabura.) Docum-
       ento de Trabajo ASE.5. Quito, Ecuador: Proyecto
       INIAP/Cornell.
Uquillas, J., P. Garrett, y C. Campbell
   1986   Gúia de Entrevistas para el Análisis Regional de
       los Sistemas de Producción Agropecuaria. (Interview
       Guide for the Regional Analysis of Farming Systems.)
       Documento de Trabajo ASE.8. Quita, Ecuador:
       Proyecto INIAP/Cornell.

15
Technological Domains of Women in Mixed Farming Systems of Andean Peasant Communities

Maria E. Fernandez

Extension and social science literature recognize that a division of labor by gender is common to most peasant production systems. In the Andes, women are responsible for the tasks of grazing animals, collecting fodder, curing animals, seed selection, sowing and weeding, among others. Men perform tasks such as ploughing, branding, purchase of supplementary inputs, and harvesting. It is commonly argued that men make all decisions related to production.

It can be argued that in Andean mixed cropping and livestock systems, there is a constant in the division of labor, such that men have greater responsibility over crops and women over animals, or vice versa. This responsibility results not only in a division of tasks, but in control over technological knowledge related to a specific area of production, and consequently in the right and the ability to make decisions about the products of that area.

In Andean mixed farming systems, women are the principal herders. In similar systems in many parts of Africa, women are the principal cultivators. The division of labor among family members by gender does not necessarily imply specific biological reasons for one or the other to assume certain production activities. The distribution of responsibility and tasks is basically a functional one. Women within a given farming system are often assigned tasks and responsibilities that are compatible with the care of small children. "Women's tasks do not take them far from the home, and frequently require less concentration than men's...the division of labour is informal and flexible." Men can and do cook, carry water, clean the house, wash clothes, or graze sheep when the women are away or are busy at another task (Harris 1985:28).

Male and female members of a social group carrying out particular tasks also control bodies of technical knowledge

and skills needed for the management of the areas of
production for which they are responsible. This control of
knowledge and skills relates to the decision-making power
men and women have over the production process and over
disposal of the products. This is not to say that within a
family unit a man or a woman has complete liberty to make
decisions within these areas of responsibility.

Small farm management might be compared to that of a
large company. The farm family resembles the management
unit with the adults forming "a board of directors."
Decisions made by this board are made for the good of the
unit as a whole. Specific areas of production are run as
sub-divisions of the enterprise. The section managers have
a great deal of liberty in their production decisions.
Managers also carry out specific tasks in others' production units as a means of maintaining checks and balances.
The entire farm enterprise encompasses multiple crops,
animals, tasks, and responsibilities. Members of the
family (nuclear/extended) and the community take on various
roles at different times under different circumstances —
from "board" member to simple laborer.

The biological parts of this system also overlap. For
example, oxen are used as farm equipment (ploughing), as
sources of fuel (manure), for reproduction, and as financial reserves. Llamas are used for transport and as
sources of manure, fiber, meat, shoes, and ropes (from the
hides and intestines). The donkey carries seed, farm
implements, and manure to the field and brings back the
harvest to the family patio. Sheep graze on the weeds of
fallowed lands or on the stubble of recently harvested
crops and provide the main source of manure for cropping.

Within the farm system, task allocation and responsibility for production decisions are overlayed and inter-
acting. While men are responsible for agriculture, women
do the seed selection and planting. While women are
responsible for livestock production, men do the branding
and care for the supplementary feeding of oxen.

The interaction of tasks along gender lines and between
them has broad implications for the transfer of knowledge
and technology. How technological knowledge is passed from
one generation to another and who controls such knowledge
within a specific farming system are closely linked with
the gender division of labor and decision-making power.

THE ORGANIZATION OF THE HIGHLAND COMMUNITIES

Aramachay is a community of 120 families situated on
the southern side of the Mantaro Valley at an altitude
higher than 3,500 meters above sea level. The soil varies

from clay-like to sandy and from black to red. The capacity for water retention increases with altitude. Natural vegetation is mainly native grasses and shrubs are more common than trees. The rainy season lasts from September to March and crops are planted between October and December in staggered fashion. Between one and two-thirds of agricultural land is fallowed at any given time. Fallowing periods range from three to seven years depending on altitude and soil quality (Mayer 1979).

Livestock include cattle, sheep, swine, poultry, guinea pigs, and donkeys. Grazing is done on fallow or communal range land, the use of which is governed by the community assembly. The total number of animals held by each family is associated with its wealth and is determined, in part by the relation between agricultural and pastoral activities. On average, a family maintains 25 sheep. Most of the livestock are either criollo (breeds adapted during and just after the Spanish Conquest) or criollo crossed with recently imported breeds (Corriedale).

Crop and livestock production interact. Most households raise sheep and cattle, and plant potatoes, barley, broad-beans, wheat, peas, olluco, mashua and quinua (traditional Andean crops). As a rule, traditional cultivation techniques are used, although fertilizers and pesticides are applied to improved varieties of potatoes destined solely for market. Households plant an average of eight crops per growing season, on a variable number of dry farming plots (sometimes more than 40), making up a total area of not more than three hectares and most often less than one.

Labor is shared by all active members of the family. Livestock production is the responsibility of women, and children aid in the grazing. Men are responsible for dipping and branding activities. Agricultural activities are the responsibility of men, although women are responsible for seed care, selection, planting, and food processing. Women also share cultivation and harvesting activities with men. At peak labor times, older men and women take over household chores and grazing activities. The family labor force averages four to six adults. By the age of 15, a youth is considered capable of an adult's work load although he or she continues to work under the guidance of a more experienced member of the family, usually of the same gender. It is not uncommon for family heads (women and men) to work seasonally outside the community either in mining or as agricultural day laborers.

THE WOMEN'S LIVESTOCK AND CROP PRODUCTION
COMMITTEE OF ARAMACHAY

The National Institute for Agricultural Research and
Extension (INIPA) of Peru and the Small Ruminant Collaborative Research Support Program (SR-CRSP) began working
in the Community of Aramachay in 1983. The community was
chosen for its representativeness of highland mixed farming
systems where the majority of Peru's rural population and
small ruminants are concentrated. Aramachay is the hub of
11 similar communities.

The objective of the project is to look for technological alternatives based on small farmers' knowledge and on
an understanding of the traditional production system.
This information, which takes into consideration ecological, economic, and organizational constraints, is being
used to select by the project researchers together with the
farmers recovered or introduced technologies that will
improve production. Although the project aims at sheep
production in this case, research on alternatives encompasses the crop component of the system due to its complementary interaction with livestock production.

The methodology described involves a multi-disciplinary
team including an animal science specialist, an agronomist,
an economist, and an anthropologist, doing participatory-
action research with organized groups within the community.
The project began activities after making a joint agreement
with the village assembly to work on production problems
identified by the farmers (men and women) themselves.

The communal assembly is made up of all male heads of
households, together with widows and single women who are
family heads (Swindale 1985). The assembly appointed a
committee of ten members to work closely with the research/
action team and to serve as a link with the larger population. In spite of a specific request by the team and the
recognition by the male members of the assembly that women
are as active in the production process as men, no women
were included in the committee. Although the thrust of the
project was in the direction of livestock production, the
problems posed by the committee were centered on agriculture. Work was begun on these problems, but a concern
remained that women -- the principal livestock producers
-- were at the margin, although they had been the active
voice in the original signing of the agreement (Fernandez
1986).

In an attempt to solve the impasse, an International
Labour Organization (ILO) project, oriented toward research
and action on women's labor and use of fuel, was invited to
help create an organization that would provide women with a

channel for productive action. A sociologist with the project, working in the village, decided that the basis for organization existed in the mother's health committee, a recognized support organization for the local health center. The leadership of this committee was made up of wives of family heads. The village assembly appointed the men whose wives formed the leadership of the committee. Nearly two years were spent supporting this organization. In spite of the enthusiasm of a small number of women, things never seemed to get off the ground. Neither the small animal production, food processing, or bread-making projects (proposed by the ILO project), nor the attempts at forming a stable organization were successful.

After a careful joint team evaluation, a different approach was decided upon. All of the women above the age of fifteen, married or not, were invited to a meeting to discuss production problems. Twenty-three women from the 120 families attended the first meeting. Twenty of them participated actively and posed their production priorities in the following order: (1) control of internal and external parasites in sheep; (2) provision of fodder during the dry season; (3) improved management of communal rangelands; (4) improved seed selection and conservation techniques; and (5) knowledge of seed density criteria.

The women expressed their desire to work as a group and decided to call themselves "The Women's Livestock and Crop Production Committee of Aramachay." Three weeks later, a meeting was called to discuss the possibility of conducting on-farm research of legumes for fallowed lands with the two objectives of increasing fodder capacity and improving soil quality for the following rotation cycle. Thirty women attended this meeting and twenty signed up to plant one-quarter to one-half yugada (the area an ox-team can plough in one day) of land for the experiment. Only two of the women left the meeting place to consult with their husbands on the feasibility of designating the plots for this use.

When the pasture specialists visited the community one week later, the experimental lands had been ploughed and four plots were used for demonstration of planting and fertilization techniques. Fourteen women participated in the group training effort and received the seed necessary to plant the parcel each had been allocated. Simple registry sheets were designed on which each woman could set out data on plant growth over time. The design of the sheets took into consideration that some of the women are functionally illiterate and that most have very little experience with written material. Ten days later the twenty plots had been planted.

A discussion on improved animal nutrition included a group analysis of the possibility of improving the communal rangelands. The women decided to make a formal request to the community assembly for two hectares of natural pasture to use in experimenting with improved techniques. To oversee this land, they would organize themselves which required building a hut nearby (the area is 45 minutes walking time from the village center) and sleeping on the site by turns. All agreed that a major hurdle to improved communal range management and conservation arose from boundary disputes with neighboring communities. The women suggested that it was imperative to resolve these disputes, and stated that if the community assembly did not take action soon, they would take the initiative to work out a settlement themselves.

At the following meeting, after reporting that the community assembly had ceded the requested communal range land for improvement tests, the women decided that this effort could be better guaranteed if it were initiated at the beginning of the 1986-87 agricultural year. They felt that more time was needed for planning and organizing among themselves and with their husbands. They decided to request a veterinarian from the nearby livestock experimental station to instruct them on types of parasites in sheep and their influence on production. These discussions are underway at present.

CASE STUDY ANALYSIS

Although this case study allows for an analysis of project and team methodology, this paper concentrates on the aspects of community organization.

The fact that women were not included in the collaborating committee did not seem to show a desire on the part of men to exclude them. Rather it seems that within a mostly male group (the community assembly), it is "natural" to appoint members from the group itself. Although men did make an effort to think of women who could act on the committee, they did not have much success. This may be due to a lack of cross-gender groups within the village as well as to the previously limited participation of women on formal committees. It appears to be an example of the imposition of inflexible roles as a result of community organizational factors rather than of a conscious exclusion of a given category of community members.

The organization of the village mother's health committee on the basis of household head (male) appointeeship, would seem to have another rationale behind it. When a man accepts an appointment for his wife, he is acting as a spokesman for the production unit. In the community assembly, the family spokesman, generally the husband or elder male member, is charged with the obligation of speaking in the family unit's interest. If this charge is not carried out well, the man's or the assembly's decision can be modified in a subsequent assembly. No one doubts that this public change in opinion is due to consultations or pressures exercised upon him/it, when the decision made is considered to be contrary to the interests of the group or its members.

There is a tendency, however, to charge women with reproduction-related rather than production-related public responsibility. This practice cannot be attributed to the local social system alone, because it is the Ministry of Health that encourages that the health support committee be made up of women.

The ability of a large number of women to prioritize their production interests at an initial meeting suggests that they are not task, but rather problem-oriented. The five problems listed are directly related to the quality of production. Only the second priority, provision of fodder in the dry season, might be interpreted as a request for a technological solution. The other four express a need for broader technological knowledge. The women are conscious of their responsibility for making decisions within the production unit in specific areas. It is also evident that the area of greatest responsibility is that of livestock production, although there is concern for agricultural production as well. It might be worth mentioning that the woman who suggested point five (seed density criteria) is a household head. In any case, the interaction between crop and livestock production spheres is evident.

When the women signed up for the experiment on fallowed lands, not only the team but the principal researcher herself was surprised. No one expected that the women could make unilateral decisions concerning "farm lands." Further inquiry revealed that fallowed lands revert to communal use for grazing. This is clear evidence that women make decisions over livestock production. It would appear that because of the type of interaction between private and communal use rights, it was not necessary to ask for assembly approval for the use of fallowed plots as it was in the case of the communally managed rangelands.

In the case of the discussion of boundary problems, it became clear that women publicly accept the right to influence their production unit representatives (male heads) in the resolution of broader political problems. Morever, in stating that they will initiate action in the event that the assembly is not effective, they clearly recognized that

they can be active members of community associations as
well as of the family production unit.

The implementation of the research on the experimental
plots and the request for formal training on the relationship between parasites and sheep production reinforce the
analysis of women's decision-making power. This case study
was chosen not for its singularity, but because it seems to
unify a series of observations made during the past three
years.

In terms of research conclusions, the information
illustrated here is only one more step in understanding the
interactions between the organizational and biological
components of the system. The limits of women's technological knowledge and skills, as well as the spheres over
which women have decision-making control, are still to be
defined.

CONCLUSION

The result of this analysis is not that women's tasks
should be alleviated or that men's understanding of women's
work load should be improved. If women can increase their
production decisions (or at least not have their present
position undermined), social change will take place from
within the production system. Women will be able to demand
equality from a position of strength and not beg for it
from a position of weakness.

In a farming system where social and productive activities of all kinds are divided along gender lines, outside
agents will cause less damage to the power balance within
the community if they respect this division. When a male
animal science specialist enters a village where the women
are responsible for animal husbandry, he still tends to
find contact with male counterparts easier. He therefore
unconsciously reinforces the right of men to give information and make decisions regarding animals which can later
be reflected by the men themselves in the respect they show
toward their wives' productive efforts. The same specialist could feel that little is known about animal health,
management, breeding, and grazing, therefore deeming the
peasant ignorant, when in truth a great deal of knowledge
exists, but among the women and not among the men.

Perhaps a complete balance between male and female
decision-making in village production units never existed.
It is also possible that the influence of other, more
male-oriented societies, has created or reinforced an
imbalance. In any case, in most situations, women are now
at a disadvantage. They have had less access to formal
schooling and less contact with outside agents and

institutions that often makes them voluntarily retire to
secondary positions when communication with researchers and
extensionists is needed. This means that if research and
validation are to be carried out with producers themselves,
when these producers are women much more time must be given
to the process of encouraging their participation and
constructing channels for the expression of their ideas and
knowledge.

NOTES AND ACKNOWLEDGEMENTS

The research for and preparation of this paper was
supported by the Title XII Small Ruminant Collaborative
Research Support Program under grants numbers AID/DSAN/XII
-G-0049 and DAN-1328-G-SS-4093-00 through the SR-CRSP
Sociology Project. Additional support was provided by the
University of Missouri-Columbia.

REFERENCES

Fernandez, M.
    1986   Participatory-Action-Research and the Farming
            Systems Approach with the Highland Peasant.
            Columbia, MO: University of Missouri Press.
Harris, O.
    1985   Complementariedad y Conflicto: Una Visión
            Andina del Hombre y la Mujer. Allpanchis. Vol. XXI
            No. 25. Cusco, Peru: Instituto Pastoral Andina.
Mayer, E.
    1979   Land Use in the Andes: Ecology and Agriculture
            in the Mantaro Valley of Peru with Special Reference
            to Potatoes. Lima, Peru: International Potato
            Center.
Swindale, A. J.
    1985   Diagnóstico de las Comunidades Alto-andinas del
            Valle del Mantaro. Lima, Peru: IVITA/SR-CRSP.

16

The Household Enterprise
and Farming System Research:
A Case Study from Taiwan

Rita S. Gallin and Anne Ferguson

The concept of the farm as a system and the examination of interdependencies among its component parts are key to the conceptual framework of farming systems research and extension (FSR/E). FSR/E is predicated on a systems orientation, but for most practitioners, the boundary of this system is carefully delineated at the farm level. Although off-farm activities are recognized as important, household decisionmaking is assumed to depend on and to revolve around cropping and livestock activities. Broader political and economic structures are usually accepted as givens and their influence on farm level decision-making is not fully investigated. The singular focus on the farming activities of the household leads to development strategies based on new or improved agricultural technologies designed to overcome local production and utilization constraints (DeWalt 1985; Shaner et al. 1982:5).

This paper draws attention to the limitations of this conceptual framework through a case study of the evolution of the farming system in Hsin Hsing, a village in Chang-hua County, Taiwan, between 1958 and 1979. A longitudinal analysis suggests that change in the agrarian structure has been mediated by a complex series of interactions between the agricultural and industrial sectors of society brought about by the government's development strategies and the broader process of the internationalization of capital. Although most village households continue to farm, few depend on farming alone. Farming and off-farm work have become inextricably intertwined in what we have termed the household enterprise. Changes in inter- and intra-household dynamics in the village are considered within this framework.

HSIN HSING VILLAGE, 1958-1979

Hsin Hsing is a nucleated village located beside a road that runs between two market towns, approximately 125 miles southwest of Taiwan's major city, Taipei. Its people are Hokkien (Minnan) speakers whose ancestors emigrated from Fukien, China, several hundred years ago. Within the village, the household is the basic socioeconomic unit. As in the rest of China, such a household takes one of three forms: conjugal, stem, or joint. The conjugal family household consists of a husband, wife, and their unmarried children; the joint family household adds two or more married sons and their wives and children to this core group. The stem family household, a form that lies between the conjugal and joint family types, includes parents, their unmarried offspring, and one married son with his wife and children.

In 1958, the registered population of the village was 609 people in 99 households or economic units (chia). Approximately four-fifths of the population was between the ages of one and forty-four years and slightly less than half was male (Table 16.1). Conjugal family households predominated, accounting for 66 percent of village households or 56 percent of the population. In contrast, only five percent of village family households, ten percent of the population, were of the joint type, while the remaining 29 percent of family households, or 35 percent of the population, were stem units.

Land tenancy was widespread among villages before implementation of the Land Reform Program of 1949-1953, a program designed by the government both to ensure political stability (Ho 1978:162) and to strengthen agriculture as a base for industrialization (Ho 1978:175; see Chang 1954 and B. Gallin 1966:93-98 for discussions of the land reform program). Prior to the land reform, 58 percent of the land was cultivated by tenant farmers, but by 1957 only 27 percent of the land was farmed by tenants. Despite this change in the tenancy/ownership ratio, most families cultivated farms far too small to support all family members. In 1957, 45 percent of the village families cultivated below .5 hectare and 84 percent cultivated below 1.0 hectare.

No significant industries or job opportunities existed locally to provide supplemental income to absorb the excess labor produced by a growing population on a finite land base. Thus, during the 1950s, almost all families were full-time agriculturalists, deriving most of their livelihood from two crops of rice, marketable vegetables grown in a third crop, and, in some cases, wages from farm labor or remittances from migrants. Men worked in the fields, taking care of tasks such as plowing, harrowing, transplanting, and harvesting — tasks women were considered incapable of performing. Men also assumed responsibility for the care of the more valuable livestock such as oxen, which provided the major draft power in plowing and hauling as well as "backyard fertilizer" to meet agricultural demands. Women managed the house and children, raised poultry, grew vegetables in small garden plots, helped with agricultural chores such as weeding fields or drying rice, dried and preserved crops, and in their "spare time" wove fiber hats at home to supplement the household income.

Wet rice cultivation, however, requires cooperation beyond the household level to meet labor demands at various stages of the rice production process (Huang 1981:32, 64-65). Extra laborers are needed at transplanting and harvesting so that these tasks can be completed within a short time span. Cooperation is also required to ensure an adequate water supply from the irrigation system. Thus, although household members formed the core of the management unit, the farm system also included kinsmen and neighbors who participated in collective labor in addition to working their own land.

This situation in the village began to change in the late 1950s and early 1960s as the growing intensity of population pressure on the land created problems of underemployment and farms too small to support household members. Increasing numbers of village males began to migrate to the larger cities of the province to seek jobs and supplemental income (Gallin and Gallin 1974). During the earliest phase of this population shift, migrants tended to be older married men. Some eventually brought their wives and children to the city. Others continued to maintain their families in the village while they worked away from home.

As a result, the population of the village in 1965 was different in some, though not all, ways from the population in 1958. The number of people resident in Hsin Hsing remained fairly constant over the seven years from 1958 to 1965 (Table 16.1). Approximately four-fifths of the resident population continued to be between the ages of one and forty-five years, but the percentage of the population aged sixteen to forty-five years decreased slightly. The percentage of males dropped from 48.7 percent to 44.9 percent, and more strikingly, only 34 percent of the sixteen to forty-four year old cohort was male.

TABLE 16.1

POPULATION OF HSIN HSING
BY TIME PERIOD AND AGE, 1958–79

Age	1958		1965		1979	
	N	Percent	N	Percent	N	Percent
1 - 15	269	44.2	237	46.8	151	39.4
16 - 44	235	38.6	166	32.8	129	33.7
45 - 64	90	14.8	78	15.4	78	20.4
65 and older	15	2.5	25	4.9	25	6.5
TOTAL	609	100.1	506	99.9	383	100.0
Sex Ratio (m/100f)	95		82		113	

Sources: 1958, Household record book, Pu Yen Township
Public Office; 1965 and 1979, field interviews.

Note: The figures for 1958 are for all people registered as members of Hsin Hsing households,
whether they were resident or only registered
there. An estimated 509 people actually lived in
Hsin Hsing in 1958. The figures for 1965 and
1979 record only people resident in the village.
In 1965, 612 people were registered as members of
Hsin Hsing households, compared to 606 people in
1979.

The emigration of males added responsibility for household farm management to women's traditional responsibilities. Women hired people to plow, to transplant seedlings,
to weed, and, if the men could not leave their work in the
city, to harvest the crops. They also paid wages and taxes
and arranged to exchange rice for fertilizer under a government "barter system." In addition, they spent a great
deal of time in the fields supervising laborers or checking
the flow of irrigation water. Women thus became primary
farm managers rather than an auxiliary labor force within
agriculture, assuming responsibilities formerly monopolized
by the men of their households.

The new sexual division of labor occasioned by the migration of men was facilitated by Taiwan's land consolidation program begun in 1962. As farm labor was extracted from the rural areas to fuel industrialization, the government began taking steps toward replacing human labor with farm machines (Ho 1978:159; Huang 1981:121-128). The physical structure of farm land was rebuilt, consolidating scattered, fragmented holdings and reorganizing irrigation systems and roads to establish the preconditions for farm mechanization. In addition to making it possible for farmers to eventually use farm machinery, this reworking of the landscape also reduced labor demands in the rice growing season by requiring less labor for travel to scattered plots, for transportation of farm tools to and harvests from the fields, and for monitoring irrigation water. The program thus made possible changes in farm management that facilitated the feminization of agriculture.

As migration out of the village continued throughout the 1960s, labor shortages became acute, farm profits decreased, and agricultural production declined. In Taiwan as a whole, production leveled off and varied by a small amount from year to year in the late 1960s (Taiwan Statistical Data Book 1979:59). This decline in production might well have continued but for certain international and national developments in the 1970s.

The spur for change in the 1960s had come originally from the influx of foreign capital in response to the government's policy of export-oriented industrialization. Industrialization brought about rapid urbanization and migration from rural areas to the cities. But industrialization had not been restricted to a few urban centers in Taiwan. During the 1960s, industry began to disperse to the countryside to gain access to low-cost labor and raw materials. By 1971, 50 percent of the industrial and commercial establishments and 55 percent of the manufacturing firms in Taiwan were located in rural areas (Ho 1979).

The move of industry to the countryside was accelerated in 1973 by the government's implementation of policies to spur rural development and the enactment of the "Accelerated Rural Development Program" to stimulate farm mechanization (Yu 1977). Large segments of the rural population had been absorbed by urban industry and the value of a farmer's production in 1972 was only one-fifth that of a non-farm worker's production (Huang 1981:3). To stem the stagnation of agriculture, the government abolished the rice-fertilizer barter system in 1972 and instituted a guaranteed rice price in 1973.

These policies, that created a climate in which farmers believed they could derive profits from the cultivation of their land, were followed by the oil crisis of 1974 and the world recession and inflation of 1974-1975. These events, in combination with the resultant changes in the world market, slowed the pace of industrialization in the cities of Taiwan (Taiwan Statistical Data Book 1979: 78). More than 200,000 urban workers lost their jobs (Huang 1981:163) as some factories shut down and others cut back production. As a result, the city began to lose the aura of El Dorado and the countryside began to acquire one of promise.

A comparison of the structure of the village population in 1979 with the population in 1958 and in 1965 suggests one result of these developments (Table 16.1). By 1979 only 383 people lived in Hsin Hsing, but the proportion of males had increased to 50.9 percent. More strikingly, males constituted 51 percent of the sixteen to forty-four year old cohort, approximately one and one-half times the percentage of males in this cohort in 1965. In part, this difference reflected a decrease in male emigration and an increase in the migration of unmarried females to urban areas. But the difference also indicated the return of earlier emigrants in response to rising costs and intense competition for jobs in the cities relative to rural areas.

Further examination of the data suggests another way in which the villagers responded to national and international developments. By 1979, conjugal family households no longer predominated in the village; only 45 percent of households (30 percent of the population) were of the simple type. Fully 18 percent of family households (34 percent of the population) were of the joint type, while the remaining 37 percent of family households (36 percent of the population) were of the stem form.

The reasons for this growth in joint family households have been documented in detail elsewhere (Gallin and Gallin 1982a; R. Gallin 1984a and 1984b). The villagers believed that this type of household provided the means for socio-economic success in a changing world. A household that included many potential wage workers, as well as other members who could manage the household, supervise children, and care for the land, had a better chance of diversifying economically than did a household of small size.

Demographic changes paralleled changes in the economic system of the area and village. Labor intensive factories, service shops, retail stores, and construction companies burgeoned in the market towns and rural countryside. By 1979, seven small satellite factories, three artisan work-shops, and 26 shops and small businesses had been established in Hsin Hsing (Gallin and Gallin 1982b).

Four-fifths (79.7 percent) of the village men and over one-half (54.4 percent) of the women in the productive ages were engaged in off-farm economic activities.

But this does not mean that off-farm employment supplanted agricultural work as the sole occupation of village households. Fully 83.6 percent of the village households continued to farm, but of these, only 6.6 percent (or 5.5 percent of the total households) depended on farming alone (Table 16.2). Households engaged in both farming and off-farm work were by far the most common (78.1 percent) in Hsin Hsing in 1979. Only 13.7 percent of village households were engaged solely in off-farm work, while 2.7 percent of households were retired from economic activities.

One reason so few households depended solely on farming was that they considered farming an unprofitable venture. On average, Hsin Hsing farmers realized less than NT$2,000 (US $55) from the rice they grew on .097 hectare of land in 1978. Moreover, family farms had decreased in size as a result of traditional inheritance patterns and the lack of an open, liquid land market after the land reform. In 1979, the average acreage tilled by a farming household was about .078 hectare per person, a considerable decrease from the .116 hectare per person cultivated in 1957.

Nevertheless, Hsin Hsing farmers continued to cultivate the land because: it was "a resource that must be used"; it was a source of food (rice); taxes levied on it had to be paid; and additional taxes were imposed if it was not cultivated. In economic terms, the farmers' marginal incomes from cultivating the land were greater than their marginal costs of not cultivating it. To achieve this greater marginal income the villagers changed their patterns of land use and farm management.

TABLE 16.2

OCCUPATIONS OF HSIN HSING HOUSEHOLDS, 1979

Occupation	Households	
	N	Percent
Farming Only	4	5.5
Farming and Off-Farm Work	57	78.1
Off-Farm Work Only	10	13.7
Retired	2	2.7
TOTAL	73	100.0

Source: Field interviews, 1979.

Traditionally, the first and second annual crops were devoted to the cultivation of rice. In 1978-79, however, only 65.6 percent of Hsin Hsing farmers cultivated rice during these crop periods; 19.9 percent cultivated vegetables or sugar cane, and 14.8 percent cultivated no rice at all. Moreover, in the third crop period, traditionally devoted to the cultivation of vegetables for marketing, more than one-third (36.1 percent) allowed their land to lie fallow (Table 16.3). These changes in patterns of land use were related to changes in patterns of farm management.

In the 1960s, the migration of men and the assumption by women of the major responsibility for the cultivation of rice had undermined the traditional pattern of work done by farm households and the patterns of cooperation among households. The introduction of power tillers, threshers, and harvesters rang the death knell for inter-household cooperative labor and led to increasing task specialization. Men with capital and technical knowledge bought the machines and moved around the countryside, performing work for individual farmers on specific days.

TABLE 16.3

LAND USE OF 61 FARMING HOUSEHOLDS
HSIN HSING, 1978-79

Crop	Households	
	N	Percent
First and Second:		
Rice Only	40	65.6
Rice and Vegetables	6	9.8
Rice and Sugar Cane	6	9.8
Sugar Cane Only	9	14.8
Third:		
Sugar Cane	15	24.6
Vegetables	24	39.3
Fallow	22	36.1

Source: Field interviews, 1979.
Note: Because sugar cane takes eighteen months to mature, the fifteen households raising this crop during the third growing season may include households that grew sugar cane in the first two cropping periods.

This situation contrasts with Gleason (1985:14) who
states that farm mechanization in Taiwan has displaced
Hokkien women from agriculture. In the case of the Hokkien
community of Hsin Hsing, farm mechanization did not displace women from agriculture. Rather, it turned over certain tasks traditionally performed by male farmers to a few
male specialists.

The introduction of the power tiller that is used to
transport materials, tools, and crops as well as to till
the land, was accompanied by the demise of the village
cattle population; only one ox remained in Hsin Hsing in
1979 and it was kept as a "pet." The disappearance of
cattle from the village increased the use of chemical fertilizer, although this was not the only factor responsible
for rising fertilizer consumption. The abolishment of the
barter system in 1972 lowered the price of chemical fertilizer and led to increased use.

Other innovations also affected farm management practices. Herbicides were introduced to eliminate weeds,
thereby reducing the time women spent in this arduous field
task. Tube wells, operated by diesel engines, replaced
human energy in obtaining irrigation water. The operation
of this equipment was monopolized by men; they had been
taught to operate it by extension agents and they believed
"it is just as easy to operate it ourselves as to teach the
women how to do it."

In short, the mechanization and chemicalization of
agriculture led to a shift from animal and human to fossil
energy. But it did not displace women from the land; some
women continued to serve as farm managers, hiring and
supervising the agricultural labor force in the absence of
their husbands who worked off-farm. Other women came under
the direction of a male manager, engaging in farming tasks
and supervising paid laborers. These tasks increasingly
became the work of older women who assumed the role responsibilities of their daughters-in-law to release them for
off-farm employment (R. Gallin 1984a).

Households without such a supportive network adopted
other strategies to promote their well-being. They used
their small amounts of land in ways to obtain the largest
amount of income from it. Some increased the amount of
sugar cane they grew. Because the government-owned Taiwan
Sugar Corporation assumed most of the responsibility for
cane cultivation, men could work off-farm while women
tended to the irrigation of the fields. Others increased
the amount of vegetables they grew, the men drawing upon
the labor of their wives and children to help them in

production. This also contrasts with Gleason's (1985) findings, since Hsin Hsing households chose labor-intensive production when female labor was less, rather than more, available.

In sum, the policies adopted by the government to foster economic growth were accompanied by profound changes in Hsin Hsing's farming system. The penetration of the industrial labor market and technological advances led to the reorganization of inter- and intra-household relationships. Increasing involvement in off-farm employment by Hsin Hing villagers was accompanied by the demise of kin and neighbor-based cooperative labor groups and changes in the sexual and generational division of labor. Agricultural tasks formerly done by farm households or related households were performed by a few paid specialists. Managerial positions previously restricted to men were held by women. Farm work originally carried out by younger women was conducted by older women. In the course of adapting to the outcomes of the political economy of Taiwan, the villagers' work in agriculture and industry became inextricably intertwined as they strove to ensure the viability, productivity, and growth of the household enterprise.

CONCLUSION

This longitudinal study of the evolution of the farming system in Hsin Hsing has a number of implications for farming systems practitioners. First, as the case makes clear, the organization of the farming system in the village at the three points in time (1958, 1965 and 1979) was highly dependent on and could not be understood apart from developments in other sectors of the economy. For example, out-migration of males in the early 1960s occasioned by the government's policy of industrialization set in motion a series of changes in the social organization of the productive process resulting in the decline of inter-household cooperative labor and the increased involvement of women as farm managers. By the mid-1970s, government agricultural policies and the spread of industry to rural areas had brought about labor shortages in the agricultural sector, resulting in production stagnation and decline. Although most households in Hsin Hsing continued to farm, in nearly all cases farming was combined with and conditioned by off-farm employment. Both farming and off-farm employment had become essential components in what is termed the household enterprise. Although this specific configuration of events may be unique to Taiwan, the underlying processes are not and have been documented in other contexts (Ferguson and Horn 1985). This suggests that there are strong arguments for viewing off-farm activities as central rather than as tangential to a farming systems analysis.

Second, this case study showed how changes in the agricultural production process were mediated by alterations in household or family organization. There was more involved here than changes in family size and composition over the household developmental cycle. The Hsin Hsing material demonstrates alterations in the prevalence of different household types over a 20 year period in tandem with changes in the national economy. In 1958, the conjugal family household was the most prevalent type in the village, but by 1979 less than half of the village households were conjugal units and the number of stem and joint family households had grown. Household structure was influenced by larger processes of economic development (industrialization, out-migration) and exerted an influence on the paths this development took. For example, in 1979 household organization was affected in part by local employment opportunities in industry and itself influenced farm management practices. Stem and joint family households, having different kinds of labor resources, were involved in different kinds of agricultural production and labor allocation practices than were conjugal family households. This case study suggests that farming systems practitioners should gather information on the range of household types. It is within this context that the division of labor by gender and age and its implications for technological development can best be analyzed.

ACKNOWLEDGEMENTS

B. and R.S. Gallin carried out the research discussed in this paper from 1957 to 1982. The first field trip in 1957–58 involved a 17 month residence in the village. This was followed by two separate studies in 1965–66 and 1969–70 of out-migrants from the area. The latest research spanned two months in 1977, six months in 1979, and one month in 1982. During these visits, the Gallins collected data using both anthropological and sociological techniques, including participant observation, in-depth interviews, surveys, censuses, and collection of official statistics contained in family, land, school, and economic records.

234

REFERENCES

Chang, Yen-Tien
    1954   Land Reform in Taiwan. Taichung, Taiwan:
    Department of Agricultural Economics, Taiwan
    Provincial College of Agriculture.

DeWalt, B. R.
    1985   Farming Systems Research. Human Organization
    44:2:106-114.

Ferguson, A. E. and N. Horn
    1985   Situating Agricultural Research in a Class and
    Gender Context: The Bean/Cowpea CRSP. Culture and
    Agriculture 26:2:1-10.

Gallin, B.
    1966   Hsin Hsing, Taiwan: A Chinese Village in
    Change. Berkeley, CA: University of California
    Press.

Gallin, B. and R. S. Gallin
    1974   The Integration of Village Migrants in Taipei.
    In The Chinese City Between Two Worlds. Mark Elvin
    and G. William Skinner, eds., pp. 331-58, Stanford,
    CA: Stanford University Press.
    1982a The Chinese Joint Family in Changing Rural
    Taiwan. In Social Interaction in Chinese Society.
    Richard W. Wilson, Sidney L. Greenblatt, and Amy A.
    Wilson, eds., pp. 142-50, NY: Praeger Publishers.
    1982b Socioeconomic Life in Rural Taiwan: Twenty
    Years of Development and Change. Modern China 8:2:
    205-246.

Gallin, R. S.
    1984a Women, Family and the Political Economy of
    Taiwan. Journal of Peasant Studies 12:1:76-92.
    1984b The Entry of Chinese Women into the Rural Labor
    Force: A Case Study from Taiwan. Signs 9:3:
    383-398.

Gleason, J.
    1985   Women's Work and Crop Diversification in Taiwan.
    Culture and Agriculture 26:2:10-16. (see also
    Chapter   , this volume).

Ho, S. P. S.
    1979   Decentralized Industrialization and Rural
    Development: Evidence from Taiwan. Economic
    Development and Culture Change 28:1:77-96.
    1978   Economic Development of Taiwan, 1860-1970. New
    Haven, CT: Yale University Press.

Huang, Shu-min
    1981   Agricultural Degradation: Changing Community
    Systems in Rural Taiwan. Washington, D.C.:
    University Press of America, Inc.
Shaner, W.W., P.F. Philipp, and W.R. Schmehl
    1982   Farming Systems Research and Development:
    Guidelines for Developing Countries. Boulder, CO:
    Westview Press.
Taiwan Statistical Data Book
    1979   Taipei, Taiwan: Council for Economic Planning
    and Development, Executive Yuan.
Yu, T. Y.H.
    1977   The Accelerated Rural Development Program in
    Taiwan. In Industry of Free China, pp. 2-16,
    Taipei, Taiwan: Council of Economic Planning and
    Development, Executive Yuan.

17
The Contribution of Women
to Agriculture in Taiwan

Jane E. Gleason

Recent research conducted by anthropologists, economists, and other social scientists has confirmed that women
in developing countries are an integral component of agriculture as producers and decision-makers. Most agricultural researchers now realize that an understanding of the
influence that women have on a farming system is essential
for increased effectiveness of efforts by international
agriculture research centers, development organizations,
and national agriculture programs.

Women in rural Taiwan are often farmers and work with
their male counterparts in almost every aspect of agricultural production. There are tendencies for women to perform certain farm tasks that men are less likely to do, and
vice versa. The presence or absence of women working on
the farm appears to have a pronounced effect on crop choice
and diversification. An abundance of female labor is generally associated with labor intensive crops, such as vegetables and fruits, while men are responsible for activities
that require machines. In Taiwan, this implies that farm
families with more male than female labor are more likely
to grow mainly rice and sugar cane. However, if a farm
husband and wife work together as a team, the variety of
crops grown is far greater. This study shows that farm
decision making is based, at least in part, on the perception that "labor" is not a homogeneous input.

RESEARCH SITES

Tainan County is located in the southern half of western Taiwan. It was the area first settled by Chinese
immigrants during the latter years of the Ming Dynasty
(1328–1644) and first decades of the Qing Dynasty (1644–
1911). For this reason it is often spoken of as the

historical and cultural center of Taiwan. Although in
recent times its importance has been overshadowed by Taipei
politically and industrially, it still stands as the most
important agricultural region in Taiwan. The low plains
along the coast supply Taiwan with a large quantity and
variety of grains, vegetables, and oilseeds, and the inland
foothill region produces a large percentage of Taiwan's
fruits.

Three townships in Tainan County, Ma-dou, Shan-hua, and
Shan-shang, were selected to serve as research sites for
this study. They were chosen on the basis of discussions
with scientists from the Asian Vegetable Research and
Development Center (AVRDC), where the research was based,
and local Farmers' Association officials. AVRDC is the
international agricultural research center responsible for
research to adapt certain vegetable crops to tropical environments. These three townships were deemed appropriate
because of their geographic proximity to AVRDC and their
differing agricultural environments.

Ma-dou is located in the lowland plains. An area
cultivated with sugar cane, much of which is owned by the
Taiwan Sugar Corporation is located on the outskirts. Upon
entering Ma-dou, the sugar cane fields give way to
asparagus and other vegetables grown along the banks of the
Zengwen River. On this side of Ma-dou there are also
orchards of pomelo and avocado, two fruits that people in
Tainan County have come to associate with agriculture in
Ma-dou County. On the other side of Ma-dou, the familiar
paddy rice fields predominate. Scattered among them are
areas of sugar cane, corn, sorghum, and a few vegetables
such as Chinese cabbage and cauliflower.

Shan-hua is similiar to Ma-dou in terms of agroenviron-
ment, and like Ma-dou, is in the Jia-nan irrigation network. Rice is the predominant crop in the summer, and in
winter, peanut, tomato, and various types of brassicaceous
crops are common. Strawberry is a newly introduced crop
that is increasing in popularity in Shan-hua, as are certain flowers such as salvia, grown for seed production. As
in Ma-dou, sugar cane is a principal crop in Shan-hua. For
purposes of analysis in this paper, data from Ma-dou and
Shan-hua are combined and presented together.

Shan-shang has a very different agricultural environment, as it is located in the foothill region of Tainan
County (Shan-shang in Chinese means "on the mountain").
From Shan-hua to Shan-shang, after passing through the
usual rice fields, the road leads up a hill and into fields
of mango, banana, and the ever-present sugar cane. In the
summer, there are also fields of water caltrop, a crop

that farmers in Shan-shang find desirable as a replacement for rice. There are occasional fields of sesame and mulberry trees used for silkworm production. On the other side of Shan-shang, the road goes further into the mountains. Most of the cultivable land in this area is devoted to mango.

The population of Ma-dou township is much greater and far more urbanized than that of Shan-shang. The population of Ma-dou in 1983 was 47,066, constituting 9,876 families. The farm population of Ma-dou was estimated at 24,816 (4,755 farm families) or 52.725 percent of the total (48.147 percent of the families). The population in Shan-shang was 8,384 persons, or 1,776 families. Of this total, the farm population was 5,525 (1,054 farm families) or 65.89 percent of the total population (59.34 percent of the families).

The amount of irrigated land in Ma-dou is 2,234.37 hectares, or 66.07 percent of the total, while the total amount of agricultural land in Shan-shang is 1,526.32 hectares, of which 1,478.17 hectares or 96.15 percent is non-irrigated, and only 48.15 hectares or 3.85 percent is irrigated.

In Ma-dou, rice is the most common crop, accounting for 1,425.65 hectares (both cropping seasons). Corn (884 hectares) and sugar cane (561.33 hectares) are the next two most important crops, followed by sweet potato (287 hectares) and pomelo (203.70 hectares). In Shanshang, the most important crop is sugar cane (464.42 hectares), followed by mango (298.80 hectares), then rice (119.20 hectares, two crops), and corn (112 hectares). Peanut and banana are also significant crops in Shan-shang.

The differences in the cropping patterns between the two townships are largely the result of a difference in water availability. In Shan-shang, a far greater portion of total agricultural land is devoted to permanent crops such as mango, banana, sugar cane, and peanut, because these crops do not depend heavily on irrigation. Since the greater part of cultivable land in Ma-dou is irrigated, rice is the major crop in the summer, and permanent tree crops and sugar cane occupy a smaller percentage of arable land. A stable water supply also allows farmers to double or triple crop. The multiple cropping index for the two townships is substantially different (Ma-dou = 142.40 and Shan-shang = 106.90), which further illustrates the differences in cropping patterns between the two areas.

DATA

Data were obtained at the Asian Vegetable Research and Development Center (AVRDC) in Tainan County, Taiwan. Thirty farm families near the AVRDC agreed to keep records of all farm activities for a full year beginning in May 1984. Thirteen of the families live in five villages in Ma-dou township, four families live in one village in Shan-hua township, and 14 live in three villages in Shan-shang. Record sheets were developed by the Agricultural Economics Department at AVRDC and were distributed and collected weekly. These sheets provided information on hectarage of land devoted to each crop and the amount of labor and capital used each week on each crop. For example, if a farmer sprays Chinese cabbage, she or he records which family member did the spraying, for how long, what the value of the labor was, how much of what chemical was used, and the value of the chemical. Labor was broken down into six separate categories: family male, family female, hired male, hired female, volunteer male, and volunteer female.

In addition to traditionally defined farm activities, farmers were requested to record daily household expenses and the amount of income from sources other than the farm. The category for household expenses included the amount of money spent on consumption items such as food and clothing, and other expenditures such as school tuition, medical supplies, and transportation fees. With regard to income from sources other than the farm, the most important category was off-farm income, that is, income derived from an off-farm job held by the husband, wife, or other members of the family living at home. The definition of an off-farm job included factory work or other types of non-agricultural employment as well as employment as a farm laborer for a daily wage. In addition, Chinese households frequently received money from grown children who do not live at home. Information related to this type of income was also included on the record keeping sheet. Farmers were also asked to record interest from savings accounts or on money that had been lent to neighbors or friends.

In spite of the high quality and completeness of the data obtained, the size of the sample may make generalizations of the conclusions difficult. However, variation in farm and family size in this part of Taiwan is not large, therefore, the data acquired from these 30 farm families adequately represents average farms in the area. Also, the farm families who participated in data collection were deemed to be "average" families for the region by the local Farmers' Association.

THE CONTRIBUTION OF WOMEN TO AGRICULTURE IN TAINAN COUNTY

Throughout the year, the farm women's main economic activity was farming on all but nine farms. Of those nine

TABLE 17.1

INCOME EARNED FROM OFF-FARM EMPLOYMENT

Farm Family	Shan-Shang Male	Female
1	1,446.87	47.50
2	0.00	0.00
3	0.00	0.00
4	1,568.75	55.25
5	0.00	0.00
6	0.00	140.00
7	15.00	0.00
8	5,029.50	1,388.00
9	1,515.00	1,500.00
10	2,500.00	782.50
11	1,543.50	2,725.00
12	0.00	0.00
13	0.00	0.00

Farm Family	Ma-Dou Male	Female
14	177.50	2,579.33
15	5,911.67	0.00
16	0.00	0.00
17	525.00	0.00
18	0.00	0.00
19	0.00	2,753.75
20	90.00	0.00
21	0.00	0.00
22	0.00	2,217.75
23	176.25	40.50
24	3,273.75	428.75
25	0.00	0.00
26	3,602.70	231.88
27	525.00	1,262,75
28	4,492.00	397.50
29	852.50	852.50
30	4,171.00	917.50

Note: Figures are in U.S. currency ($U.S.).

who did not work regularly on the farm, six farm women
(wives) had off-farm employment. Four of these women
worked in local factories and two were farm laborers.
These six women often worked on their own farms in the
evenings and many worked full-time on the farm on Sundays.
Of the remaining three, two occasionally worked with their
husbands on the farm, especially during heavy labor-using
periods such as harvest. Work on the farm tended to be
seasonal; for example, one woman worked in a factory in
Ma-dou during the summer, but worked full-time on the farm
in the winter. Conversely, another farm family planned a
cropping pattern that allowed both partners off-farm
employment in the winter. In this family, both the husband
and wife harvested sugar cane for the Taiwan Sugar Corporation.

The financial contribution made by the wives to the
family was considerable. Those who had steady off-farm
jobs contributed their salaries to the household while
those who worked the majority of the time on the farm often
obtained additional income as farm laborers for a daily
wage. Many worked temporarily in nearby factories during
slack periods on their own farms. Fourteen of the 30 farm
women in the sample earned income from off-farm sources.
Table 17.1 shows the extent to which farm wives and husbands make financial contributions to the family. Off-farm
financial contributions of the husband are often greater
than that of the wife, but this does not necessarily
reflect a difference between the husband and wife in the
number of hours worked.

Remuneration as a farm laborer depends on the type of
work. Women are hired to do hand labor such as weeding or
hand harvesting of fruit or peanuts, and men are hired for
machine operation such as spraying pesticides and land
preparation. Women generally earn US $6.50 to US $8.75 per
labor-day depending on the crop. Harvesting or weeding
vegetables usually commands a lower wage than harvesting
water caltrop or mushrooms. Men earn far more per day than
women. For spraying pesticides or pruning fruit trees, men
usually earn US $12.50-15.00 per day. The reason given for
wage differentials between men and women is that "women's
work" is easier and lighter than "men's work," and hence
deserves a lower wage. In addition, farmers recognize the
hazards of spraying chemicals and therefore this work commands a premium. Compensation is determined on a per hectare basis for land preparation using bullocks, rototil-
lers, tractors, and for mechanical planting and harvesting
of rice.

When discussing hiring practices, farmers invariably speak of certain tasks as women's work and others as men's work. Farmers never hired women to prepare land or spray pesticides, nor were farm wives ever hired by others to perform this kind of work. Farm wives were hired only for weeding, harvesting, and other types of hand work, and only on one occasion was a farm husband hired to do handwork (he harvested peanuts).

Utilization of farm family labor is not as clearly delineated by gender as is hired labor, though men are far more likely to participate in women's work than vice versa. Husbands will often help their wives with weeding or hand harvesting, but wives generally do not spray pesticides, cultivate, or prepare land unless only a simple shovel is required. Women in agriculture in Taiwan seldom use machines; farm tools for women are confined to simple tools such as a knife, scythe, or shovel. In only one of the 30 families did a wife, who managed the farm without her husband, spray pesticides on a regular basis. On one other occasion, a woman sprayed pesticides because her husband was on a travelling vacation.

However, there may be some bias in the data regarding the amount of women's work a man actually does. It is common in the countryside of Taiwan to see groups of women working together in the field with one man. For example, when harvesting tomatoes, which is women's work, a farm family will usually hire several women who pick tomatoes with the farm wife while the husband uses a wheelbarrow to move the filled tomato cartons to the side of the road for factory pick-up. He makes sure that the women have enough cartons, and he is responsible for determining which rows are to be harvested. Therefore, while the husband is technically harvesting tomatoes, and the data indicate that he is doing a woman's task, the actual tasks that he performs are different from those of the women. This is because the husband as farm manager is responsible for inspecting the fields and deciding what tasks are required. He usually assigns jobs, organizes, and oversees the work that is performed. It is interesting to note that many Taiwanese farm wives refer to their husbands as <u>tou gei</u>, meaning "boss," and frequently when asked questions about the farm, they will either say what the <u>tou gei</u> said or request that he be asked directly.

LABOR BY GENDER AND CROP

Twenty-one different crops were planted on the thirty farms. This accounts for all seasons. Women contributed somewhat more hours of labor to the production of these crops than men. The total number of labor hours recorded on all farms for all crops was 58,697, of which 53.61 percent was female labor, and 46.39 percent male labor. Only a few crops used more male than female labor. Over 70 percent of the total labor used for rice and almost 60 percent of total labor for bananas was supplied by men. The amount of male labor used for silkworm production is marginally greater than the amount women provided (Table 17.2).

The differences in female and male labor used for production of various crops is dependent on the types of tasks needed for each crop. Rice production in this part of Taiwan is almost fully mechanized which means that much of the women's work formerly associated with rice production has been eliminated. In southern Taiwan, one seldom sees groups of women transplanting rice because of widespread use of rice transplanters. Herbicide use in Taiwan has to some extent replaced the need to weed, that is also women's work. Banana production requires more male labor because the fruit is heavy and therefore men are more apt to undertake this task. Silkworm production, a process that is repeated six times a year for three weeks each time, utilizes a large amount of male labor relative to female labor because a saw is used to cut mulberry branches. For silkworm production female labor is needed primarily to prepare the silkworms for the spinning of cocoons. At that time, each farm family will hire several women for one or two days of work.

For all other crops, the amount of hand-work required for productions exceeds the amount of work that uses machines. There may, however, be a small amount of under-reporting of the hours of hired male labor used for land preparation. Since remuneration is based on hectarage, not time, farmers recorded the costs of these tasks rather than the time used.

It is clear from the data of the 30 farmers that women did not use machines and that men did some hand-work even though it is generally considered women's work. This implies that as agriculture in this part of Taiwan proceeds toward greater mechanization, more women than men are likely to be forced into other sectors of the economy. Thus, mechanization and the use of chemicals in agriculture may displace female laborers and in some cases actually generate an additional need for male labor.

An example of a possible future change in technology that will have a significant impact on female labor is the peanut harvester. Many farmers have mentioned the need for a machine to harvest peanuts because teams of women, as

TABLE 17.2

LABOR USAGE (IN HOURS) BY GENDER AND CROP

Crop	Female	Male	Total
Shan–Shang:			
Banana	1,245	1,731	2,976
Corn	1,131	428	1,559
Edible Sugar Cane	513	506	1,019
Longan	690	859	1,549
Mango	2,283	1,346	3,629
Rice	297	970	1,267
Silkworms	2,629	3,052	5,681
Tomato	2,278	1,368	3,646
Water Caltrop	2,861	2,221	5,082
Watermelon	2,736	1,612	4,348
TOTAL	16,663	14,093	30,756
Percentage	(54%)	(46%)	(100%)
Ma–Dou:			
Chinese Cabbage	233	197	430
Cauliflower	130	271	410
Corn	2,867	2,490	5,537
Flowers	815	413	1,228
Lima Bean	3,205	2,457	5,662
Peanut	2,276	721	2,997
Pepper	339	130	469
Pomelo	1,055	1,097	2,152
Radish	312	240	552
Rice	867	2,193	3,060
Spinach	213	189	402
Strawberry	243	172	415
Sweet Corn	27	25	52
Sweet Potato	720	181	901
Tomato	483	697	1,180
Watermelon	1,014	1,660	2,674
TOTAL	14,808	13,133	27,941
Percentage	(53%)	(47%)	(100%)

many as 30 women per hectare, now are hired for one day at harvest. This accounts for a high percentage of production costs. The same is true for processing tomatoes. If peanut and tomato harvesters become a reality in southern Taiwan, the employment impact will be primarily on women. Estimating how great that impact will be is beyond the scope of this paper, but it should be examined before the harvesters are introduced. Teaching women to operate agricultural machinery would obviously lessen the severity of the employment impact.

WOMEN AND CROP CHOICE

Farm families theoretically choose crops so that the production inputs available to them — land, labor, and capital — maximize revenue. Also, farmers may have objectives other than profit maximization, such as risk minimization, that affects what they perceive as an optimal crop mix.

Crop choice is dependent on a number of environmental and economic parameters which farmers cannot control. Farmers in Ma-dou and Shan-hua exhibit different cropping patterns than those in Shan-shang in part because of differing geophysical characteristics. Other factors such as government policy or marketing conditions may similiarly affect cropping patterns in a given area.

The amount and type of family labor that is available to work on the farm also influences crop choice. Women in agriculture in southern Taiwan have a clear role in determining crop choice, but not because they are explicit decision-makers in the family. That role is usually dominated by men, although women may be consulted about their crop preferences. Rather, women's role in determining crop choice is the following: they provide additional labor to the family farm that in most cases complements that of the husband since few farms are operated only by women; they provide labor that is perceived as qualitatively different from that of men and perform tasks that men are less likely to do unless as part of a man/woman team. The effect on crop choice is illustrated by an increase in the variety of crops grown, by more labor-intensive crop production, and by a clear-cut decrease in the percentage of land that is devoted to rice and sugar cane.

Data for one full year, from May 1984 to May 1985, were used to assess the effect that women had on crop choice. The sample was divided into those in which female family labor worked on the farm and those in which it did not. The sample was also split into different agricultural environments, with Ma-dou and Shan-hua comprising one type of agricultural environment and Shan-shang the other.

Differences in crop choice between families with on-farm working females and those without were very dramatic in both regions. In Ma-dou and Shan-hua, farms without female labor planted twelve crops of which five accounted for almost 95 percent of total hectarage (Table 17.3). Those five crops, rice, sugar cane, corn, sorghum, soybean, and peanut, are considered male crops either because they require more labor with machines than handwork or because they are not labor intensive. In Southern Taiwan, a male farmer who grows only rice and sugar cane in the summer and corn or sorghum and sugar cane during other seasons, is generally not faced with a labor constraint that prevents him from planting other more labor intensive crops. Farm size in this area of Taiwan is still small relative to the amount of labor available and rice or corn and sorghum production allows farmers considerable free time. The data indicate that once planting is completed, farmers only need to inspect the fields daily and occasionally spray with pesticides. Hence the decision not to grow crops other than these is partially a function of the type of labor available to the farm. Male farmers are more likely to choose labor-intensive crops when female labor is available.

In contrast to these farms, the cropping patterns of farms with female family labor included twenty-two crops, and the percentage of land devoted to the crops mentioned above was noticeably less. Rice was planted on 31.02 percent of total cultivated land, and sugar cane accounted for 10.59 percent. Other important crops in this category were pomelo, watermelon, lima bean, and tomato (Table 17.4).

Farm families in Shan-shang exhibited the same tendency as those in the other areas. In Shan-shang, 73.33 percent of hectarage of farms without female family labor was devoted to sugar cane and rice, whereas on farms with female labor the percentage was 32.28. The decrease in sugar cane was the largest, dropping from over 50 percent of the total to approximately 20 percent. Like Ma-dou and Shan-hua, there was a large increase in the variety and number of crops grown on farms with female family labor. Farms without female family labor cultivated only five crops, while those with female family labor planted 15 different crops (Tables 17.3 and 17.4).

Since the variety of crops was far greater on farms with female family labor than farms without, female family labor thus allows farms to be more diversified. A diversification index was utilized to determine whether or not the level of diversification between the two categories of farms was significantly different. The index is calculated as follows:

$$DI = \frac{1}{\Sigma P_j^2}$$

where P_j is the portion of land that is devoted to crop j. Monoculture has a DI=1, and if total area of one farm is divided equally into two parts on which two different crops are cultivated, DI=2. The greater the value of the DI, the greater the level of diversification.

TABLE 17.3

CROPS AND TOTAL HECTARAGE OF FARMS
WITHOUT FEMALE FAMILY LABOR

Crop	Ma–Dou/Shan–hua		Shan–Shang	
	Ha.	Percent	Ha.	Percent
Rice	8.51	39.60	2.06	21.13
Sugar Cane	2.73	12.70	5.09	52.20
Tomato	.12	.56	1.30	13.34
Corn	4.70	21.18	————	————
Sorghum	2.07	9.63	————	————
Lima Bean	.33	1.54	————	————
Sweet Corn	.23	1.07	————	————
Sweet Potato	.30	1.40	————	————
Cauliflower	.23	1.07	————	————
Soybean	1.23	5.72	————	————
Mungbean	.10	.46	————	————
Peanut	1.17	5.44	————	————
Silkworms	————	————	.50	5.13
Water Caltrop	————	————	.80	8.21
TOTAL	21.49		9.74	

TABLE 17.4

CROPS AND TOTAL HECTARAGE OF FARMS WITH FEMALE FAMILY LABOR

Crop	Ma-Dou/Shan-hua		Shan-Shang	
	Ha.	Percent	Ha.	Percent
Rice	12.71	31.02	4.52	12.64
Corn	5.65	13.79	2.94	8.22
Sugar Cane	4.34	10.59	7.02	19.64
Tomato	.98	2.39	1.92	5.37
Mango	.29	.71	5.30	14.83
Edible Cane	.72	1.76	.46	1.28
Watermelon	3.20	7.81	6.33	17.71
Sorghum	2.63	6.43	——	——
Peanut	3.72	9.08	——	——
Lima Bean	1.86	4.54	——	——
Sweet Corn	.15	.36	——	——
Sweet Potato	.80	1.95	——	——
Pomelo	.90	2.20	——	——
Sesame	.30	.73	——	——
Cauliflower	.10	.24	——	——
Spinach	.08	.20	——	——
Green Pepper	.23	.56	——	——
Strawberry	.10	.24	——	——
Radish	1.01	7.81	——	——
Flowers	.40	.97	——	——
Banana	——	——	2.40	6.71
Water Caltrop	——	——	2.31	6.46
Silkworms	——	——	.90	2.52
Cassava	——	——	.51	1.43
Okra	——	——	.20	.56
Orange	——	——	.35	.98
Papaya	——	——	.40	1.12
Longan	——	——	.18	.50
TOTAL	40.97		35.74	

Table 17.5 gives values for the diversification index for all farms. Two separate analyses of variance tests were computed to determine whether location or the presence of family female labor or both contributed to the differences in the level of diversification among the four groups of farm families. The results showed that location was not statistically significant but female family labor was.

TABLE 17.5

DIVERSIFICATION INDICES OF 30 FARMS

	With Female Labor		Without Female Labor	
	Farm	DI	Farm	DI
Shan–Shang:	1	4.10	5	1.97
	2	5.73	8	1.10
	3	4.54	11	1.28
	4	2.78		
	6	3.69		
	7	2.40		
	9	1.98		
	10	1.54		
	12	2.27		
	13	2.23		
Average DI:		3.13		1.45
Ma–Dou/Shan–Hua:	15	2.85	14	2.23
	17	2.86	16	1.25
	18	2.42	19	1.67
	20	1.46	21	2.12
	23	1.68	22	1.34
	24	1.73	30	1.78
	25	3.04		
	26	2.70		
	27	3.20		
	28	3.87		
	29	2.09		
Average DI:		2.53		1.73
Overall Average DI: W/FL		2.86	W/O FL	1.63

In this area of Taiwan, diversification of agriculture means that more risky crops are introduced as replacements for rice and sugar cane. These two crops are considered relatively risk free compared to vegetables and fruits, because government policy protects the prices of both crops and therefore they are subject to minor price variation. Also, farmers are knowledgeable about production problems, having grown these crops for many years. Cultivars that perform well in the specific environment of the region have been developed and adopted by all farmers. Apart from rice and sugar cane, corn, sorghum, peanut, and soybean also have guaranteed government prices. It is noteworthy that in Ma–dou and Shan–hua almost 95 percent of total hectarage of farms without female labor and in Shan–shang over 50 percent (there is no guaranteed price for rice in Shan-shang) was devoted to crops with fixed government prices. Hence the presence of females on the farm significantly increases the proportion of hectarage that is planted in crops that are subject to price volatility. Therefore, inclusion of crops other than those mentioned above into cropping patterns in southern Taiwan is to accept added risk, despite the agricultural economics literature that claims that diversification is a form of risk management.

IMPLICATIONS AND CONCLUSION

The availability of female labor in the farming system in southern Taiwan increases the variety of crops planted, as well as the level of diversification. This occurs because rice cultivation is now highly mechanized and most rice production tasks are done by men. If women are able and willing to work on the family farm, labor intensive crops such as vegetables and fruits are more likely to be grown because of the type of labor required for their production.

The data on the 30 Taiwanese families suggest that economists should not view labor as a single production input. Rather, it should be evaluated as separate male and female components. Decision-making models should therefore take into account two types of labor constraints. In addition, various crops in the system should be evaluated in terms of those requiring male labor or those requiring female labor.

Additional research is needed to determine to what degree the principles revealed in this part of Taiwan are applicable to the rest of the country and to other areas of Asia. Since Taiwan's agriculture is at a higher level of development relative to other countries in Southeast Asia, trends in Taiwan may serve to forecast developments in other areas in Asia with similiar agroenvironments, such as Thailand or certain regions of the Philippines or Indonesia. This may indicate that the availability of female labor in the farming system may be a key to the adoption of vegetables or other subsidiary crops. It is also clear that women will be the users of modern vegetable technology, and they are most likely to be affected by technical changes in their production.

NOTES AND ACKNOWLEDGEMENTS

An earlier version of this paper, entitled "Women's Work and Crop Diversification in Taiwan," appeared in Culture & Agriculture 26 (Spring):10-16.

18
Gender-Differentials in the Impact of Technological Change in Rice-Based Farming Systems in India

Bahnisikha Ghosh and Sudhin K. Mukhopadhyay

This paper examines the differences in the use of men's and women's time in various agricultural and other activities in rice-based farming systems, based on recent evidence from farm-household surveys. The discussion answers the following questions: (1) What is the total number of hours worked per day by an adult male and female on average, and what is the distribution of time spent in different activities? (2) Is there any difference in the male-female allocation of time because of socioeconomic characteristics, such as between cultivators and landless laborers or between agricultural and non-agricultural households? (3) What are the differences between the contributions of male and female workers in the various operations in rice cultivation? (4) What are the major factors explaining differences in male-female time allocation in the cultivation of rice and does technology impact significantly upon the proportion of female labor to total labor used?

Since the concept of "work" or "economic activity" remains elusive, it is difficult to analyze the gender dichotomy of the labor market. Assessing the role of women as contributors of labor inputs and generators of family income and utility is especially difficult. If "work" is defined according to conventional practice, as in the Census of India, the majority of women disappear from the "economically active" category and most of their valuable contributions are ignored. However, women emerge as significant participants in the generation of family income and economic welfare if the true extent of their activities are noted. These activities are necessary for subsistence and the ultimate welfare of the household and produce commodities that otherwise must be purchased with cash.

One way to redefine "work" or "economic activity" is to subject the whole gamut of activities that an individual

undertakes to intensive scrutiny by time allocation
methods. Since the total time available for the household
from its members is the major source of income, time
allocation studies provide economists with an opportunity
to reconsider employment concepts (Ghosh and Mukhopadhyay
1984).

Time allocation methods are used here to redefine total
activity into "economic production" activities and "home
production" activities, the latter encompassing not only
such activities as education, child care, and domestic
work, but also activities that may be called "expanded
economic" (Pradhan and Bennett 1981). "Expanded economic"
activities include the making of fuel, knitting, sewing,
etc. These activities use market goods and services,
home-grown inputs, and the time input of household members
to produce commodities (or "characteristics") that are the
"true" objects of utility. They could justifiably be
called "home production," based upon the concept of a home
production function.

This paper is based upon data collected from farm and
household surveys in six villages (Barasat, Nagarukhra,
Simulpukuria, Bhandarkona, Singa, and Chanda) in the district of Nadia, West Bengal. Data were collected on a complete enumeration basis for all individuals in the villages. The data measured detailed allocation of time into
all activities, both "economic" and "home" production, for
the six month period of the major rice season in the
region. "Economic production" activities were divided into
agricultural and non-agricultural and "home production"
activities included education, child care, domestic work,
and other activities. Households were classified into four
broad occupational categories: only agriculture; only
non-agriculture; both agriculture and non-agriculture; and
others, including the unemployed and persons not in the
labor force. To examine differences in economic status,
households were also classified into cultivators (owners
and tenants) and landless agricultural laborers. The cultivators were further divided into users of traditional and
modern technology in rice cultivation. This study, based
upon an intensive examination of the data from these six
villages, was supplemented by a regression analysis of both
these and additional Farm Management Survey data for two
other districts in West Bengal.

A BRIEF DESCRIPTION OF THE POPULATION

The six villages comprising 1,570 households have a
total population of about ten thousand (Mukhopadhyay 1984).

The sex ratio averages 951 females per 1,000 males, in contrast to the average of 911 for the state of West Bengal. The age/sex composition of these villages suggests that females outnumber males only at age groups 0-9 and 15-19, while at all other ages the reverse is true. The age distribution of the total population of these villages is similiar to that of rural West Bengal. The data available on age at marriage indicate that 75 percent of marriages occur before age 17 for women and age 27 for men. The average number of children per woman in these villages is six, about 17 percent higher for all ages than for rural West Bengal in 1981, producing a higher total overall fertility rate (TFR). The relatively low sex ratio, low age at marriage, and the high fertility rate may have important effects on the allocation of time to different activities by women in these villages.

Information on literacy rates for the population suggests that: (1) male literacy is much higher than female literacy, with 33 percent of males and almost half of the females being illiterate; (2) for both sexes, literacy is much higher among recent cohorts compared to the older population; (3) when the population is classified into landowners and landless laborers, the literacy rate appears to be much higher for the former, the differential being substantial for the females; and (4) if the landowning families are further classified into those who use the new technology and those who continue to grow only traditional varieties of rice, the high-yielding variety (HYV) rice growers appear to more literate than the others, with the differential again being especially noticeable in the case of females.

In these villages the total cultivable land is owned by about 49 percent of the households. Of these households, owner-cum-sharecroppers and pure tenants constitute a little over 11 percent, while about 36 percent own and cultivate their own land. Of the remaining households with no land, 31 percent of the families engage in only non-agricultural activities and about 14 percent are agricultural laborers. The rest, about six percent of total households, earn their living from both agricultural and non-agricultural occupations.

Modern rice technology seems to have exerted a significant influence upon farmers in these villages. Although 40 percent of all farmers still adhere to traditional rice cultivation technology, the rest have turned completely to modern technology (20 percent) or use modern in combination with traditional technology (40 percent).

GENDER DIFFERENTIALS IN TIME ALLOCATION

The total population aged 15-50 in the six villages selected for this study consisted of 2,319 men and 2,123 women. On average, a man and a woman in this age range spend about 9.47 hours and 9.53 hours per day respectively on all activities except rest, recreation, and personal needs (Table 18.1). The man spends about 6.25 hours per day on economic production activities and 3.22 hours on home production activities. In contrast, the woman spends only about 1.35 hours per day on economic activities and 8.18 hours per day on activities related to home production, including child care, education, housework, sewing, knitting, making cowdung cakes for fuel, and others.

If the activities related to home production are included, the total hours spent by women are higher than those for men in each occupational category of the population. Women spend much longer hours in home production activities than the men, which more than compensates for the hours spent by men in economic production activities. The number of hours spent by women on home production activities remains fairly steady and never falls below 5.3 hours per day in spite of variations in the number of hours spent on economic production activities; while the time spent by men on home production activities varies substantially depending on the time spent in economic production activities.

The study villages are dominated by rice production that is the most important agricultural activity and the major avenue for income, employment, and work. It is useful to examine in detail the relative importance of the time spent by men and women in rice production. Male--female differences in participation in rice production are influenced by social, economic, and technological factors. Differences exist among the various operations in rice cultivation as well as between operations conducted in the field and in the household.

In West Bengal, female participation in field work in rice cultivation is uncommon among Bengali Caste Hindus and Muslims who form the vast majority of farmers in these villages. Female participation is more important among other castes and tribes, but they are only a small percentage of the total population. It is not surprising, therefore, that the total female labor input in rice cultivation in these villages is only about ten percent, with the rest of the labor coming from the men (Mukhopadhyay 1985). Among the different operations, the proportion of female labor is highest in processing that is carried out within the

TABLE 18.1

TIME ALLOCATION OF MALE AND FEMALE POPULATION: HOURS PER PERSON PER DAY

Occupational Categories

Activities	Only Agriculture		Only Non-Agriculture		Both		Others		Total	
	Male	Female	Male	Female	Male	Female	Male	Female	Male	Female
Economic Production Agriculture plus Non-Agriculture	5.21	3.82	8.56	6.34	10.05	7.18	1.95	0.90	6.25	1.35
Home Production Education, Child care, Housework, and Other Activities	2.48	5.29	1.70	5.38	2.20	5.88	7.49	8.50	3.22	8.15
TOTAL WORK BURDEN	7.69	9.11	10.26	11.72	12.08	13.06	9.45	9.40	9.47	9.53

Source: Mukhopadhyay (1984).

household premises (Table 18.2). Such female labor comes mainly from the family. Family women and hired men engage in processing work, whereas male family members perform mostly fieldwork and supervision. Weeding, which consumes the largest proportion of total labor used in rice cultivation, is also the most important activity for women. Most weeding is done by hired female labor rather than by family members. Supervision may be undertaken both in the field and within the household premises; in the field using solely male family labor, while female family members supervise processing and storage operations done by hired labor in the household courtyard. About 98 percent of all supervision is done by male family members and makes up 44 percent of their total work burden; a small amount is done by female family members who spend only about 11 percent of their time in this activity.

TABLE 18.2

MALE-FEMALE DIFFERENCES IN LABOR INPUT IN RICE CULTIVATION

| | Percentage of Hours Spent By | | | |
| | Male | | Female | |
Activities	Family	Hired	Family	Hired
Seed Bed Preparation	6.35	1.55	0.60	0.40
Field Preparation	14.29	11.63	0.31	0.64
Transplanting	3.37	25.03	0.82	26.41
Irrigation	8.37	2.03	0.66	—
Application of Fertilizer	3.34	2.37	0.13	0.07
Weeding	8.94	26.01	4.51	35.20
Harvesting	6.37	20.04	4.00	21.30
Processing	4.58	11.34	77.25	15.97
Supervision	44.39	—	11.71	—

Source: Mukhopadhyay, 1984.

Transplanting and harvesting are important activities for hired labor, with both male and female laborers allocating about the same share of their time to these activities. Seedbed preparation, irrigation, and application of fertilizers appear to be the responsibility of male family labor.

A pattern of labor use thus emerges that emphasizes the gender-specific allocation of different operations of rice cultivation. Hired female laborers typically work in the field, spending most of their time in such tedious and laborious jobs as transplanting, weeding, harvesting, and processing while female members of owner-cultivator households remain busy with processing, storage, and supervision activities. The male member of such a household spends most of his time on his own farm in supervision and field preparation. For male hired laborers, their most important activities also appear to be weeding, transplanting, and harvesting, although these activities absorb a relatively smaller proportion of their time than of the female hired laborers.

Gender differences in time allocation among the different components of economic production and home production activities are also dependent on the economic status of households. To examine this, as well as the impact of technology on gender differentials in labor use, one of the six villages (Bhandarkona) was subjected to closer scrutiny on the basis of complete enumeration. The total number of households was divided into: (1) those cultivating 100 percent high-yielding variety rice; (2) those cultivating 100 percent traditional variety rice; (3) those cultivating different combinations of high-yielding variety and traditional variety rice; and (4) landless agricultural laborers' households. Since households cultivating combinations of high-yielding variety and traditional variety rice could not provide dependable information on the distribution of labor and other inputs by technology, discussion has been confined to only categories (1), (2), and (4); (1) and (2) are cultivators and (4) are landless laborers.

For the population aged ten and above, a female spends on an average about nine and a half hours per day in both economic production and home production activities, compared to less than seven and a half hours for a male. The distribution of these two activities shows that the bulk of women's time is spent on home production activities while the average man does not spend much more than three hours per day in such activities (see Figure 18.1). Compared with cultivators, the average number of hours worked by women in landless households increases for economic

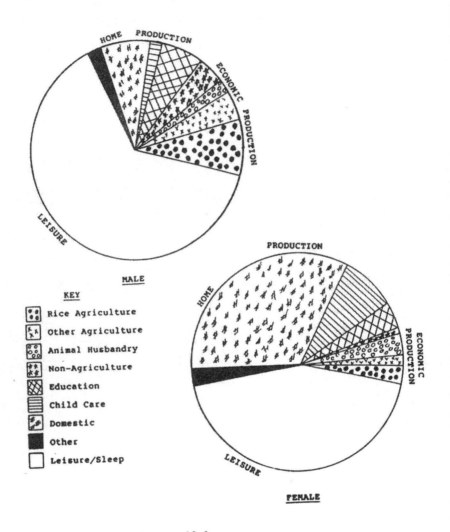

FIGURE 18.1

ALLOCATION OF TIME TO DIFFERENT ACTIVITIES:
MALE-FEMALE DIFFERENTIAL

production activities while the hours worked in home production decreases. For men, however, the number of hours worked on both sets of activities are higher for the landless laborers than for the cultivators. This may be due to the landless laborers' abject poverty and lack of any household assets. Their female household members have to scramble for survival outside the home, and have practically no resources for home production. At the margin, in such poor households, the man works somewhat more than the woman.

Disaggregation of the cultivators into high-yielding variety (HYV) and traditional variety (TV) rice growers shows that the total number of hours worked increases on an average from 6.2 to 9.2 hours respectively for the male and female traditional variety growers to 7.5 and 10 hours for the high-yielding variety growers (Table 18.3). This is due to increased agricultural activities, total economic production activities, as well as total home production activities. Women in both the high-yielding and traditional variety growers' households work longer than the men due to their greater participation in home production activities (see Figure 18.2). However, in agricultural as well as total economic production activities, the number of hours worked by the women increases from traditional variety to high-yielding variety technology.

Most of the increase in agricultural activity for women in high-yielding variety technology is in processing and supervision in rice production (Table 18.3). For men, there has been a marginal decline in the total number of hours worked in rice production as a result of the change to high-yielding varieties. The new technology produces a larger volume of output requiring more threshing, cleaning, parboiling, etc., yet the short duration modern varieties require fewer hours of work in the field. When one takes into consideration the total rice-based farming system including non-rice crops and animal husbandry, the total labor requirement increase is more pronounced. However, there is very little gender differential in this increased labor requirement for the farming system as a whole.

The workload of the women in home production activities, including domestic chores, child care, and education, also increases from 7.85 to 8.23 hours per day on an average as a result of the change from traditional variety to high-yielding variety rice cultivation. The corresponding increase for the male is only from 3 to 3.17 hours per day (Table 18.3). Most of the increase in home production

TABLE 18.3

TIME ALLOCATION BY MALE AND FEMALE POPULATION IN ECONOMIC AND HOME PRODUCTION ACTIVITIES: BHANDARKONA
(HOURS PER PERSON PER DAY)

Activities	Landless Laborers		TV Cultivators		HYV Cultivators	
	Male	Female	Male	Female	Male	Female
Agriculture: Rice						
Field Work	3.54	0.53	1.35	0.003	0.68	0.00
Processing & Supervision	0.35	0.09	0.42	0.17	0.41	0.64
Total Activity	3.89	0.62	1.77	0.17	1.09	0.84
Agriculture: Other	0.49	0.26	0.61	0.01	1.49	0.06
Animal Husbandry	0.48	0.85	0.61	1.12	0.76	0.91
Total Agriculture and Allied Activities	4.86	1.73	2.99	1.29	3.35	1.81
Total Economic Production	6.90	2.11	3.24	1.36	4.38	1.81
Total Home Production	3.28	6.68	3.00	7.85	3.17	8.23
Education	0.67	0.38	1.33	0.63	1.53	1.62
Child care	0.25	1.69	0.19	1.59	—	1.14
Domestic	2.33	4.61	1.22	5.13	1.42	5.15
Other	0.17	—	0.26	0.50	0.22	0.32
TOTAL WORK BURDEN	10.18	8.99	6.24	9.20	7.55	10.04

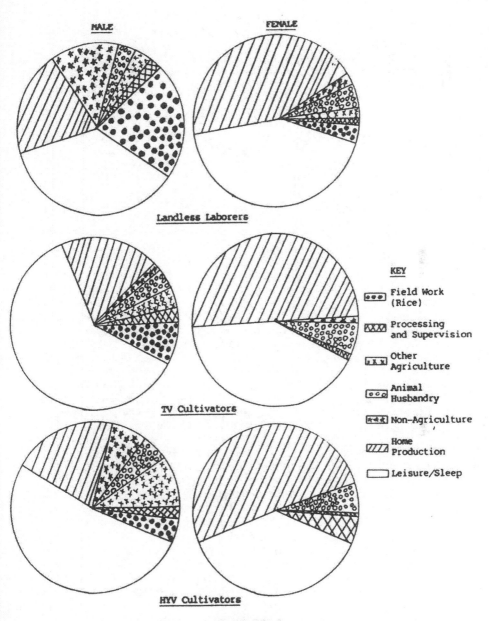

MALE FEMALE

Landless Laborers

KEY

- ●●● Field Work (Rice)
- ✕✕✕ Processing and Supervision
- ✗✗✗ Other Agriculture
- ⊙⊙⊙ Animal Husbandry
- ✦✦✦ Non-Agriculture
- ⁄⁄⁄ Home Production
- ▢ Leisure/Sleep

TV Cultivators

HYV Cultivators

FIGURE 18.2
GENDER DIFFERENTIAL IN
ALLOCATION OF TIME TO DIFFERENT ACTIVITIES:
BY SOCIO-ECONOMIC GROUPS AND BY TECHNOLOGY

activities for women is in domestic chores, while for men it is in educational activities. A result of the change to the new technology seems to be a spurt in educational activities for both men and women and the increase is actually larger for women.

EFFECT OF TECHNOLOGY

The basic hypothesis of this paper is that technological change in rice cultivation has had a differential impact on male-female labor use. This hypothesis was examined and supported generally by the survey data presented in the preceding section. A more rigorous test was undertaken on the basis of data from 52 farms cultivating traditional variety and 26 farms growing high-yielding variety rice. A log-linear functional form was postulated and variations in total female labor use and proportion of female labor to total labor use were explained with two equations with these independent variables: farm size, expenditure per hectare on seeds (representing quality and plant population), farm yard manure and compost, commercial fertilizer, insecticide, herbicide, and plant treatment, and technology. The expected effects of seed and manure upon both total female labor and its proportion in total labor are positive because these activities require more hours in animal husbandry, transplanting, weeding, and harvesting -- all female labor intensive activities. The effects of fertilizer and plant treatments are expected to be negative since these are applied more by men and tend to displace female labor that would otherwise be needed for plant care and weeding. Farm size may have an indeterminate effect through the interaction of a positive output effect and a negative substitution effect. Technology in the form of a change from traditional variety to high-yielding variety is likely to increase total labor including female labor use while its effect upon the share of female labor may still be indeterminate. The results presented in Table 18.4 conform to expectations. They demonstrate that the transition to the new technology has a strong positive impact not only on total female labor use but also on the position of the female vis-a-vis the male. It also appears that the application of commercial fertilizer and plant treatment has strong negative effects upon female labor use, both absolutely and relatively to male labor use. The effects of seed and farm size are not significant.

TABLE 18.4

LABOR USE FUNCTIONS: HOUSEHOLD SURVEY DATA (NADIA):
POOLED (LOG LINEAR)

Intercept/ Independent Variables	Dependent Variables (N = 78)	
	Total Female Labor	Proportion Female in Total Labor
Intercept	1.74	−3.79
Farm Size	−0.02 (0.13)	0.07 (0.12)
Seed	0.56 (0.32)	0.23 (0.37)
Manure/Compost	0.25 (0.16)	0.04 (0.15)
Fertilizer	−0.44* (0.14)	−0.93** (0.23)
Plant Treatment	−0.31* (0.10)	−0.32* (0.10)
Dummy (for technology)	0.37** (0.11)	0.11** (0.03)
R^2	0.42	0.28

Note: Figures in parentheses denote standard errors.
* significant at .05 level.
** significant at .01 level.

An additional test was done with data from 75 farms
cultivating traditional variety and another 75 cultivating
high-yielding variety rice in the districts of Hooghly and
Birbhum in West Bengal. A log-linear function, similiar to
the above, was fitted for three equations: (1) for the
traditional variety farmers, (2) for the high-yielding
variety farmers, and (3) for all farmers pooled together.
Variations in the share of women in total labor is
explained in these three functions with the help of six
independent variables: area under the crop, cropping
intensity, farm business income, expenditure on labor,

total value of output per hectare, and technology. It was hypothesized that the proportion of female labor in a rice-based farming system will be positively affected by cropping intensity, farm business income, and technology, whereas the effects of area, expenditure on human labor, and total value of output per hectare would be indeterminate because of the simultaneous operation of both positive and negative effects. The results (Table 18.5) show that cropping intensity, farm business income, and expenditure on human labor have all exerted positive effects upon the proportion of women in total labor in all three equations.

TABLE 18.5

LABOR USE FUNCTIONS: FMS DATA (HOOGHLY AND BIRBHUM)
(LOG LINEAR)
DEPENDENT VARIABLE: SHARE OF WOMEN IN TOTAL LABOR USE

| | | Technology | |
	Local	HYV	Pooled
Intercept	−2.60	1.07	0.75
Independent Variables			
Area Under the Crop	0.06 (0.08)	0.81 (0.81)	0.09* (0.04)
Cropping Intensity	−0.18 (0.32)	0.91** (0.23)	0.56** (0.14)
Farm Business Income	0.20** (0.05)	0.20** (0.05)	0.20** (0.05)
Expenditure on Labor	0.21 (0.15)	0.37 (0.25)	0.21 (0.13)
Total Value of Output Per Hectare	0.54* (0.25)	0.16* (0.08)	−0.01 (0.12)
Dummy for Technology	--	--	5.10** (0.07)
R^2	0.44	0.53	0.46

Note: Figures in parentheses denote standard errors.
* significant at .05 level.
** significant at .01 level.

The coefficient for value of output per hectare is also significantly positive in the two equations representing the two technologies separately whereas it is insignificant in the pooled equation. The effect of areas is insignificant throughout. The equation with the pooled data strongly demonstrates the positive impact of the high-yielding variety technology through the high and statistically significant coefficient for the dummy variable representing technology. This statistical exercise thus provides strong additional evidence in support of the hypothesis that the new high-yielding variety technology in rice cultivation has exerted a fairly strong impact on female labor use.

CONCLUSION

Micro-level farm-household data from a rice-based farming system in eastern India shows that though the mechanism of time allocation by men and women in such systems is extremely complex and is subject to social, cultural, physical, and economic factors, there is overriding evidence that in terms of total time spent on activities -- both those traditionally considered economic as well as those related to home production -- the contribution by women is often larger than that of men. This occurs even though the female labor supply is subject to additional constraints, mainly demographic and sociological. This study also shows that while the extent of the male-female labor differential varies according to social, economic, and environmental characteristics, such variations are not as significant as the impact of technological change. The new rice technology has given rise to the higher use of female labor, both absolutely and in comparison to male labor. A disaggregated examination of the components of technology suggests that the bulk of the increased workload for women is in the form of operations that are largely unrecorded, unmeasured, and unaccounted for in traditional economic literature. This is because these activities fall into the category of home production, the importance of which is yet to be properly acknowledged, at least in most of the developing world.

REFERENCES

Ghosh, B. and S. K. Mukhopadhyay
   1984   Work, Income and Women: A Macro-Micro Exercise
   in India. Presented at the workshop on Women, Technology and Forms of Production, Madras Institute of
   Development Studies, India.

Mukhopadhyay, S. K.
   1984   Constraints to Technological Progress in Rice
   Cultivation -- An Experiment-Survey Research in West
   Bengal. Sponsored by the Ford Foundation,
   University of Kalyani, India.

   1985   Labour Use in Rice Cultivation: Male-Female
   Differential in Time Allocation. Presented at the
   International Conference of Agricultural Economists,
   Malaga, Spain.

Pradhan, B. and L. Bennett
   1981   The Role of Women in Hill Farming Systems.
   Seminar Proceedings: Appropriate Technology for
   Hill Farming Systems. Nepal: The Department of
   Agriculture.

19
Women in a Crop-Livestock Farming Systems Project in Santa Barbara, Pangasinan, Phillipines

Thelma R. Paris

In April 1985, a project design workshop on women in rice farming systems was held at the International Rice Research Institute (IRRI) to organize a collaborative and coordinated effort to undertake research and action in five general program areas: development; extension; the impact of new technologies; complementary studies, such as the dynamics of agricultural household behavior, the functioning of rural labor markets, and analysis of the policy environment that affects farm and household decisions; and sensitization. The ultimate aim of this collaborative effort, to be developed under the overall umbrella of the Asian Rice Farming Systems Network (ARFSN), is to institutionalize women's concerns within the agricultural research and extension agencies dealing with rice farming systems.

This action-research project in Santa Barbara, Pangasinan is one of the ARFSN sites that intended to improve the existing farming systems through an integration of suitable crop and animal production technologies. Specifically, this project is developing ways of increasing the utilization of crop by-products and residues as animal feeds through crop-livestock research (Roxas et al. 1984).

Women's concerns will be considered at the various stages of the technology development process including design, dissemination, and extension. The following women's concerns are to be integrated within the farming systems project (Cloud 1985; Castillo 1985):

(1) analysis of women's productive activities within the farming systems, including their roles in the household and in management;
(2) identification of the factors influencing women's productivity in farming systems such as access to productive resources (information,

technology, land, labor, capital, markets) and
access to and control over the benefits of production;
(3) exploration of existing, emerging and possible
technology options conducive to the expansion of
women's productive capacity as well as human
development potential; and
(4) application of this understanding throughout
the farming systems research process.

CROP—LIVESTOCK PROJECT

The crop—livestock project in Santa Barbara, Pangasinan
began in 1984 as a collaborative project between the Institute of Animal Science, University of the Philippines in
Los Banos, the Ministry of Agriculture and Food, and the
Rice Farming Systems and the Agricultural Economics of the
IRRI. This project uses the farming systems approach,
conducting research in three main stages: identifying the
local farming system and its constraints; selecting and
improving existing technologies and techniques used in
overcoming these constraints; and then testing and adopting
these technologies under the conditions in which men and
women farmers have to work.

A team of scientists chose Santa Barbara based on its
proximity to major livestock auction markets, government
support agencies and experiment stations and its potential
for crop and livestock improvement. Two villages within
Santa Barbara represented distinct production systems and
socioeconomic environments. Malanay is an irrigated area
where two high-yielding variety (HYV) rice crops are grown,
while Carusucan is a purely rainfed area where generally
only one crop of rice is grown. Malanay can be considered
as more prosperous than Carusucan because of the availability of irrigation from the National Irrigation
Administration, availability of electricity, and proximity
to markets and trading centers.

Cropping and Livestock Technologies

Cropping systems component. Improvements in the
cropping systems component were done through component
testing (variety, fertilizer trials, etc.) and new cropping
patterns were designed based on the historical rainfall
distribution and farmers' existing practices.

In the rainfed area of Carusucan, generally only one
rice crop per year is grown. However, farmers in this
research site plant two rice varieties depending upon the
type of soils. Farmers would usually plant early maturing

high yielding varieties (IR36, IR58) and local varieties in heavy, flood prone soils. Although there is no clear cut demarcation between the land categories, the farmers are able to avoid crop damage or losses by identifying the appropriate cultivar types to be planted in their plots.

After cropping pattern trials and component technologies were further tested and refined, the agronomist recommended that legumes be grown before and after rice as long as the farmers follow the right cut-off dates in the middle part of May and November. Otherwise, when grown beyond these dates, the legumes planted will be destroyed by too much water. Legumes provide protein and income for the family; residues can be used as cattle feed, maintaining feed availability throughout the year especially during February and March.

The farmers' previous practices in mungbean production included: (1) use of hired tractors for land preparation; (2) use of low yielding varieties; and (3) unscheduled frequency of sprayings (four times) even ten days before harvesting (this may be potentially harmful to both humans and livestock). As an improvement in these practices, the agronomist recommended that (1) furrowings be done by carabao (water buffalo), (2) high yielding varieties be used, (3) drilling be in furrows; (4) no fertilizer and no weed control be done, and (5) only three insecticide applications (two, twelve, and thirty-five days after emergence) be done.

Although no cropping pattern was tested in the irrigated site, component technology trials such as variety, fertilizer rates, ratooning potential of the second rice crop, and fitting an upland crop in between the two rice crops were verified. Except for the fertilizer rates and variety trials, the rest of the trials failed because of the uncertainty in the water release by the National Irrigation Administration during the first crop. This eventually flooded the mungbean crop at flowering stage, and the potential of the main rice crop to ratoon was hindered because of the lodging caused by floods.

Variety and fertilizer components in the cultural management of the two rice systems will be widely tested in farmers' fields. These two components of variety and fertilizer can contribute to a higher grain and fodder yield and can be much more economical than what the farmers are presently using. Further component technology trials on forage grasses are being done to improve fodder yields to augment the supply for ruminants in this area.

Livestock component. Generally the animals in the area are undernourished due to the low nutritive value of the fodder, so animal nutrition, health, and breeding interventions were designed. For both areas, feeding the cattle with a minimum of two kilograms of Leucaena (leguminous tree) per day was initially recommended to improve the protein quality of the fodder. Unfortunately, an unknown jumping lice destroyed the Leucaena all over the country, negatively affecting this particular intervention.

At the irrigated site, component technology trials are now being done to try forage grasses and protein rich indigenous fodder, the feeding value of which is not known to the farmers. The breeding scheme of carabaos through artificial insemination of Murrah breed also stresses the importance of maintaining healthy animals through proper nutrition.

At the rainfed site, the feeding of leguminous residues to supplement rice straw as cattle fattener is being introduced. Rice straw is preserved after the harvesting season to last for the whole year. Since only one rice crop is predominantly grown in this area, there is more seasonality in the availability of feeds for the animals. The use of legume residues as feeds during the dry season could space out the availability of feeds.

In both sites, there are gender differences in farm production activities. Women specialize in swine, mungbeans, cowpeas, and vegetables, while men specialize in rice, cattle, and carabao. However, men's and women's activities complement each other.

"BREAKING INTO" THE FARMING SYSTEMS RESEARCH GROUP

The Women in Rice Farming Systems (or WIRFS) representatives attended as observers in one of the workshops in the project site. At this meeting where representatives of the collaborating agencies were present, WIRFS began to understand the overall objectives of the project, the various components, the existing crop and livestock practices, the constraints, the complexity in designing the technologies for both crops and livestock practices, and the on-going activities. Despite the use of a so-called multidisciplinary approach involving the animal nutritionist, livestock specialist, agronomist, economist, crop protection specialist, animal breeder, and veterinarian from the different agencies, no social scientist was formally included in the team. An economist represented the socioeconomic component of the team, but his tasks were specifically focused on analyzing the economic viability of the crop and livestock technologies. The need for a social scientist was realized when the recommended livestock intervention (minimum of two kilograms Leucaena daily feeding as cattle fatteners) was not adopted by the livestock cooperators even before the lice epidemic occurred. A social scientist then conducted case studies of livestock cooperators and noncooperators to find out the reasons behind adoption and non-adoption of the recommended intervention. Certain problems were discovered such as: most farmers did not know that the leaves can be dried in the sun and mixed with rice straw and other grasses; female technicians came to teach the women "how to plant" Leucaena but the male technicians came more often to find out about the cattle; gathering of the leaves was a "bother" to the farmers' routine; the frequency of cutting the stems for higher herbage yields and the proper way of feeding leucaena to animals was only explained to the farmers but not to the women who fed and gathered forage for the animals (Juliano et al. 1985). The meetings concerning the livestock technology involved only men since it was assumed that women did not have anything to do with these animals. After this, the scientists realized the importance of studying the "human element" and considering the "clientele" of the proposed intervention. The social scientist is now formally included in the team.

COLLECTING INFORMATION ABOUT WOMEN AND THE HOUSEHOLD

The next step was to collect information about the household and about women's roles in farm activities in the area. WIRFS obtained the socioeconomic profiles for both research sites but there were no studies or information on the participation of women in agriculture disaggregated population figures, the nutritional status of children, and the number of acceptors of family planning methods, were provided by the Municipal Health Office. Although a benchmark survey was done during the initial stage of this project to obtain information about the household, landholdings, cropping patterns, livestock inventory, crop residue utilization, livestock feeding practices, and constraints in crop and livestock production, there was a lack of information on the specific tasks and responsibilities by different household members, and on household consumption patterns. A diagnostic survey was conducted in November 1985 of the selected farmer cooperators of the project and non-cooperators as well. Questions were asked about participation in specific crop and livestock activities by gender; off-farm and non-farm activities; access to information, land, credit, and agricultural household technology; the inventory, and ownership of assets; and an animal inventory was taken. After conducting the diagnostic survey, the team stayed in the site for three days, talked with key informants, especially women, and observed on-going activities in the village.

THE ROLE OF WOMEN IN THE FARMING SYSTEM

Crop Production Activities

There is a sexual division of labor in rice production. Women pull seedlings while men do land preparation, transplanting, weeding, fertilization, spraying, harvesting, threshing, and hauling. Social custom was mentioned as one of the reasons why women in these areas do not transplant as do women in the nearby provinces. Women are more involved in making labor arrangements than in buying farm inputs. In the irrigated area, women are mostly responsible for selling products and by-products. Also, in this area, transplanters are not immediately paid after the service has been rendered, but they are given the right to harvest and thresh the crop. Since the transplanters are primarily men, very few women participate as hired labor in harvesting. In the rainfed site, exchange or reciprocal labor arrangements exist for land preparation, pulling seedlings, transplanting, and harvesting. The difference in hired labor arrangements can be explained by the presence of more landless workers who harvest in the irrigated area during the second cropping season. In the rainfed site only two percent of the households are landless so that exchange labor can be practiced by those who have farms (Table 19.1).

For those farmers who grow mungbeans, cowpeas, and vegetables, it is the women who harvest, thresh, and make marketing decisions on the selling price, marketing outlets and when to sell (Tables 19.2, 19.3, 19.4).

Livestock activities

Men are generally responsible for large animals (cattle and carabao), although women and children help with feeding, gathering forage, cleaning the shelter, cleaning the animal, taking the animals to the fields, and collecting and disposing wastes. Putting up the animals' shelter, and buying and selling large animals are mainly done by men (Table 19.5).

Swine and poultry are considered women's responsibility. Their care, maintenance, and selling are mainly done by women. Women also buy rice bran and take rice for

TABLE 19.1

PERCENTAGES OF MEN, WOMEN, AND CHILDREN IN SAMPLE HOUSEHOLDS PROVIDING DIFFERENT SOURCES OF LABOR BY SPECIFIC ACTIVITY IN RICE PRODUCTION *

Activity	MALANAY (N = 26 Households)								CARUSUCAN (N = 27 Households)							
	Family			Hired			Exchange		Family			Hired			Exchange	
	M	W	C	M	W	C	M	W	M	W	C	M	W	C	M	W
Seedbed Preparation	85	12	4	8	12				89	4	7	11	4	4	7	7
Land Preparation	73			58					85		7	7			7	
Pulling	12	38	12	8	77				4	63	4	4	48	4		26
Transplanting	46	4	4	73	8	4		4	67	11	7	59	11			
Broadcasting	81			4				4	78	11	4	4			15	
Weeding	92	12	23	8	4				78	18	15		4		4	
Fertilization	92		4					4	70	4	11	4			4	
Spraying	88			12					63		4	4				
Harvesting	46	27	8	62	27	4		4	89	33	26	26	18		11	
Threshing	38	12	15	62	19	4			22		7	74	4			
Hauling	65			38	8				82	11	11	7			7	
Gleaning/Stacking	8	8			8				15	22	11					
Other Post-Harvest	65	46	12			4			48	15	11					
Buying Seeds	38								11	7						
Buying Fertilizers and Chemicals	62	31							74	15		4				
Hiring Labor (Making Arrangements)	77	54							70	41						
Selling Products and By-Products	38	58							41	26						

* Numbers refer to the percentages of women, men, or children who perform each operation.

TABLE 19.2

PERCENTAGES OF MEN, WOMEN, AND CHILDREN PROVIDING DIFFERENT SOURCES OF LABOR BY SPECIFIC ACTIVITIES IN MUNGBEAN PRODUCTION

Activity	MALANAY (N = 7 Households) Family			Hired			CARUSUCAN (N = 18 Households) Family			Hired			Exchange		
	M	W	C	M	W	C	M	W	C	M	W	C	M	W	C
Land Preparation	43						89	6	6						
Broadcasting	57						6	50	22	11					
Weeding	14						17								
Fertilization	14						6								
Spraying	43	14					50	6	6	6	39	6	6		
Harvesting	29	43	14	14	43		61	67	22	22	28	6	6		
Threshing	14	43	14		29		44	50		22	17				
Hauling							44	28	6						
Other Post-harvest		14			14		17	6		6					
Buying Seeds							22	22							
Buying Fertilizers and Chemicals	14	14					53	17		6	6				
Hiring Labor (Making Arrangements)	14	14					22	17							
Selling Products and By-products		57	14				17	39							

TABLE 19.3

PERCENTAGES OF MEN, WOMEN, AND CHILDREN PROVIDING DIFFERENT SOURCES OF LABOR BY SPECIFIC ACTIVITIES IN COWPEA PRODUCTION

Activity	MALANAY (N = 7 Households)						CARUSUCAN (N = 14 Households)					
	Family			Hired			Family			Hired		
	M	F	C	M	F	C	M	F	C	M	F	C
Land Preparation	33						75		25			
Broadcasting	67	33					50	25	25			
Weeding				33			50			25	25	
Spraying	67						25					
Harvesting	67	67	33	33			50	75	25			
Threshing							25	25		25	25	
Hauling							50					
Other Post-Harvest												
Buying Seeds		33					25	25				
Hiring Labor (Making Arrangements)							25					
Selling Products and By-Products		33					25	50				

TABLE 19.4
SAMPLE HOUSEHOLDS USING DIFFERENT SOURCES
OF LABOR BY SPECIFIC ACTIVITIES IN SQUASH PRODUCTION
MALANAY, SANTA BARBARA, PANGASINAN

Activity	Households (N = 6)	
	M	W
Land Preparation	4	
Planting	4	
Broadcasting	1	
Weeding	1	
Fertilization	3	1
Spraying	1	
Harvesting	2	1
Buying Seeds	1	1
Buying Fertilizers and Chemicals	1	1
Selling Products and By-Products	2	3

milling to make fodder for the animals. Usually, men do
not interfere in the care and maintenance of swine.
Decisions about price and marketing, or how income from
swine sales will be spent are made chiefly by women (Table
19.5).

Income Generating Activities

In rice production, women earn income by working as
hired labor in pulling seedlings. A woman can pull seedlings at a rate of about 120-150 bundles a day for P.25 per
bundle (U.S.$1 = P19). There is no standard wage rate for
pulling seedlings per day; income earned depends greatly on
the individual woman's skill and speed. Women are
generally preferred over men to pull seedlings because they
do the job better and faster. Women earn additional income
by selling vegetables (tomatoes, squash) and legumes
harvested from small plots or on the upper rice beds.
During our three-day visit to the village in the
rainfed area the WIRFS team was lucky to witness the
processing of glutinous rice "diket" that takes place only
once a year in October. Glutinous rice is grown and
harvested two weeks earlier than the other rice varieties

TABLE 19.5

PERCENTAGES OF MEN, WOMEN, AND CHILDREN IN SAMPLE HOUSEHOLDS PROVIDING DIFFERENT SOURCES OF LABOR IN SPECIFIC ACTIVITIES IN CATTLE AND CARABAO PRODUCTION

	CATTLE						CARABAO					
	MALANAY (N = 13)			CARUSUCAN (N = 18)			MALANAY (N = 18)			CARUSUCAN (N = 17)		
	Family Labor											
Activity	M	W	C	M	W	C	M	W	C	M	W	C
Putting up Shelter	77			44	6		83	17	6	18		
Preparing Feeds		15	23	78	22	28	89	28	22	88	6	24
Feeding		38	8	72	39	39	94	28	11	76	18	59
Watering	92	23	23	83	28	44	89	17	6	88	24	59
Cleaning Animal	77	15		89	17	28	83	22	17	76	24	29
Waste Disposal	54	23	15	56	11	28	61	22	17	53	12	41
Gathering Forage	92	8	8	61	22	50	83	17	6	82	12	47
Buying Animal	69			33	6		67			41	12	
Buying Feeds	8			22			11	6		24		
Taking Animal to Market	46	15		44			17			53		

to take advantage of the high price on November 1 (All Saints Day), a special holiday. This is a major traditional income generating activity of women in this rainfed site. The rice requires arduous processing techniques and consumes much fuel. Because of the large fuel need, cattle dung is used for this purpose.

These constraints made the team think of developing technologies that are important to women such as: a high-yielding, shorter maturing glutinous rice variety; low, light machinery that can reduce drudgery and cooking time; and fuel-saving devices, all of which may enable women to earn more. Since the data about cropping patterns was presented in an aggregate form in the initial benchmark survey, the importance of glutinous rice as a source of income for women and the household in general did not become immediately evident.

Other income generating activities pursued by women include selling fish, snails, pigs, and tending small variety stores.

For men, transplanting and harvesting on other farms as well as fishing, carpentry, and construction work are the major secondary sources of income. Large animals are sold to meet major expenses such as hospitalization, to finance a family member going abroad, or to replace another animal, while the selling of small animals is done to meet immediate expenses such as schooling and daily household needs (Table 19.6).

GENDER DIFFERENCES IN ACCESS TO PRODUCTIVE RESOURCES

Access to Education, Training, Organizations, and Credit

The majority of the farmers and their wives in the sample households have about six years of schooling. Of household members above 15 years old, more men than women have reached high school. More men than women (five percent and two percent) have no education in the rainfed site. There is only one elementary school in both sites and the irrigated site is more accessible to the high school located in the municipality. Training classes tend to be gender specific. Farmer's Class, Azolla Training, and Crop-Livestock Training classes were mostly attended by men, while Nutrition and Food Preservation classes were mostly attended by women. There were no training programs for women on vegetable gardening, swine and poultry raising, or in other income generating activities which would provide additional knowledge and skills.

TABLE 19.6

PERCENTAGES OF MEN, WOMEN, AND CHILDREN WITHIN HOUSEHOLDS
USING DIFFERENT SOURCES OF LABOR BY SPECIFIC ACTIVITIES
IN POULTRY AND SWINE PRODUCTION

POULTRY:	MALANAY (N = 18) Household			CARUSUCAN (N = 17) Household		
Activity	M	W	C	M	W	C
Putting Up Shelter	20					
Preparing Feeds		7	7	20	10	5
Feeding	20	93	40	25	35	45
Watering		33	7			5
Waste Disposal	13	7	13		5	5
Buying Chickens				5		
Buying Feeds		13				
Selling				10	10	

SWINE:	MALANAY (N = 18) Household			CARUSUCAN (N = 17) Household		
Activity	M	W	C	M	W	C
Putting Up Shelter	47	7	7	27	7	6
Preparing Feeds	7	7	7	27	20	7
Feeding	13	80	13	20	87	20
Watering	13	53	7		73	
Cleaning Animal	27	40	7		87	
Waste Disposal	27	27	27	7	33	7
Gathering Indigenous Feeds	27	27	27	7	33	7
Buying Animal	20	40		13	67	
Buying Feeds	13	60	7	13	60	
Taking Animal to Market	20	53		27	60	

Formal organizations for men give farmers access to
credit, inputs, technology, and markets for rice production. For instance, a new presidential decree was recently
issued authorizing the National Food Authority (NFA) to
grant farmers' cooperatives priority in selling their
produce to the government at higher prices than the market
price. NFA also gives credit for inputs at lower interest
rates than the private banks. Women do not have access to
formal credit since they are not formally organized and do

not have collateral such as land titles. In the study
sites, share tenancy and leasehold are the predominant land
tenure arrangements and women are not involved in these
transactions. Women have more access to informal credit
particularly for food consumption purposes. They are
responsible for borrowing to meet emergencies and food
shortages.

Access to Household Technology

Women's access to improved household technology enhan-
ces their productivity. In the irrigated site, the women
have more access to labor-saving devices such as sewing
machines, refrigerators, and electric irons, since electricity is available there. Water is provided from pumpwells
near their households so access to water is not a major
problem. The sample households in the rainfed area tended
to use firewood for fuel, but because of the increasing
scarcity of bamboo and wood in the area, some households
had to resort to cow dung for fuel. In the irrigated site
some households use rice hulls for fuel with the use of
specially made rice hulled stoves.

INTEGRATING INFORMATION ABOUT WOMEN INTO THE PROJECT

Another crop-livestock workshop was held January 3,
1986 in Santa Barbara to assess accomplishments, discuss
problems, and present the future plans of the project.
This time the WIRFS members were no longer mere observers
but part of the team. The importance of incorporating
women's concerns in the project was underscored. WIRFS had
shown that women take active roles in crop and livestock
activities; that these activities provide women's major
sources of income; and therefore, women could be ignored in
the technology design, dissemination, and extension
process. The following points were emphasized in the
dialogues with the whole team.

First, some productive activities that are being modified by new technology interventions are the responsibility
of women. The introduction of mungbeans after rice means
additional household income but also requires women's and
children's labor in harvesting and threshing; therefore,
discussions about this proposed crop should include women's
concerns. Improved cultural practices and pest management,
as well as the use of mungbean residues as fodder for
animals, should be demonstrated not only to men but to
women as well.

Second, it is important to understand who will have an additional incentive to participate in the proposed intervention. Since it is the women's responsibility to harvest, thresh, and market mungbeans, the increase in productivity may provide incentives to adopt the technology. Because women have to be aware of market prices and to realize profits from production, their market and financial orientations will have to be developed. Women may be expected also to cooperate with innovations in swine raising, such as growing root crops or home grown feeds to substitute for commercial feeds, because the care of swine is women's responsibility. Women decide how to spend the income they derive from selling swine.

Third, research resources should not be concentrated only on men's activities. In both research sites, livestock interventions have always been focused on ruminants rather than on swine. Also, there has been little on-farm research conducted on vegetable and root crop production in this area. There have been no training classes on swine and vegetable production from which women may benefit. Proposed crop technologies should also consider the importance of the crop in the diet.

Finally, the value of tailoring component research to meet farmer's needs was very much recognized during the workshop and as a matter of fact was being done by the project. There was also a consensus that since an understanding of technology is crucial to adoption, it would be ideal if senior researchers themselves could explain their research results to the potential users.

INCORPORATING WOMEN'S CONCERNS

Based on the interactions among the production and social scientists in the project, several developments came out of the workshop. A class was held on January 18, 1986 to consider women's participation in crop and livestock activities in which a female livestock nutritionist explained the importance of the nutritive value of different crop residues and fodder. She clarified misconceptions about the abortive effects on pregnant cows of cassava leaves and other leguminous fodder. She also gave lessons on preservation of rice straw and legume residues, and explained the proper feed mix for the animals. She discussed the different sources of available home-grown feed that can be given to swine and how to produce earthworms as feed for chickens. In contrast to the usual practice of inviting only men to attend meetings, this time women were also invited. Twelve women out of twenty-six

people in the rainfed site attended. In the irrigated
site, only eight of the twenty-four individuals attending
were women. Some reasons cited for non-attendance were
that nobody would be left in the house to cook and take
care of the children, and attending meetings concerning
livestock was unconventional.

Since the important role of women in the dissemination
of technology has been recognized, details of the technology were explained to them. In the breeding intervention,
detecting estrus as well as maintaining the health of the
animals is crucial for the success of the intervention.
Women were invited to a class conducted by the animal
breeder to learn how to detect estrus, how to monitor
estrus cycles, and to detect animal illnesses. When the
women who attended the class were interviewed, they said
that they did not know anything about detecting estrus in
large animals but they did for swine. Misconceptions about
producing smaller offspring through artificial insemination
were clarified.

In the proposed intervention of adding legumes before
and after rice, the women can remind their husbands of the
planting cut off dates and the proper time to spray. Since
the women harvest and thresh mungbeans, they can also
preserve the mungbean hay by drying it in the sun and
storing it in a dry area.

Since glutinous rice is an important source of income
of women in the rainfed site, the farmers under the guidance of the agronomist will test a high-yielding and early
rice variety (IR65) to compare yield potential and cooking
quality with the local glutinous rice variety currently
used. Later, its cooking quality and other characteristics
essential to the rice delicacy will be tested.

The potential of developing light machinery for processing glutinous rice will be explored in October when
actual processing takes place. An agricultural engineer
will be invited to see if some light low-cost machinery can
be designed.

An animal husbandry specialist will be invited to look
into local sources of suitable feeds for swine which are
women's responsibility and major source of income.

Once the scientist recognized the importance of communicating laboratory results to the farmers, they volunteered
as extension agents in the dissemination of technology.
The female animal nutritionist who conducted classes was
also the same person who conducted research in the laboratory to establish the feeding value of crop residues for
cattle and carabao. For someone who was used to working on
problems of large commercial livestock growers and teaching

in the classroom using technical terms, reaching the farmers and imparting her technical knowledge in layman's language for the first time was a considerable accomplishment and a rewarding job for her.

These classes, involving men as well as women, begin a series that will provide farmers with an understanding of the technology, maintain interaction between the scientists and the farmers, and create feedback for a dynamic technology development process.

ACKNOWLEDGEMENTS

The author acknowledges Dr. Gelia T. Castillo, IRRI Visiting Scientist, for her valuable comments and suggestions on this paper.

REFERENCES

Castillo, G.
  1985   Women in Rice Farming Systems Research Program Proposal, (unpublished). Prepared at Los Banos, Philippines: International Rice Research Institute.
Cloud, K.
  1985   Women's Productivity in Agricultural Systems: Considerations for Project Design. In Gender Roles in Development Projects. Catherine Overholt, et al., eds., pp. 17-56. West Hartford, CT: Kumarian Press.
Godilano, S.
  1986   Year Two Results of Cropping Pattern and Component Technology Testing and Plans for Year Three, Crop-Livestock Research Site, Santa Barbara, Pangasinan, Philippines. Rice Farming Systems Department. Los Banos, Philippines: International Rice Research Institute.
International Rice Research Institute
  1985   Women in Rice Farming, Proceedings of a Conference on Women in Rice Farming Systems. Los Banos, The Philippines: IRRI. September.
  1985   Report of the Project Design Workshop on Women in Rice Farming Systems. April 10-13, 1985. Los Banos, The Philippines: IRRI.
Juliano, P. and L. Tolentino
  1985   The Leucaena Story in Santa Barbara, Pangasinan: A Case Study. Los Banos, Philippines: University of the Philippines at Los Banos.

Laufer, L.
   1985   Methodological Issues: Women in Farming Systems
          Research.   Presented at the Farming Systems Research
          Socioeconomic Workshop.   Los Banos, The Philippines:
          International Rice Research Institute.
McKee, C.
   1984   Methodological Challenges in Analyzing the
          Household in Farming Systems Research:
          Intrahousehold Resource Allocation. In Proceedings
          of Kansas State University's 1983 Farming Systems
          Research Symposium, Animals in the Farming System.
          Cornelia Butler Flora, ed., pp. 593-603, Manhattan,
          KS: Kansas State University.
Paris, T. and L. Unnevehr
   1985   Human Nutrition in Relation to Agricultural Pro-
          duction: A Project in the Philippines.   Presented
          at the Farming Systems Research Socio-Economic
          Workshop.   Los Banos, Philippines: International
          Rice Research Institute.
Roxas, D. and R. Olaer
   1984   On-Farm Crop-Livestock Systems Research in Santa
          Barbara, Pangasinan.   Report of the Crop Livestock
          Systems Research Monitoring Tour, Philippines and
          Thailand, December 10-18.   Los Banos, Philippines:
          International Rice Research Institute.

20
Gender Related Aspects of
Agricultural Labor in Northwestern Syria

Andrée Rassam and Dennis Tully

Technological change is an essential part of economic and social development (Cain 1981); however, technological change can have far-reaching effects on rural communities. Therefore, policy makers, donors and agricultural research centers are giving greater attention to socioeconomic factors in the design and extension of new technologies.

New technologies may shift the sexual division of labor, increasing men's or women's work loads, sometimes with adverse effects. They may also affect the balance of opportunities and access to economic resources. One common pattern is for the mechanization of tillage to reduce men's labor, since tillage is often a male task. However, improved tillage or increased crop area may lead to more crop production, that increases post-harvest female tasks (Spence and Byerlee 1976; Nyanteng 1985).

Where new technology reduces women's labor, this may also have adverse effects if poor women depend on the income from agricultural labor. In Java, the introduction of rice mills is said to have replaced 12 million female work hours (ILO 1981), while herbicide use in Kenya has eliminated weeding as a job opportunity for some rural women (UNGA 1978).

Hand in hand with technological change, a second major factor affecting many rural populations is the development of off-farm income opportunities. These are usually in urban areas and often most available to males. Off-farm incomes can stabilize household incomes and offset declining farm sizes; however, the absence of adult males may increase the labor burden of women, children, and old persons (Dasgupta 1977; Nash 1983; Tully 1984). Paradoxically, men's migration may reduce women's and children's labor if non-agricultural income satisfies household needs. They may be replaced by hired labor if it

is available, or else a general labor shortage may develop.
For example, agricultural decline in Oman and Yemen has
been related to extensive labor migration (Birks and
Sinclair 1980).

Data on the division of labor and income generating
activities are becoming increasingly important in farming
systems research. Labor issues need to be assessed, since
they may limit the adoption and diffusion of technological
change (Somel and Aricanli 1983). Studies by The International Center for Agricultural Research in Dry Areas
(ICARDA) Farming Systems Program have addressed gender
related issues (Nour 1985). Gender issues will also be
considered in a new project on mechanization and labor constraints in the Middle East and North Africa, that will be
conducted in collaboration with national research organizations. The results are expected to have implications for
research design, organization, and priorities.

AGRICULTURAL CHARACTERISTICS OF SYRIA

Syria is among those Middle Eastern countries where
agriculture is a major factor in the national economy. It
is a rich country with regard to its land and water resources. Fundamental changes have occurred in the last thirty
years both in agricultural technique and in the organization of production. New technologies were adopted rather
quickly in Syria. For example, land preparation is almost
completely done by tractor today and harvesting, particularly of cereals, is increasingly mechanized. When cereal
is hand harvested, virtually all is threshed by standing
mechanical threshers, rather than the animal-drawn sled
formerly used. Herbicides are used by the majority of
wheat farmers in wetter areas, and seed drills have
replaced broadcasting to some extent.

The organization of production has also changed tre-
mendously with the Agrarian Reform of the 1960s. Land was
distributed more equitably to increase the number of small
holdings. Currently, three-fourths of the holdings are
less than ten hectares, 24 percent of farms are from ten to
100 hectares, and fewer than one percent are larger than
100 hectares (FAO 1982). As part of the reform, cooperatives and credit facilities were organized to give farmers
access to inputs and new technologies.

Although farms are small, agricultural production has
shown remarkable progress in recent decades. Agriculture's
contribution to GDP grew at a respectable 4.4 percent rate
in the 1960s and at 7.2 percent in the 1970s. Syria also
increased its food production per capita by 68 percent
during the 1970s, while most Middle Eastern countries have

been unable to increase food production at the rate of
population increase. However, the industrial sector and
oil production have grown faster than the agriculture
sector, resulting in a decrease of agriculture's relative
importance in the national economy. The percentage of the
labor force in agriculture dropped from 53 percent in 1965
to 33 percent in 1983. The contribution of agriculture to
GDP decreased from 29 percent to 19 percent from 1965 to
1981 (World Bank 1985).

Overall, rural development with the increase of agricultural production, industrial labor demand (off-farm
employment), and the extension of new technology have
affected the division of labor between men and women and
between households and hired labor. This has also led, as
will be shown later on, to a predominance of women in the
unskilled rural labor force while most skilled jobs within
or outside rural areas are dominated by men.

METHOD OF THE STUDY

This paper draws on a survey of 47 land-owning households in four villages of Aleppo Province in Syria, supplemented by visits and interviews to major labor supplying
villages. Data were collected on farm labor for the
production of crops and livestock for the 1982/83 cropping
season. Twelve households were randomly selected from each
of four villages located in Northwestern Aleppo Province.
Two villages were selected in a relatively wet area,
designated "Zone 1" by the Syrian Government, with
approximately 450 millimeters mean annual precipitation.
In this area, wheat, barley, legumes, and summer crops
(melons, sesame) are grown. Two other villages with
approximately 325 millimeters mean annual rainfall were
selected in "Zone 2". The same patterns of crops are found
except that chickpeas are not grown in Zone 2.

Data from the households were collected at three different periods corresponding to the different seasonal
tasks. Both husband and wife were present at each interview session and information was collected from both.
Labor has been disaggregated by age and sex, and household
labor has been distinguished from hired labor. The number
of hours spent by each age and sex category has been calculated for each task and each crop. (For more details on
the method used in the study, see Rassam, 1986.)

LABOR INPUT IN ON-FARM ACTIVITIES

Labor input for agricultural tasks, whether provided by
household or hired labor, differs by gender depending on

the crops and the techniques used in accomplishing the tasks. Some tasks are mostly carried out by men, particularly the mechanized operations. Men's tasks include land preparation, chemical weed control, mechanical harvesting, and threshing. Among manual tasks, chemical fertilizer application and seeding are generally done by men in the villages studied. Ordinarily the contribution of women to these tasks is limited to assistance; however, in some villages where men are heavily involved in non-agricultural work, women may broadcast seed or fertilizer themselves. Activities such as spreading manure in the field, selecting seed, planting summer crops, and hand weeding are normally done by women. Seed preparation and the various steps in the harvest process seem to be shared by men and women although within these processes there is also specialization by sex. Livestock activity is divided into males' and females' tasks. For example, livestock are fed mostly by women, while animals are usually herded with the help of children and shepherds from the villages. Selling sheep is a male task while selling poultry is a female task. Dairy products can be sold by either sex.

To simplify the presentation the contribution of children under the age of 13 (and usually over 10) who supply approximately seven percent of labor hours is omitted. Their productivity in major tasks, such as harvesting and weeding, is estimated by farmers at about half that of an adult. Children tend to work with their mothers, especially in hired labor; the correlation between children's and adult women's hours in hired labor from outside the village is .87. Henceforth, the percentages of various categories of labor presented will be based on the total adult labor hours.

In general, men's and women's contributions to agricultural labor (in terms of hours of physical work and including both family and hired) are almost equal (Table 20.1). Household labor provides 61 percent of the total work hours in agricultural operations; females provide 57 percent of this. Hired labor used in total agricultural production is equally divided by gender.

The work provided by each sex depends on the degree to which the production is mechanized. For example, in cereal crops where most of the operations are mechanized, the contribution of hired males is higher than that of hired females (33 percent versus 5 percent). The opposite is found in legume crops where most operations, particularly harvesting, are not mechanized. Women's and men's contributions are 36 percent and 12 percent respectively. Tables

TABLE 20.1

CONTRIBUTION OF MALES AND FEMALES AS PERCENTAGES OF THE TOTAL TIME SPENT IN ON-FARM AGRICULTURAL PRODUCTION

Labor Contribution	Cereal Crops	Legume Crops	Summer Crops	Tree Crops	Total Crops
Household					
Male	23	18	44	47	26
Female	39	34	42	28	35
SUB-TOTAL	62	52	86	75	61
Hired					
Male	33	12	9	24	20
Female	5	36	5	1	19
SUB-TOTAL	38	48	14	25	39
GRAND TOTAL					
Male	56	30	53	71	46
Female	44	70	47	29	54
Percentage of Area Allocated to Each Crop	50	25	19	6	100
Percentage of Hours Spent in Each Crop	30	40	22	8	100
Mean Hours Per Hectare	46	135	94	99	—

Source: From villages surveyed in study.

20.2 and 20.3 show the contributions of men and women by crop and by activity as well as the proportion of hours devoted to each task.

The cereal harvest is a good example of current trends, because it is partly mechanized and partly manual, and both sexes are involved in manual harvesting. Only one farmer manually harvested the entire cereal crop but an additional twelve farmers harvested some cereal by hand, so it is possible to make a comparison between groups by technique (Table 20.4). Even when the cereal is harvested by a combine, there is also associated hand labor — primarily gathering up the straw for use as feed.

TABLE 20.2

CONTRIBUTION OF MALES AND FEMALES AS PERCENTAGES OF HOURS SPENT IN LEGUME AND CEREAL PRODUCTION

AGRICULTURAL ACTIVITIES	LEGUME						CEREAL					
	Percent Hours Spent by Task	Percent Total Adult Input		Household Only			Percent Hours Spent by Task	Percent Total Adult Input		Household Only	
		Male	Female	Male	Female			Male	Female	Male	Female
Tillage Operations	3.7	100	0	32	0		10.1	100	0	32	0
Seeding	1.5	86	14	30	14		5.1	86	14	19	14
Herbicide Use	—	—	—	—	—		0.8	95	5	20	5
Fertilizer Use	0.9	81	19	49	19		11.0	79	21	31	21
Hand Weeding	16.1	14	86	13	67		20.6	5	95	5	95
Pest Control	2.5	71	29	51	29		8.5	73	27	54	27
Harvesting	58.4	15	85	11	26		26.2	38	62	10	39
Transport	5.3	74	26	36	26		12.0	84	16	34	16
Threshing	6.6	57	43	31	43		4.6	62	38	35	38
Winnowing	1.6	89	11	48	9		0.4	83	17	47	17
Cleaning	2.4	34	66	34	66		0.4	27	73	21	61
Bagging	1.0	44	56	44	56		0.2	66	34	66	34
TOTAL	100.0	30	70	18	34		100.0	56	44	23	39

Source: Villages surveyed in study.

TABLE 20.3

CONTRIBUTION OF MALES AND FEMALES AS PERCENTAGES OF HOURS SPENT IN SUMMER CROP AND TREE CROP PRODUCTION

AGRICULTURAL ACTIVITIES	SUMMER CROP						TREE CROP					
	Percent Hours Spent by Task	Percent Total Adult Input		Household Only		Percent Hours Spent by Task	Percent Total Adult Input		Household Only	
		Male	Female	Male	Female		Male	Female	Male	Female
Tillage Operations	11.1	100	0	24	0	16.0	100	0	23	0
Planting	20.0	18	82	18	57	22.0	34	66	19	62
Thinning & Weeding	27.0	37	63	37	63	2.0	21	79	21	79
Pruning	24.0	76	24	75	24	35.0	70	30	57	30
Hoeing, Irrigating & Fertilizer Use	3.0	91	9	91	9	24.0	92	8	78	8
Pest Control	2.0	32	68	32	68	1.0	17	83	6	83
Harvest and Transport	13.0	45	55	45	55	—	—	—	—	—
TOTAL	100.0	53	47	44	42	100.0	71	29	47	28

Source: Villages surveyed in study.

TABLE 20.4

MEAN HOURS OF LABOR IN CEREAL HARVEST: DIFFERENCES
BETWEEN FARMS USING MECHANICAL AND MANUAL TECHNIQUES

Labor Variables	All Mechanical	Part Manual
Household		
Male	7.9	18.2(NS)
Female	20.3	76.2**
TOTAL	28.2	94.4**
Hired		
Male	19.2	20.8(NS)
Female	4.7	60.2*
TOTAL	23.0	85.9**
All		
Male	27.9	39.9(NS)
Female	23.3	140.4**
TOTAL	51.1	180.3**
Ratios		
Female/TOTAL HOUSEHOLD	.77	.79(NS)
Female/TOTAL HIRED	.03	.40**
Female/TOTAL ALL	.57	.81**
Household/All	.43	.53(NS)

Note: ** significant at .01 level.
      * significant at .05 level.
      NS not significant.

In households that hand harvest part or all of their
cereals, there is more work for both men and women, but
especially for women. This is also true for hired labor;
the men's contribution is not significantly different for
hand and mechanical harvesting, but the women's contribution is much higher where hand harvesting is done. Inter-
estingly, while mechanization decreases the female labor
proportion of total labor in this task, it does not significantly affect the female proportion of household labor.
The mechanical operations are largely done by hired persons, and thus do not affect the household ratio. The mean
female proportion of hired labor is extremely small where
harvesting is done by combine.

295

Thus, mechanization of the cereal harvest substantially reduces women's labor inputs, from a mean of 180 hours per family to a mean of 51. This is in spite of larger crop areas associated with combine harvesting. Approximately half of the reduction comes from household labor, while the other half comes from hired labor. The gross amount of hired male labor is not significantly affected by mechanization. These figures do not show that in the mechanized harvest, hired male hours include more machine operator hours and fewer manual laborer hours in comparison to the manual harvest. The number of hired male machine operators increases, substituting for both male and female household and hired labor.

LABOR INPUT IN OFF-FARM ACTIVITIES

This section describes the gender differences in off-farm activities, as they relate to agricultural or non-agricultural tasks. The discussion will be divided into two parts. Emphasis will first be placed on the four survey villages with few landless families, then villages with a higher percentage of landless households will be discussed.

Working off-farm depends largely on farm size, crop productivity, and access to work opportunities. In the survey villages, 63 percent of income comes from crops and livestock, versus 37 percent from off-farm activities. Thus farming is the more important activity. However, its importance varies in relation to the productivity of agriculture. Off-farm income provides only 29 percent of income in the two wetter villages (Zone 1) compared to 44 percent in the drier, less productive area (Zone 2). Working outside the village is more frequent in the drier villages; from the twenty-four households in Zone 2, 42 percent of the family heads work in non-agricultural activities compared to 22 percent in Zone 1. The farmers from Zone 2 usually work as laborers, mainly in the construction industry. The off-farm activities of Zone 1 farmers consist of running a business in the village, driving a taxi, or teaching in the village school.

No women from the four survey villages work in non-agricultural activities outside the village. A few women work off-farm, but their tasks consist of agricultural labor within the villages, mainly planting summer crops or harvesting legume crops. Their work is limited to a few days per year and is not an important component of household income.

In many villages having less land, agricultural labor by women is a major operation. Men in such villages mostly pursue off-farm activities, while the women work more in agricultural labor. For example, in one village of 5,000 people, 37 percent of the households are landless. Approximately 300 men commute 50 kilometers to Aleppo every day to work as laborers in a government construction company. Another 100 men and some women work in construction in the village, while others work driving tractors, pickup trucks, etc. Approximately 300 women work regularly in agricultural labor, as well as about 100 men (usually either unmarried young men or old men). By combining summer crop planting, weeding, and harvesting of various crops, approximately six months of work are provided over the course of the year. This source of income is not as regular or as well-paid as urban work, but it clearly forms a larger portion of household income than in villages with more land.

The overall discussion of labor input by sex has attempted to explain patterns of variance within the data, particularly to determine factors affecting the male and female labor inputs, and the relative importance of household and hired labor. The following analysis is restricted to hours spent in crop production.

The most important variable affecting all labor categories is holding size. Total land area is significantly correlated with all categories of labor, including male household ($r = .66$), female household ($r = .56$), male total ($r = .86$), female total ($r = .77$), hired total ($r = .84$), and total labor ($r = .85$). Thus all categories of labor increase with higher land areas. Within the range of sizes of landholdings observed, the relationship of area to labor appears to be linear. In larger holdings, it might be expected to see an effect of substitution of capital for labor, but this is not observed on these small farms.

In view of the strong effect of the size variable, it is not surprising that there are also significant correlations among the labor variables, such as total male labor with total female labor ($r = .83$) and household with hired labor ($r = .41$). As requirements increase, men and women both increase their labor and hire more labor as well. It is interesting to note that the female proportion of hired labor is negatively correlated both with household labor ($r = -.29$) and with the female proportion of household labor ($r = -.29$); this substantiates the point that female labor is hired mostly for manual jobs that can be carried out by unskilled family members if they are available.

Because of the strong effect of holding size, in considering other variables it is essential to consider land
simultaneously. For example, where the male head of household has a steady job, this appears to have a strong effect
on labor inputs, with male household labor hours averaging
136 if he has a job and 406 if he does not. Large differences in means are found for other labor variables as well.

However, households with jobs hold an average ten hectares compared with 19 hectares for those without. That
is, on farms with small holdings, the male head of
household is more likely to seek off-farm work. When
holding size is entered into the analysis of variance as a
covariate, there are no significant differences in any
labor variables between families where the male head has a
job and those where he does not.

However, factors other than holding size, such as demographic variables, do have an effect on labor allocation.
Using mutiple linear regression, four variables were found
to explain most of the variance in the labor variables:
holding size, number of adults in the family, number of
family members absent on a daily basis (either working, in
military service, or away in school), and "excess of
females" (number of adult females minus number of adult
males).

For comparative purposes, regression statistics are
shown for all labor variables with all four independent
variables (Table 20.5). There are interesting differences
among the results.

First, it should be noted that household labor is
primarily linked to household size, and female household
labor is also related to the number of females. Male
labor, as one should expect, is reduced by male absentees.
Holding size is also important; all other things being
equal, household members work more if their farms are
larger. But in this case that variable appears to be less
important than demographic factors. However, holding size
is the only variable significantly related to total hired
labor. Large farms hire more labor than small ones. This
fact is also shown by the ratio of household to total
labor; larger holdings are associated with a greater proportion of hired labor. It appears that labor is hired for
large farms where family labor is not sufficient. Family
size increases the relative contribution of household
labor. However, from the total labor inputs, it appears
that farmers do not hire as much labor as they would use if
family labor was available. Family size and the number of
absentees have an effect on total labor expended. Thus it

TABLE 20.5

REGRESSION RESULTS ON LABOR VARIABLES

Dependent Variables (Labor)	Independent Variables (Beta Values)				
	Holding Size	Number of Adults	Number of Absentees	Excess of Females[a]	Adjusted R^2
Household					
Male	.36**	.60**	−.42**	−.08(NS)	.54
Female	.22(NS)	.61**	−.18(NS)	.27*	.50
TOTAL	.31**	.67**	−.33*	.11(NS)	.60
Hired					
Male	1.03**	−.45**	.25*	.08(NS)	.70
Female	.69**	.07(NS)	−.10(NS)	−.27**	.55
TOTAL	.92**	−.15(NS)	.05(NS)	−.14(+)	.70
All					
Male	.70*	.33**	−.25*	−.04(NS)	.77
Female	.49**	.51**	−.19(NS)	.09(NS)	.68
TOTAL	.61**	.45**	−.23*	.03(NS)	.77
Labor Ratios					
Female/ TOT. HOUSEHOLD	−.25(NS)	.05(NS)	.27(NS)	.49**	.24
Female/ TOTAL HIRED	.25(NS)	−.05(NS)	−.20(NS)	−.45**	.18
Female/ TOTAL ALL	−.50**	.67**	−.18(NS)	.07(NS)	.15
Household/ ALL	−.49 **	.75 **	−.26(NS)	.20(NS)	.20

Note: [a]Adult females minus adult males.
 ** significant at .01 level.
 * significant at .05 level.
 (+) significant at .06 level.
 (NS) not significant.

appears that either sufficient hired labor is not available or family labor is valued more cheaply than hired labor.

As expected, the female proportion of household labor is related to the number of females in the household. So also, it appears, is the proportion of the hired labor force that is female; where household females are numerous, hired females are fewer. This is related to the division of labor, since hired females are involved in manual labor that can also be accomplished by household females.

The substitution effect is also apparent in the differences between male and female hired labor. Male labor is hired where family size is small, indicating a higher rate of mechanization by small holders where hiring males increases when family males work off-farm. Where females are more numerous, less female labor is hired. The female proportion of total labor expended is positively associated with household size, although the female proportion of neither household nor hired labor is significantly related to size. Possibly this indicates that large families are more likely to grow legumes, since they can contribute to the harvesting themselves, but they will still have to hire from the predominantly female labor force. The female proportion of total labor is negatively associated with holding size, while on large farms, mechanization predominates and labor intensive crops may be avoided.

CONCLUSION

Male and female time contributions to crop production are approximately equal; however, males are more often involved in mechanical operations and other activities in new technology. Females are more involved with crops requiring hand labor, such as legumes. Hired labor for mechanical operations is predominantly male, while that for manual operations is predominantly female.

Overall, the trend in Syrian rural areas has been towards more mechanization and more off-farm employment of males. Thus the male rural labor force has been reduced, and those remaining have been increasingly involved in using new technologies. The female labor force has also been reduced as families have moved to cities and female work opportunities in rural areas have also declined. Continuing mechanization, particularly of legume and summer crop production, will continue to reduce female agricultural activities, including both household and hired labor.

The next task is to assess the effect of these changes on demographic trends, nutrition, income, equity, and agricultural productivity.

REFERENCES

Birks, J.S. and C.A. Sinclair
    1980   International Migration and Development in the Arab Region. Geneva: ILO.

Cain, M.
    1981   Overview: Women and Technology-Resources for our Futures. In Women and Technological Change in Developing Countries. R. Dauber and M. Cain eds., pp. 3-8, Boulder, CO: Westview Press.

Dasgupta, B.
    1977   Village Society and Labor Use. Delhi: Oxford University Press.

FAO (Food and Agricultural Organization).
    1982   Regional Study on Rainfed Agriculture and Agroclimatic Inventory of Eleven Countries in the Near East Region. Rome: FAO.

ILO (International Labor Organization)
    1981   Women, Technology and the Development Process, In Women and Technological Change in Developing Countries, pp. 33-47, Boulder, CO: Westview Press.

Nash, J.
    1983   Implications of Technological Change for Household Level and Rural Development. Working Paper 37, Office of Women In Development, East Lansing, MI: Michigan State University Press.

Nour, M.
    1985   Concern for Clients: Agricultural Research at ICARDA. In The Users' Perspective in International and National Agricultural Research, pp. 35-43, A Background Document Prepared for a CGIAR Inter--Center Seminar on Women and Agricultural Technology, March 25-29, Bellagio, Italy.

Nyanteng, V.K.
    1985   Women and Technological Innovations in Rice Farming in West Africa. In The Users' Perspective in International and National Agricultural Research, pp. 149-161, A Background Document Prepared for a CGIAR, Inter-Center Seminar on Women and Agricultural Technology, March 25-29, Bellagio, Italy.

Rassam, A.
    1986   Farm Labor by Age and Sex in Northwestern Syria:
            Implications for Two Proposed Technologies. In
            Proceedings of Kansas State University's 1985
            Farming Systems Symposium, Farming System Research
            and Extension: Management and Methodology. C.
            Butler Flora, ed., pp. 272-287. Manhattan, KS:
            Kansas State University.
Somel, K. and T. Aricanli.
    1983   Labour-Related Issues: An Assessment with
            Respect to Agricultural Research in the Middle East
            and North Africa. Discussion Paper 12. Aleppo,
            Syria: ICARDA.
Spence, D. and D. Byerlee.
    1976   Technical Change, Labor Use, and Small Farmer
            Development: Evidence from Sierra Leone. American
            Journal of Agricultural Economics, December, pp.
            874-880.
Tully, D.
    1984   Culture and Context: The Process of market
            Incorporation in Dar Masalit, Sudan. Ph.D. thesis,
            University of Washington, Pullman, Washington.
UNGA (United Nations General Assembly)
    1978   Technological Change and its Impact on Women.
            Minutes, 26 October, A/33/238.
World Bank
    1985   World Development Report. New York, NY: Oxford
            University Press.

21

The Woman's Program of the Gambian Mixed Farming Project

Margaret Norem, Sandra Russo,
Marie Sambou, and Melanie Marlett

LOCATION AND SOCIOECONOMIC BACKGROUND

The Gambia is an independent republic within the British Commonwealth, located on the west coast of Africa
between 13.30 degrees and 13.49 degrees North latitude and
between 16.48 degrees and 13.47 degrees West longitude. It
is one of the smallest countries in Africa, 250 miles long
and 15-30 miles wide, with a total land area of 10,367
square kilometers. The Gambia lies within the valley of
the Gambian River. Senegal surrounds the Gambia on three
sides, while the fourth is bordered by the Atlantic Ocean.
The highest point above sea level in the country is 90
meters.

The Gambia has a seven month dry season and a five
month rainy season between June and October. The heaviest
rainfall months are July and August and average annual
rainfall is 900-1100 millimeters (Dunsmore et al. 1976).

Current occupation of land has been determined by
historical possession; the original members of a patri-
lineage moved to unoccupied land and established exclusive
rights to it. Each village has an identifiable area of
land that falls under the village headman or Alkalo. Any
compound has the right to clear unclaimed land outside village jurisdiction and claim it for the village. Land is
transferred from generation to generation along male lines.
Women obtain rice land primarily from their husbands but
also receive some from their parents (Dunsmore et al.
1976).

Five major ethnic groups in The Gambia account for 95
percent of the population (Alers-Montalvo et al. 1983): the
Mandinka (42.3 percent), the Fula (18.2 percent), the Wolof
(15.7 percent), the Jola (9.5 percent), and the Serahuli
(8.7 percent).

Marketing outlets consist of local markets, village
traders, the Gambian Produce and Marketing Board, and

"other sources." Credit is available primarily from cooperatives, but also from Rural Development Projects and banks.

CROPPING SYSTEM

Certain crops are farmed in The Gambia according to the sex of the farmer. Groundnuts are traditionally a male crop although women are beginning to grow more. Millet and sorghum are also men's crops, but women help with harvesting. Alluvial rice and vegetables are grown by women. Maize is grown by both sexes. Seasonal migrants from other parts of The Gambia or other countries, referred to as "strange farmers," are given land to farm or cash crop on in return for working on the compound head's fields two or three days a week. These strange farmers return to their own villages at the end of the cropping season (Dunsmore et al. 1976). Land preparation is usually done by animal traction although clearing by hand is still quite common. Much of the cultivation is done by manual labor.

BACKGROUND FOR MAIZE PROGRAM AND MIXED FARMING PROJECT

In 1979 the Department of Agriculture in The Gambia introduced a cereal technology package to encourage an increase in the production of millet, maize, sorghum, and upland rice. This cereal technology package demonstrated the advantage of using good seed, fertilizer, timely operations, insecticides, and improved farming implements (Kidman and Owens 1985). Maize was introduced through this program as a crop with export potential.

The Mixed Farming Project (MFP) began in The Gambia in 1981. The objective of this project was "to improve the well-being of the rural people through intensified integration of crop and livestock production within existing Gambian farming systems." One component of this project was the maize production program. Maize was selected for emphasis since it is an easy crop to grow and a good cash crop; the government also offered a guaranteed price. Another advantage of maize is that it is an early food crop, requiring only 60-70 days to mature. Maize can be safely stored in the dry season. NCB, a Nigerian maize cultivar, was chosen for the MFP trials based on the results of maize cultivar trails conducted from 1975-1982.

The MFP maize program emphasized improved farm management practices to increase yields. Seminars were conducted for training extension personnel. The selection of farmers to participate in the program was based on their

established credit. Some women were also selected to participate in the MFP maize program that eventually led to
the development of the women's component of the MFP.

Life expectancy of Gambians is 32 years for males and
34 years for females (Dunsmore et al. 1976). Out of every
1,000 births, 285 children die before reaching two years of
age. Lack of food and malnutrition cause many of these
deaths. The MFP Women's Program attempted to address these
problems by introducing more nutritious crops and teaching
the preparation of these crops. Hand mills were introduced
to villages and cooking demonstrations conducted. A maize-
cowpea intercropping package was introduced to the women's
societies in 1985.

WOMEN'S SOCIETIES AND THE MFP

In numerous Gambian villages, the women have organized
women's societies. The extension service of the Department
of Agriculture encourages the formation of these societies
because they provide a means of bringing women together to
be taught improved agricultural methods. Women's society
members typically range in age from 15-40. These societies
are an ideal network for channeling education to village
women. The women's societies use cropping projects as a
means of obtaining funds for social and charitable purposes
by selling crops harvested from the society field. Each
society has a savings account to save money to purchase
inputs for the next growing season and to contract labor
for field preparation. Seed is reserved from the harvest
to plant in the following year.

Women's societies also hire themselves out to work on
other fields to receive additional income for the society.
The MFP hired women's societies to transplant grasses up-
country and to weed fields at the experimental station at
Yundum. Women's societies may also be called on for non-
profit community service. For example, if the village is
building a school, the society members may be asked to help
with construction or cook for those doing the work.

The 1982 MFP introductory maize program involved a few
women's societies. In 1983, 42 women's societies grew
maize, and by the 1984 season, 78 women's societies grew
maize. Extension workers and MFP staff instructed the
women about maize cultural practices such as time of planting and harvest, rate and spacing of planting, and fertilizer application.

Prior to cooking, maize must be shelled, cleaned,
dehulled, and then pounded into a meal or flour (Marlett

and Sambou 1985). This preparation is tedious and time consuming and can deter women from using maize. The MFP distributed 50 hand shellers and 40 hand mills throughout the country to facilitate maize processing. Use of these tools was demonstrated by extension workers and MFP personnel. There were some problems with the mills. After dehulling, maize had to be dried before grinding which lengthened the processing time. Another drawback was that maize had to be ground two or three times to obtain the desired texture. Nevertheless, using this tool was considered superior to hand pounding the maize.

In conjunction with the Extension Aids Training Unit of the Department of Agriculture, the Women's Program of MFP initiated maize cooking demonstrations in December, 1984. The objective of these demonstrations was to introduce Gambian women to methods of food preparation using maize as the staple ingredient. Forty female agriculture demonstrators from throughout the country were taught to conduct cooking demonstrations. The training sessions were two days long; the first day covered economic and nutritional benefits of maize and the second day was the actual cooking demonstration. Eleven recipes were prepared at each demonstration, including a maize weaning food consisting of a fine maize meal, groundnut butter, milk, sugar, pumpkin, and bananas.

Fifty-four villages were selected for cooking demonstrations; those that had grown maize in 1984 were given priority in this selection. Villages were required to provide firewood, cooking utensils, and maize. The MFP provided D17,000 (US$4,857.00) for cooking ingredients, night allowances, and training of the demonstrators. Representatives from surrounding villages were invited; approximately 100 women were instructed at each demonstration. The cooking demonstrations were well received by the villages and evolved into festive celebrations.

Demonstrators were requested to conduct follow-up cooking demonstrations at the compound level. Cassette tapes with cooking instructions in Mandinka, Wolof, and Fula were distributed to women's societies to assist with these demonstrations.

1985 MAIZE-COWPEA INTERCROPPING PROGRAM

The MFP Women's Program designed a project to introduce maize-cowpea intercropping to the women's societies during the 1985 growing season. The objectives of the intercropping project were to improve the human diet, improve stover quality for livestock, increase the rural family income, and introduce a valuable agronomic practice.

Cowpeas (Vigna unguiculata) are a summer crop which grows best with 750–1000 millimeters of rainfall. It is a drought tolerant crop and has been reported to do well in the Sudan with only 400 millimeters of rainfall (Skerman 1977). Cowpeas are quick growing and compete well with low growing weeds.

Cowpeas are a valuable cash crop and could provide a source of income for women farmers as well as improve the nutritional balance of local food production. They are high in protein and valuable for both human nutrition and animal forage. Intercropping maize with cowpeas provides one crop high in carbohydrates complemented by one high in protein. In Tanzania, Jeffers and Triplett (1979) considered the benefits of intercropping cereals and legumes one of the most important measures of the success of intercropping. Although yield is an important measure for evaluating the success of a crop, it does not reflect nutritional value or land use that are important benefits of an intercropping system (Francis 1978).

In eastern Gambia, Weil (1980) describes the soil as low in clay, sandy, free draining and low in fertility. Decline in soil nitrogen content naturally accompanies agricultural development. In The Gambia, groundnuts are rotated with millet, but maize is monocropped on the same field each year. Average fertilizer application is two bags per hectare (Eckert and Fulcher 1984). Intercropping legumes and grains may be a viable alternative to nitrogen application in agricultural systems such as The Gambia where nitrogen is unavailable or unaffordable since legumes potentially may fix some of their nitrogen and consequently conserve soil nitrogen for the benefit of other crops (Nambian et al. 1983).

The technical package for the women's societies intercropping program (Marlett and Sambou 1985) was designed on the basis of intercropping studies conducted by the MFP maize agronomist on the experiment station in 1984. The technical package was distributed to extension workers in each division who were responsible for explaining the package to the societies and assisting the women in its implementation. Society selection for participation was based on each society's financial status and village location. Twenty-eight of the 78 societies that had grown maize in 1984 were able to pay back their loan to the FAO fertilizer revolving fund credit scheme and these 28 societies were the ones selected to participate in the intercropping

program. All five agricultural divisions in the Gambia
were represented by the societies selected.

Fields were cleared by hand or by animal traction.
Each society paid for the clearing from its savings.
Cowpeas and maize were intercropped on one hectare.
Women's societies provided their own maize and the MFP
provided them with local cowpea cultivars. Several local
varieties of cowpeas were distributed to the women's
societies. Planting began in early July after the rains
began. Once the heavy rains began, 15-15-15 compound
fertilizer was broadcast over the field at a rate of 200
kilograms per hectare with 90 centimeters between rows and
25 centimeters between plants. Two weeks after planting
maize, fields were weeded and cowpeas were planted between
the maize rows. Harvesting of maize was in October.
Cowpea harvests varied with the cultivar. To obtain yield
data for maize and cowpeas, five rows were selected at
random and 10 meters of each row was harvested from each
society's field.

RESULTS

Society fields were visited by MFP personnel (Norem,
Sambou) throughout the growing season and progress was monitored and problems discussed during these visits. However, too many societies were involved for MFP to assume
complete responsibility for monitoring of the society
fields. Consequently, it was essential that MFP personnel
instruct agricultural extension personnel thoroughly about
the technical package. In some villages the agricultural
extension personnel were more effective than in others and
spent more time assisting the women's societies. The
extension workers changed the intercropping design in some
villages. In others, the technical package was abandoned
completely and cowpeas and maize were grown separately.

The women's society fields were frequently the last
fields assigned by the village Alkalo and were often in
poor condition. Women planted society fields after their
own plantings were concluded so typically planting was
late. Time available for field maintenance was very
limited; women usually worked in their society field on
Wednesday, their day of rest. Women worked together as a
group, singing while they worked.

In the 1985 growing season, abundant rainfall resulted
in high weed infestations in many of the fields. The presence of cowpeas between maize rows suppressed weed growth
in some fields, but in others the weeds were not deterred

by the cowpeas. The women were aware of the need to weed their fields more frequently but simply had no time for the additional task.

One disadvantage of growing cowpeas is that they are plagued with insect pests. Thrips (Megalurothrips sjostedti) are a major insect pest in cowpeas during the flowering stage. They infest cowpea blossoms, causing them to abort. If untreated, the cowpea plants continue developing vegetatively but lose all their blossoms before pod development can occur. Thrips are very small and can go unnoticed to the untrained eye. The MFP personnel emphasized the importance of monitoring for thrips during their visits to the society fields. Agricultural demonstrators and society members were advised to monitor their fields for insect pests and contact Crop Protection Service (CPS) when pests were observed. CPS was provided with a list of participating villages and was to provide chemicals for spraying as needed.

To participate in the program, each women's society had been required to provide its own maize seed and pay the MFP D100 (US $28.57). This is a significant sum of money and its payment indicated society interest in the program. However, of the 28 women's societies which participated, eight did not harvest either maize or cowpeas from their fields (Table 21.1). We will examine what factors contributed to these failures, including time constraints and pest damage.

Several societies planted maize but then had no time to maintain the fields. Fields became densely overgrown with weeds and the cowpeas were never planted. One society explained that a severe rice shortage had forced the women to spend much of their time searching for rice to purchase and as a result they did not have time to work in the society field. A common problem for several women's societies is that they were assigned fields two kilometers or more from their village. As they became busier in their own fields they did not have time to travel the distance to maintain the society fields. When fields were so distant, it was also difficult to find a society member or child to guard the field for insects or monkeys. One society reported that its distant field was destroyed by monkeys.

Sixteen of the societies did harvest maize. Yields varied from very high to quite low. The women had experience with maize from previous growing seasons and were aware of the labor investments needed for a good crop.

Results of the cowpeas were unsuccessful with only eleven villages reporting harvests. As previously mentioned, in several fields cowpeas were not even planted

TABLE 21.1

MAIZE/COWPEA YIELDS OF PARTICIPATING VILLAGES

	Village	Maize (kg)	Cowpeas (kg)	Comments
URD Division:	Giroba-Kunda	150	—	
	Mansa-Jang	976	270	
	Kisi-Kisi	900	380	
	Tinkinjo	1100	400	
	Tuba	1800	130	
	Taibatu	—	290	
	Bajon Koto	—	230	
	Madina Koto	—	496	
	Jendeh	100	340	
	Koli Bentang	—	—	
	Limbambulu	2300	240	
MID Division:	Bati	400	8	
	Njoben	2400	—	Planted both crops in same row at same time.
	Boiram	2500	—	Planted both crops in same row at same time.
	Saruja	830	—	180 cm. between maize rows.
	Cha-Kunda	1000	35	
LRD Division:	Kanni Kunda	—	—	Livestock grazed field.
	Madina Sancha	600	—	
	Jassong	—	—	Extension Agent resigned.
WD Division:	Kerewan	—	—	Did not intercrop.
	Busumbala	—	—	Cowpeas not planted.
	Beeta	285	—	Very late planting.
	Kassange	—	—	Results not known.
	Jorem Bunda	50	—	
NBD Division:	Kanuma	—	—	Cowpeas not planted due to weeds.
	N'Jawara	158	—	
	Sallikere	—	—	Field 2 km. from village.
	Kerewan	—	—	Field too far from village.

due to dense weeds. In many fields the thrips infestations went unobserved or had developed too far before CPS was contacted, while sometimes CPS could not respond to calls due to fuel shortages.

Another problem was that the cowpea seed distributed was from local cultivars obtained throughout the Gambia and the Casamance region of Senegal. The seed had been purchased at the local markets and the growth characteristics, including day length requirements, were not known. Possibly several of these cultivars were not planted at the right time of year to produce pods. Using cultivars of known characteristics would have enhanced the chances of obtaining a cowpea harvest, but even the use of a known cultivar would not have guaranteed success because of the severe insect problems accompanying all cowpea cultivation. (On the experimental station at Yundum, many known cultivars produced only vegetative growth due to thrips infestations.)

DISCUSSION

This method of intercropping maize and cowpeas was new to Gambian women, but growing cowpeas was not new to them. Cowpeas are a desirable food crop and are frequently grown in backyard gardens. Local cultivars are used and pesticides are not available. However, in a backyard garden the cowpeas are close to the compound and easier to monitor for insects. Pests are also easier to control since fewer plants are grown.

Some women's societies were asked about local customs for controlling insect pests. Practices for controlling the cowpea weevil (Callosobruchus maculatus), a severe problem in harvested seed, included steaming seeds or storing them in oil, ash, or chilies to repel the weevils. A survey conducted by CPS (Sagnia 1981) revealed several local methods that may be effective in controlling insect pests on cowpeas as well as on other crops. Broadcasting wood ashes on plants, particularly vegetables, to protect them from insects is widely practiced in the Gambia. Insects can be attracted to a site away from the field using bran, fresh cow dung, or green baobab fruit. They are then killed by burning or burying. Burning blister beetles is reported to produce an odor that discourages other blister beetles from the area. Neem tree berries (Azadirachta indica) are dried, pounded, and used as seed dressing or to ward off field pests.

These local methods of pest control are important to identify and evaluate. Instructing village women to use particular customs such as these may be more valuable than teaching them to rely on pesticides that are frequently unavailable and often arrive too late to be effective.

CONCLUSION

In retrospect, it is apparent that local Gambian customs for cowpea production should have been better understood prior to implementing this program. The technical package should have included more information on insect pests and extension agents should have received training regarding these pests. Ideally, a pesticide spraying program should have been set up in advance rather than relying on contacting CPS once pests were observed. Finally, a known cowpea cultivar with proper day length requirements should have been distributed. Unfortunately, this is the final season for the MFP so an important follow-up program is not possible. Hopefully these suggestions will be useful to other projects carrying on future intercropping work in The Gambia.

Despite the failure of the cowpea crop the work with the women's societies should be regarded positively. The village women were recognized as farmers and singled out for instruction. The women seemed eager to learn and were always receptive to MFP personnel. The women's societies in The Gambia are an ideal network that is already established and could easily be used by developmental projects for WID (Women in Development) efforts. In 1986 the Department of Agriculture of the Gambia established a permanent Women's Program. Marie Sambou, an MFP staff member, was assigned to this group.

REFERENCES

Alers-Montalvo, M., F.O. Dumbayo, G.D. Fulcher, B. Gai and J. Haydu
    1983   Farming Activities in The Gambia: A Preliminary Report. Technical Report No. 2. MANR/GOTG/CID/CSU. Fort Collins, CO: Colorado State University.
Dunsmore, J.R., A.B. Rains, G.D.N. Lowe, D.J. Moffatt, T.P. Anderson, and J.B. Williams
    1976   Land Resources Division #22, The Agricultural Development of The Gambia: An Agricultural, Environmental and Socioeconomic Analysis.

Eckert, J.B. and G.D. Fulcher
    1984   Annual Administrative Report No.4 for The Gambi-
    an Mixed Farming and Resources Management Project.
    MANR/GOTG/CID/CSU. Fort Collins, CO: Colorado State
    University.
Francis, C.A.
    1978   Multiple Cropping Potential of Beans and Maize.
    Hortscience 13:12-17.
Jeffers, D.L. and G.B. Triplett, Jr.
    1979   Management Needed for Relay Intercropping Beans
    and Wheat. Ohio Agricultural Research and Develop-
    ment Center. Ohio Report 64:5:67-70.
Kidman, D. and S. Owens
    1985   The Commercialization of Maize in The Gambia.
    Technical Report No. 3. MANR/GOTG/CID/CSU. Fort
    Collins, CO: Colorado State University.
Marlett, M. and M. Sambou
    1985   Food Production/Consumption Linkage. Technical
    Report No. 4. MANR/GOTG/CID/CSU. Fort Collins, CO:
    Colorado State University.
Nambian, P.T.C., H.R. Rao, M.S. Reddy, C.N. Floyd, P.J.
    Dart, and R.W. Willey
    1983   Effect of Intercropping on Nodulation and N2
    Fixation by Groundnuts. Experimental Agriculture
    19:79-86.
Sagnia, S.B.
    1981   Pest Control Methods Used by Gambian Farmers.
    Crop Protection Service, Ministry of Agriculture,
    Gambia. Personal Communication.
Skerman, P.J.
    1977   Tropical Forest Legumes. FAO: Rome.
Weil, P.M.
    1980   Land Use, Labor and Intensification among the
    Mandinka of Eastern Gambia. Presented at the 23rd
    Meeting of the African Studies Association, October.
    Philadelphia, PA.

22

Intra-Household Dynamics and State Policies as Constraints on Food Production: Results of a 1985 Agroeconomic Survey in Cameroon

Jeanne Koopman Henn

Nearly all African governments, even those with relatively vibrant traditional food sectors such as Cameroon, are worried about maintaining self-sufficiency in food (Cameroun 1980). With population in Africa growing at three percent a year and urbanization proceeding at twice that rate if not faster, such worries are seen by many observers as well founded (Daddieh 1985; FAO 1978; USDA 1981; World Bank 1981 and 1984). Much of the international policy discussion on Africa's food problems centers around markets, prices, and technological change (Commins, et al. 1986). While acknowledging the importance of the state's role in providing price and marketing incentives for producers, this paper shifts focus from the state to the peasant farmer and to the intra-household relations that condition his and her ability to respond to state policies aimed at increasing food output.

Too few policy makers and analysts have stopped to ask: Who are the nation's farmers? What are the economic, ecological and social conditions under which they work? How is the mix of export, national market, and subsistence crops or livestock produced by peasants affected by state policy? To analyze policy from the perspective of African farmers, it is necessary to understand peasants as economic agents whose farming and non-agricultural enterprises are embedded in household relations of production and distribution.

This paper examines issues of food production in southern Cameroon, a region that produces food both for subsistence and for the market as well as producing the bulk of Cameroon's cocoa exports. This region shares basic characteristics with large parts of West and Central as well as East and Southern Africa in that women are responsible for most subsistence and marketed food while men concentrate on the production of an export crop and also engage in a wide

316

variety of supplementary cash earning activities, including
production of marketable food crops.

The basic constraints on most African food production
systems result from scarcity of the classic factors of
production: capital, labor, and land. While national and
international policy discussions acknowledge the existence
of these constraints, policy makers still have too little
understanding of how various constraints affect household
decision-making. This paper analyzes the interactions
between the labor and capital constraints in peasant households in southern Cameroon and the differences in the
nature of these constraints for men and women. Drawing
upon the results of a recent set of farm surveys conducted
from July through October 1985, the hypothesis is examined
that, other things being equal, food production is subject
to a binding labor constraint. If the labor constraint is
binding, neither men nor women farmers will be able to
increase food production when new marketing opportunities
increase farmgate prices for food. A change in market
access and farmgate prices is represented in the study by
contrasting the situation of two ethnically and ecologically similar villages that have very different conditions
of access to urban food markets, one requiring 1.5 hours of
walking to reach the nearest transport point and a second
served by a major new highway to urban markets.

The paper examines the gender dimensions of farmers'
capital constraints by comparing men's and women's separate
income streams and the current gender differences in access
to credit and spending on productive inputs. The next sections describe the villages studied, the research methods,
and the results. The conclusion addresses the implications
of the findings for state policy.

AGROECONOMIC SURVEY OF THE ETON FARMING SYSTEM:
LEKIE DEPARTMENT, CAMEROON

The villages studied, Bilik Bindik and Mgbaba II, are
located 50 and 70 miles respectively from Cameroon's capital city on the northern border of the tropical rain forest. The region has two rainy seasons that support two
annual food cropping cycles. The laterite soils are said
to have been extremely fertile thirty years ago when low
population densities permitted food fields to be left fallow for at least ten years. Today about 60 percent of the
farms that average five hectares, are planted with cocoa.
Fallow periods in this shifting agriculture system have
been reduced to an average of three to four years. People

complain that the land is poor but neither fertilizers nor
pesticides are used on food crops.

The village of Bilik Bindik has enjoyed excellent
access to the food markets of the major cities of southern
Cameroon since a major highway through the village was
opened in 1982. In contrast, Mgbaba II has been almost
totally isolated from both national and local food markets.
The farming system of both villages is typical of farming
systems throughout southern Cameroon's forest region (see
Westphal 1981; Guyer 1984). Cocoa is the only major export
crop and is grown primarily by men. Women own cocoa plantations (a monocropped cocoa farms is called a plantation)
only through inheritance from husbands who have circum-
vented the traditional practice of excluding women from a
man's heirs. Fifteen percent of the women in this study
grew cocoa, but their holdings were only one-fifth the size
of the average male owned farm.

Most farms have two basic types of food fields: an
intercropped "peanut field" that all women plant twice a
year with minimal male help, and a "forest field" where men
intercrop cocoa seedlings with plantain and bananas, and
women plant the tuber, cocoyam. Some women also plant a
small field of vegetables in a stream-watered area during
the January-February dry season. About three-fourths of
the total value of food consumed in rural households is
produced in these three food fields, the greatest part by
far in the women's peanut fields.

The peanut field is oriented toward subsistence,
although surpluses are sold and certain elements of its
complex crop mix (peanuts, corn, melons, leafy vegetables,
onions, tomatoes, cassava, plantain, banana and cocoyam)
may be expanded to respond to market opportunities. The
spring peanut field is usually about one-half an acre in
size whereas the fall field is always smaller (averaging
about one-fourth acre) due to the shorter growing season
and the demands on a woman's time resulting from her
husband's cocoa harvest. Since the crops grown in the
peanut field have considerably different maturation rates,
each peanut field produces output for about two years.

Men often help their wives or mothers clear a peanut
field, but they rarely help with any other cultivation
task. By contrast, the clearing and planting of a forest
field is usually initiated by male landholders because they
will eventually establish a cocoa plantation there. Men
normally claim rights of disposition over the plantain and
bananas they plant in the forest field: most both sell and
consume them, while a few give them to their wives to sell.
Because plantain and bananas are perishable and difficult
to transport, only 65 percent of men sampled in the village

with poor market access intercropped these foods with mature cocoa trees whereas 90 percent of the men in Bilik Bindik took advantage of their ability to sell to passing vehicles or traders by growing plantain with cocoa.

Market access also determines the extent to which a dry season field is planted: 39 percent of women interviewed in Bilik Bindik cultivated dry season vegetables whereas only one woman (representing 4 percent of the sample) planted this field in Mgbaba because she had not cultivated a peanut field due to illness.

RESEARCH METHODS

The research instrument was a single structured, two to three hour interview questionnaire with each married adult in a random sample of forty households: twenty-one in Bilik Bindik and nineteen in Mgbaba (23 percent and 17 percent samples of total households in each village). The final sample included thirty-four men and forty-seven women; six households were headed by widows and seven were polygynous.

With the aid of a young woman from each village who interpreted between Eton and French, all interviews were conducted personally while residing in each village for five to six weeks. The questionnaire specified field types and categories of farming tasks, asking respondents to recall the number of days or weeks spent on each task and to indicate how many hours per day that task usually required. This allowed the estimation of annual labor hours by task. Annual production was estimated from reports of seasonal or monthly output. Sales were reported for each major food crop as well as for minor crops. Both quantities sold and earnings were specified. The questionnaire also asked about non-agricultural labor times and returns in order to estimate total labor times and to facilitate the evaluation of constraints on agricultural labor attributable to competing activities.

Although the validity of single questionnaire data to estimate annual production and labor time is often questioned, confidence about the relative validity of the data is due to the fact that the results for men's annual cocoa cultivation hours and women's annual food production hours are within seven and eleven percent respectively of results obtained in southern Cameroon by a research team which used daily interviews throughout the calendar year (Leplaideur 1978 and 1981). Women farmers' impressive ability to recall the details of their farm operations the consistency of their responses was also striking.

THE GENDER DIMENSIONS OF LABOR AND CAPITAL
IN PEASANT FARMING

Table 22.1 presents the sample estimates of women's and
men's annual working hours. Two sets of data are presented
for men: one for all thirty-three men in the sample and
one for the twenty-seven men who neither participated in
full time wage work nor declared themselves retired. The
most striking contrast in Table 22.1 is between men's and
women's total labor hours. Men's total weekly labor averages thirty-two hours, while women's is over sixty-four
hours. Even though much of this enormous disparity results
from differences in domestic labor hours, (thirty-one hours
a week for women and four for men), an important difference
was also observed in agricultural labor hours (twenty-six a
week for women and twelve for men). Even full time male
farmers work only fourteen hours a week in agriculture, ten
of which are devoted to cocoa production and four to food.

Most of women's agricultural labor is spent producing
food for family consumption -- over sixteen hours a week.
In contrast, male farmers spend three hours a week on subsistence food production, one hour helping to clear women's
peanut fields and two hours cultivating plantain in the
forest field. Whereas most labor surveys report that
African men do the "heaviest work" in food production, of
cutting trees and clearing new fields, this survey found
that men did less clearing labor than might have been
expected. While men did about half the labor necessary to
clear a forest field (a project undertaken every three to
four years), women spent twice as much time as men clearing
the biannually cultivated peanut fields.

Women also spend considerably more time than men producing food for the market. Ninety-four percent of the
women but only 24 percent of the men sold food. Food sales
were women's major source of cash income; average annual
earnings in 1984-85 were $433. Cocoa is most men's primary
source of income. Men's average net earnings from the 1984
cocoa season ($760) far outstripped their average earnings
from food ($134).

However, men's earnings from food are very high when
evaluated on the basis of the $3.80 hourly return they
receive for less than one hour a week of market oriented
food work. By comparison, women received an average of
only $0.71 per hour when allocating nearly eight hours a
week to cultivating marketable food plus four more hours
processing cassava for marketing. While the small percentage of men actually selling food and the very limited
range of food crops produced by male farmers require
that their estimated return to marketed food work be

TABLE 22.1

MEN'S AND WOMEN'S AVERAGE LABOR HOURS

	All Men (N = 33)		Full Time Male Farmers (N = 27)		Women (N = 47)	
	Annual	Weekly	Annual	Weekly	Annual	Weekly
Food Production Hours	168	3.2	196	3.8	1253	24.1
Family Food	133	2.6	154	3.0	850	16.3
Marketed Food	35	0.6	42	0.8	403	7.8
Cocoa Production	448	8.6	529	10.2	113	2.2
Total Agricultural Hours	616	11.8	725	14.0	1366	26.3
Hunting or Fishing for Family Food	58	1.1	70	1.3	52	1.0
Cassava Processing	0	0	0	0	208	4.0
Marketed Palm Wine Production	259	5.0	316	6.1	0	0
Subsistence Palm Wine Production	121	2.3	145	2.8	0	0
Trade	25	0.5	31	0.6	41	8
Full Time Wage Work (12% of Men Participate)	243	4.7	0	0	0	0
Other Income Generating Work	137	2.6	126	2.4	48	0.9
Domestic Labor	207	4.0	221	4.3	1640	1.5
TOTAL LABOR HOURS	1666	32.0	1634	31.5	3355	64.5
Income Generating Labor	1145	22.0	1044	20.1	815	15.7
Unpaid Subsistence Labor	520	10.0	590	11.4	2540	48.9

Note: Food Production Hours and Cocoa Production Hours are subcategories of Total Agricultural Hours. Food Production Hours are divided into production for family consumption and production for sale.

treated as preliminary data, the economically illogical contrast between men's minimal effort for a high hourly return and women's considerable effort for a much lower return requires further analysis. To understand this unexpected result, it is necessary to examine men's and women's nonfood labor and related incomes.

Male farmers devoted an average of 8.6 hours a week to cocoa cultivation with net personal earnings of $1.70 an hour. Cocoa planters also mobilized approximately the same number of hours as they personally worked from family and friends, mainly for harvest processing where labor is both exchanged and "paid" for with food and drink. The average cost for harvest processing labor in 1984 was only $0.31 an hour or about 18 percent of the planter's own net return. The planter's wife's post-harvest cash or clothing "gift," that acknowledges the two hours a week she works on his cocoa plantations, has a monetary value of $0.54 an hour. Women who cleared plantations or harvested cocoa for unrelated planters made about $0.60 an hour.

Men engaged in a much wider range of income earning activities than were available to village women. The most frequently encountered male sideline activity was the tapping of oil palm trees to produce a drink known as palm wine. Two-thirds of the men in the sample collected palm wine for one or more months a year, but only half sold part of their output. Most was consumed with family and friends on the day of collection. Earnings for men who sold palm wine varied considerably from person to person but the average was about a quarter of men's average cocoa earnings. At $0.36 an hour, however, hourly earnings from palm wine were considerably lower than the return to cocoa or to men's marketed food. Further, hourly returns to marketed palm wine production were only half as high as women's hourly return to marketed food production.

Returns to the 33 percent of men sampled who supplemented their cocoa earnings with artisanal work such as house and furniture construction also varied considerably; the average was about $1.00 an hour. Hourly earnings for the two men who worked as manual laborers outside their village and for the single male peasant who developed a modern poultry raising project were at a similar level. Finally, the four men in the sample who had full time jobs outside the village earned only $0.79 an hour, less than half the average planter's hourly return to cocoa and only slightly higher than the return to women's marketed food. The comparatively low return to wage employment is all the more surprising considering that three of the men had skilled jobs and an average educational level of 6.7 years of schooling, nearly double the male average. These men have apparently chosen wage labor (that they combine with both income-producing and subsistence work on their village farms) because the opportunity to earn a cash income by working a total of 2,350 hours a year allows them to earn an annual income of $2,150, considerably more than the $1,417 the average planter earns while working 1,634 hours a year on paid and unpaid tasks.

The preceeding review of male peasants' complex mix of non-agricultural activities has shown that most men supplement their cocoa earnings with work that brings them an hourly return that is at least 30 percent higher than women's return from marketed food. On this basis, men would not be expected to enter food production. On the other hand, the few men in the sample who did produce food for the market earned a very high rate of return for producing plantain and bananas. Why then do men spend no more than a mere thirty to forty hours a year producing these marketable food crops? Economists, if they could find no more compelling constraint, might conclude that the typical male peasant places a very high value on his leisure. Sociologists would highlight the importance to men of avoiding food production activities which traditionally have been defined as "women's work."

For women there seems to be no question of placing a high value on leisure. After devoting forty-nine hours a week to family food production and domestic labor, the average woman extended her labor time by nearly eight hours a week to cultivate marketable food and by four more hours to process raw cassava into the more marketable cassava flour.

Half the women interviewed also engaged in trading activities. Twenty percent were village food traders who took other farmers' food to urban markets, and about thirty percent carried on petty trade in beer, wine, soft drinks and cigarettes within the village. Food marketers made $150 to $250 a year; petty traders earned only about $25 to $50 a year.

Summing up the labor picture, male farmers spent 11.5 hours a week on unpaid labor and twenty hours on income generating work. With a 31.5 hour total work week, it is implausible to suggest that most male villagers face a binding labor constraint. Indeed, men with wage work outside the village work forty-five hours a week. Women, on the other hand, work forty-nine hours a week on unpaid tasks and nearly sixteen more hours on income-generating work. The sheer length of women's current labor week prompts the policy relevant questions: Is women's labor constraint binding? Will peasant women be able or willing to work more than 64 hours a week if the government introduces new incentives to increase food output? The next section addresses these questions by examining men's and women's responses to an increase in producer prices for food resulting from an improvement in access to urban food markets.

COMPARISON OF WOMEN'S AND MEN'S MARKETED FOOD PRODUCTION IN VILLAGES WITH GOOD AND POOR MARKET ACCESS

In late 1982 the completion of a major highway through the village of Bilik Bindik produced significantly improved conditions for the marketing of food. Once the road opened, food could be sold directly to traders and private' individuals passing through the village in cars and trucks. The road also provided easy access to public transport linking the village to two major urban food markets. By contrasting the labor times, production, and sales of farmers in Bilik Bindik with those of farmers in Mgbaba a sense of the importance of marketing constraints on production can be gained. There, food sales were made to village women who could organize the headloading of produce to a road located 1.5 hours away at which irregularly appearing public transport then carried the trader to urban markets. The village subsample data can also be used to test the hypothesis that most village women are too overburdened with domestic labor and the production of food for family consumption to be able to devote additional labor time to production of marketed food.

Women in Bilik Bindik reported increasing their food production and processing labor after the road opened. The survey results corroborate these statements showing that women from Bilik Bindik spent 4.6 more hours a week producing food than women in market-isolated Mgbaba. Women's total work week was nearly sixty-eight hours in Bilik Bindik versus sixty-one hours in Mgbaba. Statistical comparison of means tests found both the difference in time spent in food production and the difference in total labor time significant at the ten percent confidence level. Women in Mgbaba worked less than five hours a week producing food for the market while women in Bilik Bindik spent 10.75 hours. The effects of the additional labor on women's incomes, enhanced by the lower marketing costs in Bilik Bindik were dramatic: women from Bilik Bindik made an average net income of $570 from sales of processed and unprocessed food, while women from Mgbaba made only $225 (1).

Women's response to improved marketing conditions clearly demonstrates that their labor constraint is not yet binding. Men's response is more difficult to interpret.

Male farmers in Bilik Bindik actually spent less time in
food production than men in the market-isolated vilage of
Mgbaba: 2.7 hours a week in Bilik Bindik and 3.8 hours in
Mgbaba. While this result seems to defy economic theory,
further analysis shows that men were in fact acting to
maximize their individual gains. When one looks solely at
labor to produce marketed food, men in Bilik Bindik spent
nearly one hour a week while men in Mgbaba spent barely
twenty minutes a week. It should be noted that in the Eton
region, the only food crops men produce in their forest
fields and cocoa plantations are the highly perishable, low
labor-intensity crops: bananas and plantain. When the
road opened through Bilik Bindik, men began to plant much
larger quantities of plantain and especially bananas that
were then sold in bulk to wholesale traders. At the same
time, these men also cut back on the amount of time they
spent helping their wives produce food for the family.
Women in Bilik Bindik, therefore, were obliged to make up
for disappearing male labor in the subsistence sector (2).

In sum, in response to greater ease of marketing and
rising farm-gate prices, both men and women increased output of marketed food, but men did so in a very limited way,
producing essentially only two crops while simultaneously
reducing their traditional input into family food production. Even though heavily burdened with other types of
work, women still increased the amount of time they spent
producing food for the market by about six hours a week.
The total amount of time men with good market access worked
to produce food for the market remained less than one-tenth
of women's average market-oriented labor time. The policy
conclusion is quite obvious: unless men experience unusu-
ally high prices for specific crops which are not culturally defined as "women's crops," they cannot be expected
to increase production of marketed food in response to
improved marketing conditions or moderate price increases.
Women will respond to the same incentives by significantly
increasing labor time on food production.

Women may not yet face a binding labor constraint under
these conditions for the moment, but it is clear that the
female labor constraint will surely begin to bind in the
not too distant future. Women farmers already describe
themselves as overworked. Several complain of living in a
nearly perpetual state of exhaustion. Thus, for women's
food systems to make significant progress in increasing
output, there must be considerably more research devoted to
their particular agronomic, ecological, and technological
problems. When research finds new solutions, new questions
will come to the fore. Will Africa's food farmers be able
to invest in yield and output increasing technology? More

specifically, will women farmers be in a position to allocate more capital as well as more labor to food production?
To analyze these questions it is necessary to consider the
budgets of peasant households.

ACCESS TO CAPITAL AND CREDIT:
EXAMINING MEN'S AND WOMEN'S BUDGETS

In the great majority of African families, men and
women keep separate budgets. This fact is well known among
scholars studying African households, but is rarely noticed
or perhaps deliberately ignored by international agencies
studying the constraints on peasant agriculture (World Bank
1981). The ubiquity of separate budgets has enormous
implications for women farmers' access to capital as this
section will illustrate.

Women in the market-isolated village of Mgbaba made an
average income in 1984-85 of $402, 65 percent from produc-
ing and processing food for urban markets. Bilik Bindik
women, who enjoy much better marketing conditions, made an
annual income of $787, 78 percent from food sales. In con-
trast, men in Mgbaba earned $1,140 in 1984-85, 55 percent
from cocoa and a negligible 1.6 percent from food sales.
In Bilik Bindik men earned an average of $1660, 52 percent
from cocoa and 14 percent from sales of plantain and bananas. Clearly, income from food sales is extremely important to women's budgets but marginal to men's. As the
labor time data emphasized and the income data will confirm, women both respond to and gain from economic incentives to increase food production. Even with higher
earnings from food sales, however, women's access to capital from their own earnings is much more limited than
men's.

Both men and women contribute to essential family
expenses. For example, on average women in the two villages spent 17 percent of their monetary incomes on food,
while men spent eight percent. Women allocate a much
greater share of increases in their earnings to food purchases than do men. Women in Bilik Bindik actually spent
more money on food than men even though men's incomes are
twice as high as women's. The same phenomenon was observed
for medical expenses and for purchases of essential household inputs such as kerosene and soap. Examination of the
data in these categories shows that as women increase their
incomes, men actually decrease or considerably moderate
their spending on family necessities thereby shifting more
of the burden of family maintenance costs onto women.

The effect of these intra-household economic relations
on men's and women's individual access to capital is

significant. Women farmers are restricted to a low
productivity technology in part because they do not
perceive themselves as economically capable of purchasing
fertilizers or insecticides. When asked if they would use
credit for agricultural inputs to change their farming
methods, many women replied, "how could I pay it back?"
Women worry both about the risks involved in experimenting
with unproved technology as well as the nagging fear that
if their incomes actually do improve due to new methods of
farming, their budgets might well be strained by new family
expenditure reponsibilities. Only a few of the younger
widows in the sample were unreservedly enthusiastic about
the prospects of receiving credit.

The reality to date is that here are almost no institutions for providing credit to women food farmers. Credit
for agricultural inputs, housing and schooling are provided
to men who are members of cocoa marketing cooperatives and
thus have a predictable cocoa income each year from which
the cooperatives can relatively easily deduct loan payments. Outside the male-oriented cocoa cooperative structure, only wealthy peasants have been able to obtain loans
from the national agricultural credit institutions. At the
present time in southern Cameroon, credit for agricultural
inputs relevant to food production is not effectively
available to the great majority of men or women farmers.

The interactions between labor and capital/credit constraints affecting food production are less inhibiting for
men than for women. Men dispose of both more leisure time
(that could be allocated to food production) and loan
collateral (male-owned land, cocoa plantations and predictable incomes from cocoa production) than women. Male
peasants could readily benefit from a policy of directing
food production credit through existing village cooperative
structures. For women the problems are more intractable.

Women have very little free time to allocate to food
production. While the provision of improved rural roads
has been shown in this paper to stimulate increased production and sales of food by women, labor time and capital
constraints combine to limit the progress that is possible
from improving market access alone. For women to increase
food output beyond the amounts that women in Bilik Bindik
are currently producing for the market, new technologies of
production must be tested, new inputs made available, and
new sources of credit provided. Policy makers cannot
ignore the intrahousehold relations of both subsistence and
marketed food production that attribute most of the time
consuming subsistence and domestic labor obligations essential for family survival to women. If rural women are to
dispose of more time to allocate to producing food for

urban markets, labor-saving technologies for the provision
of water and fuel, for the processing and storage of food
products, and for the production of food must be developed.
Women must also be provided with credit.

DISCUSSION AND CONCLUSION

This paper began with a series of questions for policy
makers: Who are the nation's farmers? What are their conditions of production? How does state policy affect who
produces what? This paper has shown that in southern
Cameroon, as in a great many peasant production systems in
Africa, women are the major food farmers, intra-household
relations of production and distribution seriously constrain women's access to capital and land, and state policy
has done little, outside the attempt to provide essential
road access to urban markets, to loosen the constraints
currently limiting the output of women food farmers. Further, by contrasting women's impressive response to the
loosening of the food marketing constraint with men's relatively moderate increase in food marketing under the same
conditions, this study suggests that if African governments
are seeking policies to promote both rural food self-
sufficiency and increased national provision of food to
urban markets, the specific constraints limiting women's
food production must be identified and alleviated.

Africa's food systems will have a dual character of
subsistence and market production for a long time to come.
If rural subsistence production, that currently feeds at
least 80 percent of the population in most countries, is
not to be jeopardized as policies are developed to increase
the national provision of food for urban markets, women's
food farming must be protected. Policies that induce men
alone to enter into marketed food production can have negative repercussions on subsistence food production. Men
have the power to adjust to changing market conditions by
reallocating household subsistence and expenditure obligations within families. The research reported here shows
that men may well shift family food production tasks onto
women in order to give themselves more time to engage in
income generating enterprises. They also tend to shift
responsibility for purchasing food, medicines, fuel, and
women's and children's clothing onto wives who are able to
increase their own cash earnings. Women nevertheless have

shown themselves capable and willing to respond to policy incentives that make it possible for them to increase food marketing.

If state policy-makers react to information about the complexities of the constraints on women's farming by directing policy incentives at men, the results of the research reported here indicates that it may be very expensive to induce men to produce the broad range of traditional food crops currently consumed in both urban and rural households. Rural men can choose from several income earning activities that result in a higher return to labor than that earned by women farmers. Thus, the prices of most foods would have to rise considerably if men are to produce them.

This paper suggests that food policies, technological innovations, and credit policies directed at alleviating women's labor and capital constraints are likely to be more cost effective approaches to meeting national food needs than are policies directed at attracting more male labor into the food sector. This is not to say that policies targeting women farmers, such as research on current agronomic and technologial constraints in the food sector, will not also be of benefit to men who may want to increase food production. It is merely to assert that policies targeted at women are more likely to result in a significant production response at lower cost than policies that have little likelihood of improving the conditions under which women farm.

NOTES AND ACKNOWLEDGEMENTS

(1) Net income from food sales was estimated by attributing one half of women's average transport costs by village to the marketing of food and deducting this amount from income from unprocessed and processed food sales. Other costs of production, principally the cost of the woman's hoe, that is virtually the only tool or input used in both subsistence and marketed food production, are minimal and were not netted out.

(2) Labor time data from the survey clearly indicates that men in the poor market access village (Mgbaba) spent more time producing food for family use by helping with clearing and some other tasks on women's peanut fields than men in the good market access village (Bilik Bindik). Production data shows that women in Bilik Bindik produced no less food

for family consumption than women in Mgbaba, even though the labor data indicates that they did not allocate more time to subsistence food production.

The research upon which this paper is based owes a great deal to the invaluable and untiring help of my inter- preters, Matilde Lema of Bilik Bindik and Cressance Milaga of Mgbaba. I am also extremely grateful to the eighty men and women who patiently and carefully answered my questions as well as to the many residents of Bilik Bindik and Mgbaba who so graciously shared their homes, meals, friendship, and knowledge with me. Also invaluable was the support of the Social Science Research Council which sponsored the research.

REFERENCES

Cameroun, Republique Unie du
    1980   Bilan Diagnostic du Secteur Agricole de 1960 à
            1980. Yaounde: Ministère de l'Agriculture.
Commins, S. K., M. Lofchie, and R. Payne, eds.
    1986   Africa's Agrarian Crisis: The Roots of Famine.
            Boulder, CO: Lynne Rienner Publishers.
Daddieh, C. K.
    1985   The Future of Food and Agriculture, or the
            Greening of Africa -- Crisis Projections and
            Policies. In Africa Projected. Timothy M. Shaw and
            Olajide Aluko, eds., pp. 151-169. London:
            Macmillan.
FAO
    1978   Tenth FAO Regional Conference for Africa.
            Arusha, Tanzania. September. Rome: FAO.
Guyer, J. I.
    1984   Family and Farm in Southern Cameroon. African
            Research Studies No. 15. Boston, MA: Boston
            University Press.
Leplaider, A.
    1978   Les Travaux Agricoles Chez Les Paysans du Centre
            Sud de Cameroun: Les Techniques Utilisés et Les
            Temps Necessaires. Paris: IRAT.
    1981   Modele 3C: Cameroun-Centre-Sud-Cacao-Culture ou
            Simulation de Comportement Agroéconomique des Petits
            Paysans de la Zone Forestière Camerounaise Quand Ils
            Choisissent Leur Systeme de Cultures. Paris: IRAT.

330

United States Department of Agriculture (USDA)
    1981   Food Problems and Prospects in Sub-Saharan
       Africa. Washington, D.C.: USDA.
Westphal, E. et al.
    1981   L'Agriculture Autochtone au Cameroun.
       Wageningen: H. Veenman and Zonen B.V.
World Bank
    1981   Accelerated Development in Sub-Saharan Africa.
       Washington, D.C.: World Bank.
    1984   Toward Sustained Development in Sub-Saharan
       Africa. Washington, D.C.: World Bank.

23
Intra-Household Gender Issues
in Farming Systems in Tanzania,
Zambia, and Malawi

Jean M. Due

This paper examines various aspects of the Farming
Systems Research and Extension (FSR/E) methodology and the
need to incorporate gender issues by using the results from
projects in Tanzania, Zambia, and Malawi. A conceptual
framework for the diagnosis and analysis of intra-household
dynamics in FSR/E (Feldstein et al. 1987) was used to guide
the review of these project results. In the course of this
analysis, the paper also highlights the contributions of
social scientists in FSR/E.

Hildebrand and Poey (1985) suggest two areas for the
contributions of social scientists. The first is unen-
cumbered social science that in its own right contributes
new knowledge and understanding. Examples are research on
the economies (national, regional, international) and
societies into which new technology will be introduced.
The second is social science research that has continuous
interaction with research of agricultural scientists, is
restricted to specific variables being addressed by both
groups, and in which there is active communication among
all team members. Examples are the farming systems,
culture and households into which new varieties will be
extended. Illustrations of social science research
presented here fit nicely into these two categories. The
results come from research into development of new high
yielding, disease, insect, and drought resistant beans in
Tanzania and Malawi as part of the Bean/Cowpea CRSP
(Collaborative Research Support Project) and from farming
systems research undertaken in Zambia.

The discussion of the results from the three projects
is organized into four sections that correspond to the
stages of the FSR/E methodology: diagnosis; design;
testing and evaluation; and recommendation, adoption and
dissemination.

332

DESIGN

In the original design of the Tanzania Bean/Cowpea CRSP
research it was agreed that traditional bean cultivars
would be tested for resistance to disease, insects, and
drought. Then the best of these varieties would be crossed
with other materials that tested highly from international
institutes, surrounding countries, or United States seed
banks. The breeding program focused on breeding high-
yielding varieties (HYV) for specific agroclimatic conditions in Tanzania but also included other desirable traits
such as cooking quality, seed coat, seed color, storability, and yield stability. Plant breeding is a long-term
process and the breeding program began before the diagnostic data were in; however, the breeders were nationals who
were cognizant of local conditions.

As the breeding program progressed and new data were
fed into the program from the diagnostic phase and FSR,
priorities in the program changed to include the cooking
quality, seed coat, and storability aspects as well as
narrowing the potential crosses to those with best performance based on desired characteristics. Each year the pool
was further narrowed so that the most promising lines could
be tested on farmers' fields in both high altitude-rainfall
and low altitude-rainfall areas. A shortage of seed
limited the number of farms on which the new varieties (TMO
101 and Kabanima) could be tested initially.

DIAGNOSTIC PHASE

Resources

It would seem important to have information on the
level, quality, and distribution of resources (land, labor,
capital, and other) of the community and households as to
ascertain the constraints facing farmers in the area.
Agroclimatic, soils, and census data provide a base of
information for the research domain. Visits with farmers
and agricultural and government personnel in the area can
determine if this is the appropriate locale of the research
domain. Visits of the research team to the area are important and in Tanzania one area in which only one-third of
the farmers grew beans was determined by such a visit.

After visits to prospective research areas and rapid
rural appraisal, a more detailed farming systems baseline
study was undertaken to learn about the total farming
system, the manner in which beans fitted into the system,
distribution of size of farms, production for consumption
or sale, and so forth. Farms were small, averaging 8.4

acres. The major inputs in agricultural production were
land and labor; little fertilizer or chemicals were used in
these farming systems. A few farms hired labor or oxen or
mechanized ploughing; a few farmers owned oxen. Capital
equipment owned per household was limited to hoes, axes,
machetes and simple tools; cash reserves were also small
but knowledge of the farming system was impressive, having
been acquired over the centuries.

In most areas of Tanzania land is available for
increased production. Technically the land is owned by the
state but is allocated to individuals by the community or
ethnic group. Since the Arusha Declaration of 1967 when
rural households were moved from scattered sites to ujamaa
villages, village officials control the allocation of land
within the village but, in general, continue traditional
land allocations. Households can obtain more land outside
the village unless the village is surrounded by other villages. In one area of this research land was available, in
the other (the high altitude-rainfall area) acreage could
not be expanded.

Capital for off-farm inputs (such as fertilizer or
chemicals) was theoretically available through the Tanzania
Rural Development Bank if the borrowing was undertaken in
the name of the village. In general the Bank does not
allocate credit to individual farmers. Individuals can
borrow from friends and relatives, usually at zero rates of
interest. However, more important than the official credit
constraint to individual farmers was the scarcity of foreign exchange to import farm inputs. The economy of the
country is depressed and foreign exchange reserves deple-
ted. The farming systems study showed that few farmers
borrowed through the bank during the study period and that
almost no fertilizer or chemicals were available in the
area during the 1980s. Credit, when obtainable, was available only for inputs in kind; it was not available for
tools, hiring labor, oxen, or tractors for ploughing, or
for consumption expenditures. Capital was available for
livestock purchases only on state farms and institutions.
The rural development bank does make loans to villages for
mechanized machinery and petroleum when foreign exchange
permits; however, villages in this region had not qualified
for mechanized machinery loans during this period.

Probably the most significant information discovered
regarding inputs was that women choose the bean seeds for
planting. This finding has important consequences for
extension of new varieties developed; if there is a perception that male extension workers cannot visit female

farmers, adoption of new varieties will be delayed or
distorted.

Labor

In the diagnostic phase, visits with government and
other officials did not indicate that female labor was
involved in agricultural production. However, on the drive
from the university to the potential research sites it was
obvious that females were significantly involved. When
labor data by gender were included in the baseline study of
the farming system, it was found that females contributed
one-half the labor on major agricultural crops and more
than half on minor crops. In addition, females contributed
more than one-half the labor on household tasks, child
rearing, and fetching water and fuel. Labor allocations on
the major crops by cropping practice and by gender are
shown in Table 23.1.

Females contributed more labor than males to planting,
weeding, and harvesting of all the major crops and 59 and
67 percent of total labor for beans and rice, respectively,
in this area of Tanzania. Even in bean areas where crop
acreage has been expanded with mechanized or ox ploughing,
the percentage of labor contributed by females did not
decrease markedly (Due et al. 1985).

Barnes-McConnell (1985) collected daily data from farm
households in bean growing areas of northern Malawi; she
found that over the year males and females in the household
each provided about half of the total labor used (with
child care omitted), with females contributing slightly
more (52.3 percent) than males (47.7 percent). Her data
show that females contributed 42.5 percent of the agricultural labor per year, 73 percent of the domestic requirements, and 42.7 percent of the economic inputs (buying,
selling, brewing beer, etc.). While the total was a team
effort and comparable across gender, there were large differences in gender contributions both among and within
categories.

All farming system studies in tropical Africa show
tight labor constraints at certain periods of the crop
calendar. One of the most interesting insights in Barnes-
McConnell's data is the relationship between labor demands
and health. Families spontaneously reported illnesses and
deaths, with the most illnesses toward the end of the rainy
season, (February through April), and the most deaths
within the March-June season. The reduced food availability, heavy work load, and dampness and water problems
inside the house explain this relationship.

TABLE 23.1

PERCENTAGE OF LABOR DAYS CONTRIBUTED BY FEMALES BY OPERATION BY CROP, SAMPLED FAMILIES, KILOSA, TANZANIA, 1980

	Maize	Sorghum	Rice	Cotton	Beans	Sunflower	TOTAL
Land Preparation	44	37	61	39	55	34	46
Planting	52	41	77	48	60	40	56
Weeding/Thinning	51	43	65	51	59	40	52
Spraying	0	0	0	0	0	0	0
Harvesting	54	46	71	51	69	43	58
Marketing	17	16	50	31	0	12	21
TOTAL	48	40	67	39	59	39	48

Source: Due and Anandajayasekeram (1984: 589).

Few farming systems studies interview women unless they are "heads" of households. In a study in Zambia, Due and colleagues at the Rural Development Studies Bureau of the University of Zambia (1985) interviewed women in farm households to obtain the relative hours of labor allocated to agricultural and other tasks during the farming and other seasons. The results astounded male colleagues. Females contributed more total hours per day (8.5) than males (7.4) in agriculture as well as in non-agricultural tasks (5.0 hours for females compared to 1.1 hours for males).

In some parts of tropical Africa, men and women farm fields jointly, in other parts women have some separate fields and farm other fields jointly with men; men may work fields communally with other males; or work parties may do some crop operations (weeding or harvesting) by gender. In other situations labor is allocated by crop with men taking primary responsibility for the export/non-consumable crops (cotton, coffee) while women take primary responsibility for food crops. Knowledge of these labor patterns is important for the agricultural scientists as they plan innovations.

Hired labor for crop production may be either male or female with female labor usually less well paid than male. Although hired labor is not widely used on most smallholder farms, it is used as acreage is expanded. In Tanzania work parties do weeding and harvesting on some farms but most of the labor is family labor. Work parties may be paid in food, refreshments, or cash.

In one high altitude-rainfall area of Tanzania where beans are grown on steep hillsides with 66 percent for the market and 34 percent for family consumption, 22 of 60 sampled families hired some labor but the average length of time per household was only three days. In another area where farms were larger and more hired ploughing was done, only 26 days of male labor and nine days of female labor was hired by 85 sampled smallholders. Hired machinery accounted for 26 percent of the total farm operating costs for these families but hired labor accounted for only 22 percent; female labor on average was paid 71 percent of male labor per day.

Incentives

Crop and Livestock Production. There are high incentives for these smallholder families to enter into agricultural production because that production provides over 90 percent of the family food supply. If that food supply fails, there may be little food available in the market as

surpluses are moved to the urban areas. Eighty percent of
the families reported that their first priority in farming
was the provision of family food; the second was income for
consumption and other purchases. During this time of a
depressed economy in Tanzania, wage goods were extremely
scarce and prices high. Thus the incentive to produce to
feed the family was very high, but price incentives
provided by the government through the official agency, the
National Milling Company, were low. Most governments in
tropical African countries have kept farm prices low to
placate vocal urban consumers (World Bank 1981, 1984;
Eicher 1982; Due 1986). Most of the farm families
interviewed in the Tanzanian baseline farming systems
studies sold their food crops at local markets rather than
through official government agencies, as prices in the
local markets were three to five times higher. In
addition, farmers were paid promptly, which does not happen
through official marketing channels. There is also active
trade in "over the border" markets in most countries that
does not enter official export statistics.

In the major farming systems studies in Tanzania, it
has been found that in general families consumed half of
total crop production and sold half, that percentages consumed and sold varied by crop, that all of the cotton and
coffee was sold, and that in some areas (the high altitude
bean growing areas where land is a constraint) all of the
maize was consumed. But in other areas, about 50 percent
of beans, maize, and sorghum were consumed and 50 percent
sold.

Most of these smallholder families in Tanzania have
some livestock and some of the families have considerable
cattle, sheep, and goats which are herded away from the
farm. Reported consumption of livestock (including poultry) was low; livestock provided only 23 of 2,295 calories
consumed per person per day for the sample families. But
livestock sales were a substantial addition to income from
crop sales, providing between three and 39 percent of
average total crop sales on sampled farms.

Non-Crop or "Off-Farm" Income. Most of the farming
systems studies undertaken have not generated data on
earnings of non-crop and livestock income, that is, income
obtained from selling small quantities of the minor crops
(vegetables, fruits, etc.) in the markets, selling bever-
ages along the road, brewing beer, and employment off the
farm. In tropical African rural areas, these avenues of
income were thought to be too low to include. But our data
have shown that important small sums are generated in this
manner, especially by females. Beer brewing is the most
important source of non-crop income for women; it is, of

course, another use of crops that are not usually sold
through the market. Women also engaged in working on other
farms and in selling small quantities of fruits and vegetables. In general women receive little child support (even
in female-headed households) or gifts from friends or relatives. Men worked in construction, butchering, tailoring,
farming, and road work to supplement their farm income.
This so-called "non-crop income" provided 60 percent as
much income as crop and livestock sales in the Arusha area
where there were more opportunities of off-farm income and
11 percent of crop and livestock sales in other areas more
remote from urban areas. Men earned more income from
employment off the farm while women earned most of their
income from beer brewing and selling fruits and vegetables.

The Zambia data (when farm women were interviewed)
showed these off-farm earnings to be even more important
than in Tanzania. In Zambia these income sources contributed 32 percent as much as crop and livestock sales; women
earned more income than men (K125 compared to K89 for men
per smallholder family); women earned 58 percent of the
off-farm income and men earned 42 percent.

Decisions on Use of Cash Income. Little is really
known of the intra-household decision-making on allocation
of cash income in smallholder families in tropical Africa.
This is not surprising as little is really known about
these decisions in North American families, and decision-
making varies by family and by consumer good. However,
some data were generated in the Zambian farming systems
studies when the farm women were being interviewed. These
farm women responded that husbands and wives jointly made
decisions on the use of husbands' cash earnings in 58
percent of the families with husband and wife present; in
40 percent of the cases the decision was made by the
husband alone and in only two percent by the wife alone.
When the same question was asked regarding wives' cash
income, the decision was made jointly in 50 percent of the
cases, by the wife alone in 40 percent and by the husband
alone in 10 percent.

So What? Results?

The whole emphasis of FSR/E is one from the "bottom up"
-- that is from the farmer's point of view. However, it is
clear that most farming systems researchers assumed that
the "farmer" was a man and that it was sufficient to interview or discuss constraints only with men. Often host
country colleagues were shocked when the extent of female
labor force participation was documented. These data were
obtained not from rapid rural appraisals but from more

detailed farming systems research; the rapid rural appraisals did not imply female labor was important, but that
women grew a few vegetables near the house. Agricultural
and university officials talked about "women's gardens" as
their contribution to the farming system.

When male farmers were asked about the major constraints to increasing agricultural production, the
responses, never included a shortage of labor. Drought,
rains that were too early, too late, or too little,
insects, pests, etc. were the major constraints mentioned.
Yet when female farmers were interviewed in Zambia, a very
different list was forthcoming. The women were asked what
types of development would most assist them in farming.
The overwhelming reply was improvements in farming,
followed by availability of credit, clinics and wells.
Twenty of the fifty-four responses on farming improvements
were for labor-saving devices -- oxen ploughs, tractors,
and farm equipment for hire. This is not surprising when
one documents the hours that women contribute to the
farming system! Smaller numbers requested higher farm
product prices, lower input prices, loans to purchase
cattle and grinding mills, and assistance in drainage.
Credit was listed separately as credit is not available to
women through official government channels in Zambia unless
they are widowed or divorced.

The farming systems mode emphasizes the complete farming system rather than individual crops. Thus, even when
the research focused on breeding higher yielding bean cultivars, those new cultivars must fit within the present
farming system including the tight labor constraints at
certain seasons. The farming systems mode also emphasizes
the household-farm model of smallholder production with
production for both consumption and sale. Innovations also
must fit within other input constraints in the short run;
if this is an area where fertilizers and chemicals are not
available most years, the high-yielding varieties must be
higher yielding without those inputs. It is difficult to
persuade agricultural scientists to test under "zero input"
conditions!

It is also difficult to persuade the agricultural
scientists to test the HYVs under intercropping conditions
in which most smallholder beans are grown in Tanzania. It
is much easier to obtain yield data under monoculture. As
a result of the farming systems studies, the focus of the
breeders has changed; data on family preferences of beans
for consumption, sale and storage have been incorporated
into the breeding program. It is now understood by all the
researchers that at least one HYV is needed for the cooler,
high altitude areas and one for the hotter, lowland areas.

On-farm trials, therefore, were undertaken in each area on representative smallholder farms.

When research on a major innovation is being undertaken jointly with host country scientists, it can be assumed that much of these farming systems results are already known by the national researchers involved and that the farming system-baseline data are generated primarily to increase the knowledge of the expatriate colleagues. That assumption is not necessarily valid; farming systems data are necessary for both groups. New innovations that fit into or relax the labor constraints will be more rapidly adopted than others which accentuate labor constraints. The CRSP researchers have decided that short maturing varieties of beans/cowpeas will better fit the labor constraints and food needs of smallholder families in many areas than longer maturing varieties.

TESTING-EVALUATION

A shortage of seed, petroleum, and spare parts for the transport vehicles severely constrained the number of farmers which could be included in on-farm trials of TMO 101 and Kabanima cultivars approximately fifty miles from Sokoine University of Agriculture.

Several months before planting in each farm testing domain, university personnel visited the villages and the village chairperson (a government official) and met with groups of farmers to inform them of the request to test this potentially high-yielding variety on some of their fields. At that time farmers were enthusiastic about this opportunity. However at planting time when the team again returned to the villages, the farmers were less interested in participating due to increasing dissatisfaction with government agricultural policy. Unfortunately, the same day the university faculty members went out to the villages to make final arrangements for the on-farm trials, a government official was meeting with farmers to urge them to plant at least a minimum acreage of cotton, a crop with a very low price and yield and very high labor require-ments. Farmers were angry at the insistence. The univer-sity personnel had to follow this official on the podium! Because the generally hostile attitude toward the govern-ment official transferred to the university faculty (also paid by the government), the university personnel decided it was best to ask for volunteers to try this new potentially high yielding variety. Since there was seed sufficient for only twelve farmers, twelve volunteers were selected and the planting instructions and seed left with the village chairperson and the extension agent (both

government officials). Time did not permit university
personnel to stay and plant that day, nor could they return
to visit the sites for a month.

When the researchers returned, they found that the
volunteers in one village included four "farmers" who
earned most of their income from non-farm sources (one bar
owner, one butcher) and in both villages the farmer volunteers included a higher percentage of those with larger
land holdings than the average for the village as a whole.

The new varieties out-yielded the traditional varieties
by about 70 percent, but the variation in yield from farmer
to farmer was very high. Since yield stability is important to these smallholders, this aspect of the high-yielding varieties will be investigated. In order to increase
the seed pool for the next year, all available seed was
purchased from the farmers; therefore cooking comparisons
could not be ascertained at that time. Other data relating
to farmers' preference of plant structure, disease resistance, insect resistance, etc., were also obtained.

RECOMMENDATION, ADOPTION, AND DISSEMINATION

Further testing and breeding will be done before final
recommendations will be made. However, serious questions
are being raised about the ability of the present government seed multiplication service to provide adequate bean
seeds and the present extension service to extend high-
yielding varieties. These issues will be discussed with
government officials. A marketing study is being planned
to estimate the impact on the present marketing structure
if significantly increased production of beans is realized.

The FSR/E project has still not addressed the needs of
female-headed households which are estimated to make up 15
percent or more of smallholder farm households in many
parts of Africa. Mollel (1986) investigated relative crop
acreage, crops grown, net income, and visits of extension
agents to contact, noncontact, and female-headed households
in the T & V area from 1980 to 1984. Some results of his
data are shown in Table 23.2.

It is evident that the female-headed families were
unable to prepare as much land and plant as much acreage as
the joint families (with both husband and wife present);
thus their ability to sell surplus production for cash was
decreased. These women compensated partially by allocating
their labor to non-crop income generation — brewing beer
and selling small quantities of fruits and vegetables in
the local markets. The net cash income per family of the
female-headed households was only 54 percent as high as

TABLE 23.2
COMPARISON OF CROP ACREAGE, INCOME, AND EXTENSION VISITS
BETWEEN CONTACT, NON-CONTACT, AND FEMALE-HEADED HOUSEHOLDS,
TANGA REGION, TANZANIA, 1985

	Male Contact N = 32	Farmers Male Non-contact N = 34	Female-Headed Household N = 32
Means of:			
Acres in Crops	2.8	2.1	1.4
Crop Sales[a]			
Less Farm Cash			
Operating Expenses	2774	1365	319
Off-Farm Income	962	2294	1685
Net Cash Income	3736	3659	2004
Visits by Extension Agent:			
No visits	5	19	23
1 & 2	8	9	4
3 & 4	9	4	1
More than 4	10	2	4

Note: [a]All figures are in Tanzanian shillings.

Source: Mollel, 1986.

that of the contact farmers and 55 percent as high as that
of the non-contact farmers.

The female-headed households were visited much less
frequently than joint families by extension agents; 72
percent of the female-headed farmers received no extension
visits compared to only 16 percent of the contact farmers
and 56 percent of the non-contact farmers. Only five (16
percent) of the female-headed households had one to four
extension visits per year while 50 percent of the contact
and 38 percent of the non-contact farmers were visited one
to four times. With less information, female-headed households were less able to try the recommended practices of
improved seed, early planting, spacing, weeding, and fertilizer application.

CONCLUSIONS

These experiences in Tanzania, Zambia, and Malawi demonstrate that farming systems research and extension efforts that do not have a person on the team who is sensitized to gender issues will miss important information. In these examples the importance of female labor in the farming system, of women making decisions about crops planted, seeds chosen, and income allocated would have been overlooked without the inclusion of gender in the analysis. In addition, female-headed households will continue to be invisible in most FSR/E in Africa unless gender issues are incorporated into the methodology.

REFERENCES

Barnes-McConnell, P.
  1985   Bean/Cowpea CRSP Malawi Project, 1985 Annual Reports from the Social Science Component. East Lansing, MI: Michigan State University. Mimeo.
Due, J. M.
  1986   Agricultural Policy in Tropical Africa: Is a Turnaround Possible? Agricultural Economics 1:1:19-34.
Due, J. M. and P. Anandajayaskeram
  1984   Contrasting Farming Systems in Morogoro Region, Tanzania. Canadian Journal of African Studies 18:3: 583-591.
Due, J. M. and T. Mudenda
  1985   Women's Contributions to Farming Systems and Household Income in Zambia. Women in International Development Papers No. 85. East Lansing, MI: Michigan State University and The Ahfad Journal, Women and Change (1986) 3:2:52-61.
Due, J. M., M. White and T. Rocke
  1985   Beans in the Farming Systems in Two Regions of Tanzania, 1980-82. Technical Report No. 4, Department of Rural Economy, Sokoine University, Morogoro, Tanzania and Department of Agricultural Economics, University of Illinois at Urbana-Champaign, AE-4602.
Eicher, C. K.
  1982   Facing Up to Africa's Food Crisis. Foreign Affairs 61:1:151-174.

344

Feldstein, H., S. Poats, K. Cloud, and R. Norem
    1987   Intra-Household Dynamics and Farming Systems
        Research and Extension Conceptual Framework.
        Population Council/FSSP-University of Florida
        Working Document.
Hildebrand, P. E. and F. Poey
    1985   On-Farm Agronomic Trials in Farming Systems
        Research and Extension. Boulder, CO: Lynne Rienner
        Publishers.
Mollel, N.
    1986   An Evaluation of the T & V Extension System in
        Tanga Region, Tanzania. Unpublished M.S. Thesis,
        University of Illinois at Urbana-Champaign.
World Bank
    1981   Accelerated Development in Sub-Saharan Africa:
        An Agenda for Action. Washington, D.C.: World
        Bank.
    1984   Toward Sustained Development in Sub-Saharan
        Africa. Washington, D.C.: World Bank.

24
Institutional and Policy Parameters Affecting Gender Issues in Farming Systems Research in Tanzania

Manasse Timmy Mtoi

National policy objectives, institutional arrangements, and the organizational structures of rural communities all influence household decision-making processes. They have an impact on the overall community structure; hence, they affect the process of formulating recommendation domains in Farming Systems Research (FSR). Inclusion of women into recommendation domains is very important since women account for about 60 to 80 percent of the labor force in many developing countries and they play a dominant role in smallholder production, often performing more than half of the household and family duties (Sacks 1983; Boserup 1970; Lamming 1983; Mascarenhas and Mbilinyi 1983). In Tanzania with an estimated population of 21 million (1985) and GNP per capita of about US $300, women account for about 50 to 65 percent of the agricultural labor force (Kamuzora 1978).

Policy objectives are defined as instances of government action in a particular area, whether the action is based on explicitly outlined policy objectives or is a reflection of the government's implicit policy in a particular area. Two types of policy affecting rural farm communities in Tanzania can be identified:

(1) those that aim at expanding the options available in peasant decision-making. Principally, these involve endogenous factors such as land, labor, capital, and management and
(2) those that change the external factors such as marketing infrastructure, pricing policy, restrictions and controls, and agricultural extension.

This paper examines the economic implications of the traditional division of labor by gender in the FSR project in Morogoro District, Tanzania along with the risks associated with the transfer of excess female labor during critical labor periods to male labor during the same period in an alternative farming system. The framework of analysis looks at how FSR address the distinction between real and expected roles of men and women, restrictions on labor transfer (mobility), and access to credit and marketing opportunities. Data pertaining to this study were collected in 1982 through 1984 from the FSR project area in the Eastern Uluguru Mountains of Morogoro District (1). A Quadratic programming model was used as a tool of economic analysis.

POLICY AND INSTITUTIONAL PARAMETERS REGARDING THE ENDOGENOUS FACTORS

The institutional parameters include those policies or interventions aimed at increasing the efficient use of existing farm resources namely land, labor, capital, and management. The FSR project's major role is to introduce new technological packages that will improve the existing production system. Figure 24.1 outlines a conceptual framework and relationships of institutional, cultural and policy factors posited to have an effect on gender issues in the FSR process.

Land Ownership and Use

The Tanzanian "village act" requires that each village member, male or female, be allocated separate plots of land to cultivate specific crops that are designated in the by-laws of each local village. Centuries ago, land was controlled and allocated to household or to extended family heads by patriarchal leaders of clans, but today all land is owned by the State and allocated by village government to village members. In many areas of the country, the practice of village governments is to allocate land to male household heads, who then allocate land to wives and unmarried sons and daughters as well as to young, newly married sons. In the FSR project area of Eastern Uluguru Mountains, user rights still predominate and the original user, who is often a male, has some hold over the land even if it is under fallow.

Major crops grown in the area include coffee, maize, cocoyams, bananas, rice, cassava, and pineapples. Table 24.1 shows the system of land use in the area.

TABLE 24.1

MONTHLY LABOR REQUIREMENT FOR FARM OPERATIONS FOR
MAJOR CROPS IN EASTERN ULUGURU MOUNTAINS[a]

Crop	Mean area (ha)	Total per hectare Work-days	JAN	FEB	MAR	APR	MAY	JUN	JULY	AUG	SEP	OCT	NOV	DEC	Total Work-days	Contribution by Gender	
																Male	Female
Coffee	.44	110.0	5.5	4.9	3.3	–	–	12.7	7.7	–	3.9	4.9	–	5.5	48.4	20.8	27.6
Maize	.64	72.5	9.6	7.2	–	–	6.4	6.4	–	–	–	–	8.4	8.4	46.4	8.4	38.0
Cocoyams	.56	141.0	29.4	–	6.2	–	–	–	6.2	–	–	6.2	–	31.1	79.0	60.4	18.6
Bananas	.36	106.5	2.3	4.9	3.5	–	3.7	3.5	3.1	3.2	–	4.1	4.0	5.1	38.3	28.1	10.2
Rice	.52	145.0	–	11.4	11.4	8.8	8.8	2.6	2.6	–	–	12.7	2.7	4.6	75.4	10.0	65.4
Cassava	.64	117.0	14.8	17.8	8.0	–	–	–	–	–	8.0	8.0	11.7	6.6	74.9	47.9	27.0
Pineapples	.40	118.3	10.6	5.5	4.5	4.5	–	–	–	–	–	–	9.6	12.6	47.3	47.3	–
Labor Requirements			72.2	51.7	36.8	13.3	18.9	25.1	20.0	3.1	11.9	35.8	16.4	73.8	409.7	222.9	186.8
Labor Available			51.7	51.7	51.7	51.8	51.8	51.8	51.8	51.8	51.8	51.7	51.8	51.8	621.0		
Surplus/Deficit[b]			(20.5)	0.0	14.9	38.5	32.9	26.6	31.8	48.7	39.9	15.9	55.4	(22.0)	211.3		

Note: [a]Labor requirements estimated from a sample of 46 farmers.

[b]Figures in parentheses represent deficit labor.

Source: Farming Systems Research Project, 1984.

FIGURE 24.1

CONCEPTUAL FRAMEWORK OF INSTITUTIONAL, CULTURAL
AND POLICY FACTORS AFFECTING INTRA-HOUSEHOLD DYNAMICS IN FSR

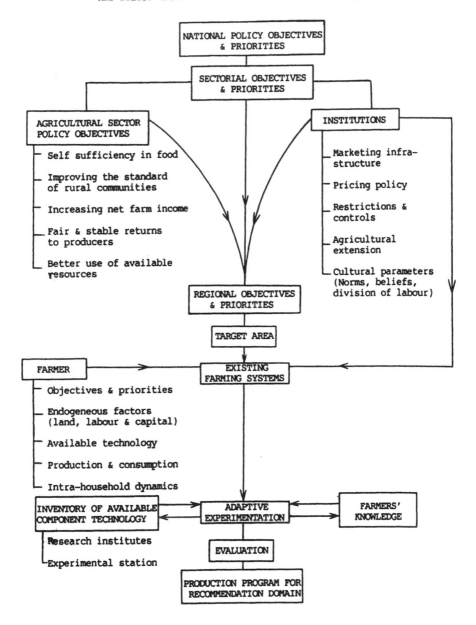

Traditional Division of Labor by Gender

Numerous detailed comparative studies of the patterns of the division of labor and the relative influence of different family members on decision-making related to the farm can be found in the literature (Due and Anandajayasekeram 1982; Lamming 1983; Mtoi 1984; Ngalulu 1977). Labor/time studies in Tanzania have shown that the major labor input of women peasants is into food production as compared to production oriented solely to the market that is dominated by men (Kamuzuro 1978; Ngalula 1977) Table 24.2 shows tasks performed by different sexes in the farming system of Eastern Uluguru Mountains.

Men tend to specialize in cash crops that include cassava, pineapples, bananas, and cocoyams. Women are mainly responsible for food crops, namely rice and maize. In the 1983/84 cropping season, men controlled about 80 percent of the average gross income of US $754 even though women contributed about 45 percent of the total farm labor requirements (Table 24.1). The exclusion of women from the cash economy tends to lower their social position vis-a-vis men. Cultural beliefs and norms exacerbate the situation to a certain extent. For example, some families in the FSR project area of Morogoro District expressed some cultural restrictions on female labor mobility from food crop production to cash crop production and marketing. However, male labor on other activities such as fetching water, collecting firewood, and cooking is considered completely taboo.

Economic Implications of Division of Labor by Gender

A risk quadratic programming (QP) model was used to test the significance of female labor on expected risk and productivity in the farming systems of the Eastern Ulguru Mountains (2). The QP-risk analysis compares two systems of farm production, the traditional system and the alternative system. The alternative system is an emergent system of production following adaptive experimentation results on cassava and pineapples from 1981 to 1983 in the FSR project area.

The families surveyed in the area had an average of 5.4 members with a range of family size from two to 11 members. Traditionally, labor consisted of family labor and hired labor. Family labor included all working days available as estimated by the farm family. Furthermore, labor was divided into three periods: November to March, April to

TABLE 24.2

TASKS PERFORMED BY GENDER, BY CROP AND BY ACTIVITY
IN THE FARMING SYSTEMS OF EASTERN ULUGURU MOUNTAINS, 1984

Crop	Women	Men
Cassava	Planting Weeding	Land Preparation Weeding, Harvesting Transporting, Marketing
Pineapples		Land Preparation Planting, Weeding Harvesting Transporting, Marketing
Bananas	Weeding	Land Preparation, Planting Scaring Wild Animals Harvesting, Transporting Marketing
Coffee	Weeding Harvesting Processing Transporting Drying	Land Preparation Marketing Planting
Cocoyams	Weeding	Land Preparation Planting, Harvesting Transporting, Marketing
Rice	Land Preparation Marketing Bird Scaring Harvesting, Weeding Transporting Processing, Cooking	Land Preparation Planting
Maize	Land Preparation Marketing Weeding Scaring Monkeys Harvesting Transporting Processing Cooking	Land Preparation Planting
Others	Collecting Firewood Fetching Water Other House Chores	House Construction Road Building

June, and July to October to reflect different opportunity costs of both male and female labor. Although labor restraints can be formulated in the model for every month of the year, this implies a rigidity in the timing of farm operations and consequently of labor use that may be unrealistic. Thus, labor restraints were formulated for those periods of the year in which labor allocation by gender is most critical. The remaining non-critical periods were also included to provide a complete accounting for labor in the farming system.

The number of days available for farm work per year was adjusted by taking into consideration days that were solely committed to non-farm activities and public holidays as shown in Table 24.3. The total average effective adult equivalent days available for farm work in the 1983/84 period was found to be 573. Female labor comprised about 45 percent of this labor.

The alternative farming system required certain labor utilization with respect to two cropping enterprises (cassava and pineapples). Though labor requirements were slightly higher in land preparation and planting, requirements for weeding were less because of spacing and mulching practices. Table 24.4 summarizes labor requirements in the alternative system of production and the resultant yields. All labor required in producing pineapples is contributed by males, while about 37 percent of the total labor requirements for cassava production is contributed by females based on the traditional division of labor. Pineapples and cassava yields increased by about 32 percent.

The QP-risk model involved the transfer of ten adult equivalent days of excess female labor for land preparation and planting during the November-March peak period to add to the male labor in the same period. The expected net farm income would increase from US $230.56 to US $394.44 if the alternative system is undertaken. This is an increase of about 71 percent when compared to the traditional system. If female labor mobility is restricted in the alternative system, the expected net farm income would decrease about 14 percent, from US $394.44 to US $338.89. Table 24.5 and Figure 24.2 indicates these changes would be effected at approximately the same levels of risk (measured in terms of standard deviation of net farm income in US dollars). The analysis further suggested that female labor transfer would ensure relatively higher efficiency ratios of expected net farm income to farm resources in terms of land, labor, and capital (Mtoi 1984).

TABLE 24.3

ESTIMATION OF EFFECTIVE ADULT-EQUIVALENT WORK-DAYS
AVAILABLE PER MONTH FOR FARM WORK

Month	Total Days	Days Unavailable for Farm Work					Effective Adult Equivalent Work-Days	
		Sundays and Wednesdays[a]	Fridays[b]	Public Holidays (no.)	Rainfall[c] Days	Net Days	Males	Females
November	30	8	2	1	1	18	26.17	21.58
December	31	8	4	2	3	14	20.35	16.79
January	31	8	4	1	4	14	20.35	16.79
February	28	8	2	1	1	16	23.36	19.19
March	31	8	1	–	4	18	26.17	21.58
April	30	8	1	1	5	15	21.80	17.99
May	31	8	1	1	3	18	26.1	21.58
June	30	8	2	–	–	20	29.07	23.98
July	31	8	2	1	–	20	29.07	23.98
August	31	8	1	2	–	20	29.07	23.98
September	30	8	1	–	–	21	30.54	26.18
October	31	8	1	–	–	22	31.98	26.38
TOTAL	365	96	22	10	21	216	314.00	259.00

Note: [a]Wednesdays are committed to communal activities such as building dispensaries, schools, etc.
[b]Fridays are committed to marketing.
[c]Rainfall days are based on 1971–80 average from Morogoro District.

Source: Farming Systems Research Project, 1984.

TABLE 24.4

LABOR REQUIREMENTS AND YIELD ON THE ALTERNATIVE SYSTEM OF PRODUCTION

Category	Yield/ha (kgs.)	Labor Requirements (work-days)					Expected Contribution by Gender	
		Land Preparation	Planting	Weeding	Harvesting and Transporting	Total	Male	Female
PINEAPPLES: Technology: .75m x .75m Trial Managed by:								
Researcher	5119	28.15	25.03	32.36	33.08	118.62	118.62	—
Researcher-Farmer[b]	4912	28.15	25.03	32.36	32.59	118.13	118.13	—
Farmer[c]	3890	24.44	25.78	44.44	25.19	119.85	119.85	—
CASSAVA: Technology: 1m x 1m Trial Managed by:								
Researcher	5870	39.38	26.68	42.97	54.21	163.24	102.38	60.86
Researcher-Farmer	5438	35.57	24.13	42.71	38.30	140.71	95.23	45.48
Farmer[c]	4458	26.68	15.49	35.30	36.20	113.67	80.53	33.14

Note: [a]Based on expected roles of men and women in the farming systems of Morogoro District.

[b]Based on farmer's traditional spacing of 0.92m x 0.92m.

[c]Based on farmer's traditional spacing of 1.14m x 1.14m.

Source: Farming Systems Research Project, Morogoro District, 1984.

TABLE 24.5

EXPECTED RETURNS AND STANDARD DEVIATION
OF THE TWO SYSTEMS OF PRODUCTION

System of Production	Expected Net Farm Income		Standard Deviation	
	Minimum	Maximum	Minimum	Maximum
	(U.S.$)		(U.S.$)	
Traditional System	41.67	230.56	7.6	10.88
Alternative System A (without female labor transfer)	150.00	338.89	7.90	10.30
Alternative System B (with female labor transfer)	172.22	399.44	8.01	11.16

The substantial increase in expected net farm income can be economically and technically achieved in the farming system of the Eastern Uluguru Mountains. This is because the alternative system does not significantly contradict the existing system in terms of additional requirements for farm resources. The difference in the two systems lies in the application of basic agronomic practices such as plant spacing and configuration, timing of critical farm operations, and mulching.

POLICY PARAMETERS REGARDING EXOGENOUS FACTORS

Policy parameters in relation to exogenous factors define the socioeconomic environment in which the farming systems operate. Generally these policy parameters are designed to steer production in those directions deemed desirable by the national policy-makers by providing guidelines for production and by creating a supporting infrastructure. Thus they are likely to influence the endogeneous factors and men's and women's responsiveness to technological change.

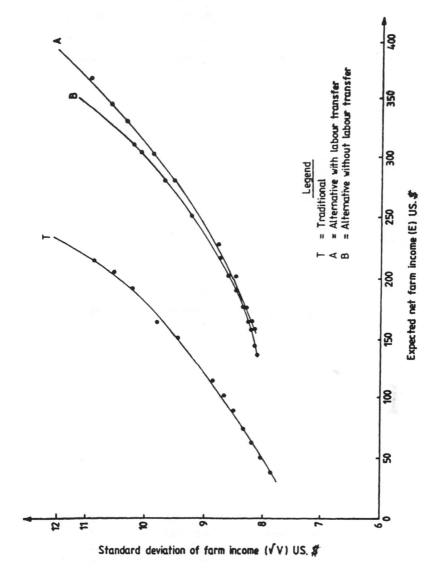

Standard deviation of farm income (√V) US. $

Expected net farm income (E) US. $

Legend

T = Traditional
A = Alternative with labour transfer
B = Alternative without labour transfer

FIGURE 24.2 E, √V Relationships generated by Traditional and Alternative Farm Production Systems

Marketing Infrastructure

Meaningful farming systems research requires an efficient marketing infrastructure that includes chemical inputs, availability of credit and adequate transport of products. Access to technological packages that require the significant use of purchased inputs are restrained by the limited cash resources of small farmers and farmers require credit facilities to adopt them. Many small farmers fail to adopt packages requiring large amounts of purchased inputs; others default on concessionary loans used to purchase inputs. The successful adoption of an innovation in a particular farming system will depend on whether or not such use is rational for an individual farm household in the light of its constraints and its priorities and in terms of the complex set of gender issues related to labor availability and decision-making.

Pricing Policy

In Tanzania, prices for export crops are set essentially by estimating probable future export prices and subtracting marketing costs from probable export earnings to arrive at unit producer price. Prices for the preferred staples, for oil seeds, and for beans are set with allowance for quality differentials. There are economic as well as political implications concerning these prices both at micro and macro levels. The large number of price controlled crops, and the great versatility of agro-economic zones over which they apply, create a high probability of unintended production consequences in some areas for some crops. Macro-level analysis on crop production indicates that Tanzanian peasants are price responsive (Marketing Development Bureau 1976).

At the micro-level, price formulation usually influences the shift of farm resources to crops that enjoy relatively higher prices. This is likely to affect coffee, rice, and maize in the farming system of the Eastern Uluguru Mountains. Obvious problems are likely to appear in a case where the FSR program concentrates on a low priced crop.

Administrative Restrictions and Controls

There is a range of administrative controls that include minimum acreage by-laws, food movement restrictions, etc. An example of such a control is the enforcement of a minimum acreage by-law for coffee, an export crop, in the

FSR project area. Pricing signals, and in some cases agronomic conditions, have worked against the growing of coffee and in favor of the more remunerative and less risky cultivation of pineapples as a potential cash crop. Adherence to the by-laws usually results in planting the required coffee acreage, but subsequently leaving the plot unattended. This is a tremendous waste of resources, particularly of male labor involved in land preparation and of the land itself. Information collected from farming system researchers in the nation might be used to inform policy makers of the wasteful effects of this procedure.

Agricultural Extension Services

The cultural environment in rural Tanzania makes it difficult for male researchers or extensionists to make effective contact with women farmers. Only four out of the 23 participating farmers in the FSR project in the Eastern Uluguru Mountains are women. However, because of women's labor participation and long involvement in agriculture, women have a store of indigenous technical knowledge that researchers could tap.

There is a lack of a well organized and articulated extension systems to take advantage of FSR programs in the country. De Vries (1981) recognized the necessity to link extension programs with FSR efforts. Though extension is coordinated at a national level, extension program within regions could be organized on a zonal or farming systems basis thus creating a good link between extension planning units and FSR units.

CONCLUSION

This FSR study has revealed several important facets that need to be incorporated in FSR and national programs in order to improve research efforts. Since there are separate economies of men and women in rural households, any improvement in the cash crop sector would mean a potential increase in the income of males at the expense of females who usually have to help in the cash crop fields without corresponding income benefits for their labor.

The study has also demonstrated that improvements in labor efficiency during critical labor deficit periods are likely to raise productivity. Farmers need to show flexibility in terms of labor specialization by gender.

It is prudently necessary to appreciate the organizational structures of rural households, if results from FSR are to bring advantage to the community at large. The

358

starting points for such an understanding would seem to be
the pattern of resource ownership and use, the character of
the productive system, the general intra-household
dynamics, and the national policy objectives.

NOTES AND ACKNOWLEDGEMENTS

(1) The FSR Project in Morogoro District, Tanzania is
supported by grants from the International Development
Research Centre (IDRC) of Canada and the Sokoine University
of Agriculture. The project started in 1981 and is expected to wind up in 1988.
(2) QP as developed by Markowitz (1959) has been
suggested as a useful model to consider expected returns
under conditions of risk and uncertainty in farm production. It minimizes directly the variance of the objective
function given a set of linear resource constraints.

The author is grateful to Dr. I. J. Minde, Dr. G.I.
Mlay, Mr. S.A. Shayo, and Mr. N.S. Mdoe of the Department
of Rural Economy, Sokoine University of Agriculture,
Morogoro for making useful suggestions and comments on
earlier versions of this paper.

REFERENCES

Boserup, E.
    1970   Women's Role in Economic Development. NY: St.
        Martin's Press.
De Vries, J.
    1981   Agricultural Extension and Farming Systems
        Research. Presented at a Conference on Farming
        Systems and Farming Systems Research in Tanzania
        held at AICC Arusha, April 14-16.
Due, J. and P. Anandajayasekeram
    1982   Women and Productivity in Two Contrasting
        Farming Areas of Tanzania. University of Dar es
        Salaam, Bean/Cowpean CRSP Project Report #-228,
        July. University of Illinois at Urbana Champaign:
        Department of Agricultural Economics.

Kamuzora, C.L.
    1978   Constraints to Labor Time Availability in
        African Smallholder Agriculture: The Case Study of
        Bukoba District in Tanzania. University of Dar es
        Salaam, ERB Seminar Paper No. 15.
Lamming, G.N.
    1983   Women in Agricultural Cooperatives. WCARRD
        Follow-up Program. Human Resources, Institutional
        and Agrarian Reform Division. Rome: FAO.
Marketing Development Bureau
    1976   Price Policy Recommendations for 1977/78.
        Agricultural price Review. Dar es Salaam, Tanzania,
        July.
Mascarenhas, O. and M. Mbilinyi
    1983   Women in Tanzania: An Analytical Bibliography.
        Stockholm, Sweden: Scandinavian Institute of
        African Studies, Upsala.
Mtoi, M. T.
    1984   Comparative Analysis of Resource Allocation
        Under Risk and Uncertainty in Morogoro, Tanzania:
        Unpublished M.S. Thesis, University of Guelph,
        Canada.
Mtoi, M. T. and A. Tarimo
    1984   Bringing National Policy Objectives to Bear in
        the Course of the Locally Oriented Farming Systems
        Research Process. Paper presented at the Regional
        FSR Workshop held at Egerton College, Kenya, August
        19-23.
Ngalula, T.K.F.
    1977   Women as a Productive Force in the Tanzanian
        Rural Society: A Case Study of Buhungwa Village in
        Mwanza District. Unpublished M.A. thesis,
        University of Dar es Salaam, Tanzania.
Sachs, C.
    1983   The Invisible Farmers: Women in Agricultural
        Production. Totawa, NJ: Rowman and Allan Held.

25

The Impact of Modern Changes in the Chitemene Farming System in the Northern Province of Zambia

Mary N. Tembo and Elizabeth Chola Phiri

PHYSICAL FEATURES

Northern Province is one of the nine provinces of Zambia. It is bordered by the Republic of Malawi to the northeast, the Republic of Zaire in the northwest, Luapula Province to the west, Central Province to the south and Tanzania to the north. The Province has a total area of 157,727 square kilometers of which 96,288 square kilometers are arable land. Although there are a few patches of good soils, the soil is generally leached and acidic because of high rainfall; but with good management many crops can be successfully grown. Substituting a permanent system of arable cropping for the traditional method of shifting cultivation will require liming to correct soil acidity. Heavy use of nitrogen and other fertilizers will be needed to maintain good crop yields. These soils are as suscep-tible to erosion as most other sandveldt soils if soil conservation practices are not used.

The annual rainfall exceeds 1000 millimeters. The cloudy and humid environment makes conditions less suitable for Virginia tobacco and maize, although both crops can be grown. Pineapples and some fruit trees grow well where irrigation can be provided and so do coffee and tea on deep soils. Pastures can be improved by reseeding, fertilizers, and irrigation. Dairy farming is possible on improved pastures near large urban areas.

POPULATION

The population of the area is slightly over 677,894. More than 60 percent of the people practice subsistence farming, and live on cassava, millet, beans, and local vegetables as their staple foods. People living along the

shores of Lakes Bangweulu, Mweruwantipa, and Tanganyika
also fish for a living (Department of Agriculture 1983:83).
In Zambia, particularly in the study area, females predominate in the population (53 percent female compared to 47
percent male), and a marked proportion of households are
headed by women (IRDP 1983). This situation has important
implications for available household labor resources and
decision-making processes at the household level. The
predominance of women in the population is not reflected in
rural development program membership. For instance only 24
percent of the participants in the Integrated Rural Development Programs were women, although these programs have
recognized the predominance of women in the Northern
Province (ASSP 1983).

TRADITIONAL AGRICULTURE IN NORTHERN ZAMBIA

Chitemene shifting cultivation is the traditional
agricultural system of the Bemba in Northern Zambia. In
this system, crops are grown in the ash that remains after
burning stacks of branches collected from trees growing in
an area five to eight times the size of the cultivated
garden. Finger millet, groundnuts, and sorghum are planted
for three years consecutively. After some years of cultivation, the plot is left fallow until the trees are grown
again. Chitemene shifting cultivation makes available the
nutrients that were previously stored in the vegetation and
utilizes the natural forest or woodland as the agricultural
fallow crop.

Traditionally both men and women performed specific
tasks within the agricultural cycle. Men lopped or felled
trees and turned fresh soils. Together with the women they
scared birds, harvested, and built granaries. Women turned
the soils, planted, weeded, scared birds, harvested, and
carried the produce to the villages for storage. They also
cultivated additional crops such as pumpkins, sweet
potatoes, groundnuts, and other nuts. Clearly, women performed more tasks than men in the agricultural sector.
Men, however, were more active in hunting, fishing, smelt-
ing, and inter-regional trade. Although single women in
these societies had access to land independently, brothers
or uncles could control the labor of their children
(Muntemba 1977). Due to the low level of technology (axe
and hoe), fields were limited in size. Shortages of staple
grains were offset by a variety of other foods such as
nuts, pumpkins, and sweet potatoes. If these, too, were in
short supply, women gathered wild relishes such as caterpillars, mushrooms, vegetables, and fruits when available.

THE COLONIAL PERIOD

During the colonial period, agricultural policies largely promoted the interests of the white settlers who occupied the most accessible and best agricultural lands. The indigenous population was seen essentially as a source of cheap labor for the copper industry and associated construction activity. A native tax, payable in cash, forced the Africans to take up employment in the mines. Africans lacked the capital to take up farming oriented to supply food to the growing urban and mining sector. That was left entirely to white settler farmers. The marketing network for African produce and the rural road system were deliberately left undeveloped (ILO 1979). Having nothing else to sell, Africans were forced to sell their labor (Sklar 1975).

As copper mining intensified in the region stretching from South Africa to the Congo (Zaire), men from the Northern, North-Western, Western, and parts of Eastern Provinces migrated in great numbers. By the 1930s, some villages could not practice the traditional land rotation system because of the absence of young men. Their absence affected the performance of other tasks such as felling or lopping trees for fuel and ash fertilization and scaring birds. Women who had started to perform most agricultural tasks, could not fell or lop trees. Less fertilization of soil occurred. Those who tried to keep up the chitemene system began cutting down entire trees because it was easier for women and children to do, but it inhibited regrowth of the trees. This had serious consequences for the fertility of the soil.

Until 1945, the agricultural rural policy consisted almost entirely of encouraging the production of cash crops such as maize, cotton, tobacco, and groundnuts. Financial, technical, and marketing assistance was limited to the South, Central, and Eastern parts of the country that were the major agricultural areas (Muntemba 1977).

In the 1950s there was an outbreak of locusts in the Northern Province and most vegetation was destroyed. Many people had no food that year. At this time cassava was introduced in the area as a crop to turn to in times of emergency. Cassava is also less labor intensive and less nutritious.

From 1960, the government wished to keep the non-producing region as a labor reservoir. For these reasons, male out-migration grew while visits by migrant workers to their homes became few and far between. Many workers were becoming permanently urbanized (Muntemba

1977). Thus, although land remained "abundant" in the Northern Province, problems connected with labor acceler-ated. In general, agricultural improvements were oriented almost entirely, though unintentionally, towards men.

By tradition, women had always had access to land. With the opening of large farms to the production of cash crops, land was allocated to male heads of households. The result was that men could practice modern methods of agri-culture and since fertile land was very scarce, husbands became reluctant to allocate plots to women for traditional crop production apart from the family plot (Muntemba 1979). Female-headed households were adversely affected.

POST-INDEPENDENCE PERIOD

Initially, the independent government was committed to "rural development" through the money economy (Klepper 1979). The government also hoped to increase agriculture's contribution to the gross national product, (Muntemba 1977). From the early 1970s, world copper prices started to slide. Agriculture, therefore, was seen as a means of earning foreign exchange. For these reasons, peasants were exhorted to increase their cash crop production, particu-larly of cotton, sunflower, and tobacco. They were also urged to grow maize and other food crops to feed the nation and to export and to increase their groundnut production for oil extraction. To aid them in increasing productiv-ity, the state heightened its support and improved the agricultural prices of cash crops. More male peasants were given training in efficient methods of production, but the women were not.

1981 TO THE PRESENT

Women's capacity to produce and to be aided in agricultural production has continued to deteriorate. The 1980 census shows that population along the line of rail provinces (Central, Copperbelt, Southern, and Lusaka) has increased, mainly at the expense of Northern and Luapula Provinces. Migration has continued even in these years, further depriving Northern Province of its much needed labor force.

The use of oxen, where available in the North, reduces the number of hours needed for cultivation, but does not reduce the time needed for weeding, sowing, or harvesting, that are women's tasks (Safilios-Rothschild 1985). A study undertaken in Mpika, Northern Province, shows that in a random sample of 112 farm households, women spend an

average of 6.6 hours per day in agriculture during the
farming season while men spend 5.7 hours per day. In
addition, women spend an average of 4.1 hours per day in
household activities while men spend only 0.4 hours. Thus
during the farming season women contribute 53 percent of
total hours of labor in agriculture while men contribute 47
percent (Due and Mudenda 1984). In another study conducted
in five villages of varying socio-economic characteristics
in Mubanga, Chinsali, it was found that 26 percent of the
women and only 15.7 percent of the men were working
full-time on the farm. Fourteen percent of the men and
10.5 percent of the women worked part-time on the farm.
Some of the men who work on the farm only part-time spend
most of their time fishing (49 percent), while the rest
attend to other chores of village life such as looking
after cattle and repairing thatched roofs. Furthermore,
women are engaged in all agricultural tasks and produce all
crops (sorghum, maize, millets, groundnuts, sweet
potatoes). Very few tasks such as weeding and harvesting
are shared with men (IRDP 1983). Therefore labor saving
technology for women is necessary to improve their
participation in both traditional and cash crop production.

The use of outside labor is linked directly to farm
sales and eventually to farm income. Thus, women's agricultural workload is heavier among low-income farm households in which labor cannot be hired (Allen 1984). The
rainy season, too, contributes to heavy demands for labor.
There is only one rainy season in Zambia in which a farmer
has to grow all necessary crops in a very short time.
Weeding is often the greatest constraint to farm expansion.

Even in remote areas where people can be said to live
entirely at subsistence level, villagers receive money
occasionally, usually remittances from relatives, or from
piece work in other people's fields, or sales of their own
agricultural produce. Most of this money is spent quickly
to buy soap, sugar, paraffin, and candles, or it is used to
buy school uniforms. Villagers also buy home brewed beer
and millet from each other or buy fish from passing
traders.

The main obstacles to regular sale and exchange of food
are the lack of agricultural surplus production and the
fact that most people do not know how to save or invest
money. Villagers sell their stocks of millet, groundnuts,
and beans for a short-term financial gain, but incur a
considerable nutritional loss. Half a basket of unshelled
groundnuts — enough to supply a household of six people
for two to three weeks — brings K2 or K2.5 (K5.86 = US
$1). The cash economy can seriously affect food

self-sufficiency when men neglect agriculture, especially the cultivation of food crops, in favor of the possibility of earning money. Since these people are not used to the money economy, the cash earned even from farming is not properly utilized for the benefit of the family.

IMPACT ON NUTRITION

Traditionally, most constituents essential to a balanced diet have been present in the Bemba environment, with the important exceptions of sufficient animal protein, fat, and salt. There is an absolute shortage of millet reckoned either by the amount available annually or from the obvious seasonal scarcity, and the peoples' reliance on this one staple makes this a dangerous lack. The most valuable supplementary foods such as the pulses only last part of the year, and the supply of green vegetables is limited to a few months. Milk is never obtainable. Except in the case of meat and milk, these deficiencies appear to be due to difficulties of production, storage and exchange rather than to any particular environmental constraints (Richards 1958).

Today the situation is much the same, except that salt is now widely available and more importantly, cassava has largely replaced millet as a staple food. Millet is preferred to cassava because of its "lasting power" (it takes longer before you feel hungry again); but the production of millet is constrained by labor shortages mentioned earlier. Since women now have the easily produced cassava as a replacement staple food, they tend to use millet more for brewing beer that can be sold to gain cash.

A study undertaken in Kasama estimates that most people in this area use at least 70 percent cassava meal to make nshima, the dough-like staple, with millet meal making up the other 30 percent. From January to May they may have to rely on cassava alone. Nutritionally this has affected children who benefited from the more nutritious nshima made out of millet.

The amount of cassava flour used in the preparation of nshima varies, but on the average it is about 400 grams/ day/person. Such cassava nshima gives a nutrient supply of 1,368 calories, 2 milligrams protein, 220 milligrams calcium, 8.0 milligrams iron, 0.16 milligrams thiamion, no vitamin A, no vitamin C, 0.16 milligrams riboflavin, and 3.2 milligrams nicotinic acid. (The requirements of an adult man moderately active according to the Food and Agriculture Organization (FAO) requirement table (1978) is 3,000 calories, 37 grams of protein, 5.9 milligrams of iron,

400–500 milligrams of calcium, 1.2 milligrams of thiamin, 1.8 milligrams of riboflavin, 19.8 milligrams of nicotinic acid, 30 milligrams of vitamin C and 750 milligrams of vitamin A). The same amount (400 grams) of millet meal made into nshima will give a nutrient supply of 1,392 calories, 26.4 grams protein, 1060 milligrams of calcium, 16.0 milligrams of iron, no vitamin A, no vitamin C, 1.36 milligrams thiamin, 0.64 milligrams of riboflavin, 2.8 milligrams of nicotinic acid.

Although maize is probably more common than in Audrey Richard's day, it is mainly eaten on the cob during February to April and is still not an important crop. The taste of maize is appreciated for nshima, but maize is considered to have less "lasting power" than millet. Moreover, the pounding of maize means hard labor for women in areas where there is no hammer mill or where they have to walk long distances to the hammer mill. Though maize production has increased in the area, most of it is sold. Few households produce maize primarily for home consumption. Sweet potatoes along with cassava, also were grown widely during the locust raids in response to direct government pressure (Richards 1939). Today, sweet potatoes are found everywhere but in insignificant quantities. They are available from April to August and serve often as a quick meal or snack (at breakfast or lunch time). Lack of suitable soil and the labor to protect the crop from birds has resulted in the disappearance of sorghum.

Animal protein sources are not plentiful in Northern Province. There is hardly any game left and hunting is illegal without a license. Sometimes people eat caterpillars and insects or on special occasions a chicken. According to the Food Consumption Survey (FCS), all these animal protein sources (game included) together constitute 5 kilograms/person/year for Northern Province; this averages 2–3 grams of animal protein/person/day.

Dried fish is a more significant source of animal protein, especially in the dry season. This fish is eaten once or twice a week, supplying an average of 10–15 grams of animal protein/person/day. Fresh fish are available in the Chambeshi valley and along the shores of the lakes, although fish have been destroyed due to intensive fishing with poison. Groundnuts and beans, both women's crops are the main vegetable protein sources giving some five to ten grams of vegetables protein/person/day, according to the Food Consumption Survey (FCS) data for Northern Province.

Leafy vegetables are the most common relish, besides mushrooms and pumpkins (also women's crops) that are

TABLE 25.1

ADEQUATE NUTRITIONAL STATUS OF CHILDREN

Farmer Category	Nutritional Status Adequate (Percent of Children)	Degree of Malnutrition (Percent of Children)		
		Mild	Moderate	Severe
Subsistence (N = 46)	70	26	2	2
Emergent (N = 73)	52	41	6	1
Small-Scale Commercial (N = 30)	50	44	6	0

Source: National Food and Nutritional Commission of Zambia.

nutritionally less relevant but important. These are usually dried and eaten during the lean months.

Fruit trees are fairly common, especially mangoes and bananas, but fruits are mainly considered to be food for children or for women. They are not eaten in important quantities except during the mango season (December-January). In the Chambeshi valley, fruits such as guavas and pawpaws are available throughout the year.

The emphasis on cash crops has also played a role in lowering the nutrition status of the area. Until recently, agriculture research in the Ministry of Agriculture and Water Development (MAWD) has focused on cash crops. Pricing policies and transportation policies of the marketing agencies have also favored cash crops. This has led to the expansion of cash crop production at the expense of traditional crops.

In 1985, the National Food and Nutrition Commission carried out a study in Mpika area to assess the Nutritional impact of the Integrated Rural Development Programme activities. The weight/height indicator (a current nutritional status indicator) showed that throughout the program area (including Chinsali) as farmers move from subsistence farming into the market economy the nutritional status of the children (6–60 months) in the households declines (Table 25.1).

Women's Access to Agricultural Extension Services

Although the expressed objectives of agricultural
extension are the reduction of poverty and the increase of
agricultural output, to date the focus has been primarily
on the latter goal and hence services have been directed to
well-off farmers who can afford modern agricultural inputs.
Advice has concentrated on maize, sunflower, tobacco, and
cotton, together with a package of cultivation methods and
inputs designed to encourage an intensive approach to production (Muntemba 1977).

Little attention has been paid to the traditional crops
such as cassava, sorghum, and millet, that are grown by
women. Thus, women farmers have had little access to
agricultural extension staff because they have been viewed
as subsistence farmers despite the fact that a considerable
number of women cultivate food as well as cash crops and
most sell some of the surplus. In some areas, there are
cultural barriers to women's access to the largely male
agricultural extension staff. In 1981, the available data
showed that there were 1,449 agricultural extension staff
members at the provincial and district level, 9.6 percent
of whom were women (Safilios-Rothschild 1985).

Although the dissemination of agricultural information
and improved agricultural practices is very important for
both wives and female heads of household, a major effort
should be made first to increase the number of women agricultural extension workers in districts in which the incidence of female heads of households is high as can be seen
in Table 25.2.

TABLE 25.2

PERCENTAGE OF FEMALE-HEADED HOUSEHOLDS BY PROVINCE

Province	Percentage of Female-Headed Households
Chilubi	50.5
Chinsali	37.4
Kasama	36.5
Luwingu	36.4
Isoka	35.4
Mporokoso	35.4

Source: 1980 Population Census.

Women's access to agricultural training centers is limited as long as training is offered at residential centers. Women farmers who are heads of household cannot take several weeks off to attend classes offered away from their village. A study in Northern Province reports that only 5 percent of the women had attended farmer training courses (Gaobepe and Mwenda 1980). In a district like Samfya in which 41 percent of farm households are headed by women, no women were reported to have attended the courses offered at the Samfya Farmer's Training Centre throughout 1983.

CONCLUSION

The women in the Northern Province have suffered from national policies that have been biased towards the urban industrial sector. The rural sector of the Province has been seen as the provider of labor for the mines and of food for the workers. The effects have been to deprive agriculture of able-bodied men and to spur the growth of female-headed households. Governmental encouragement of cash cropping has introduced the necessary infrastructure, market, extension services and credit, but closer examination shows that these have been directed to men and away from women who are primarily responsible for subsistence food production. Finally, agricultural pricing policies directed to providing cheap food for the urban sector has deprived the rural areas of potential increased income from cash crops.

The available data indicate that women in Zambia, particularily in the Northern Province, play an important role in food production. That role varies according to their marital status, the type of prevailing agriculture, and their non-farm activities. Wives have different constraints and needs than female heads of households and require different strategies and approaches in order to effectively participate in and benefit from agricultural programs. Furthermore "de facto" female-headed households created through male migration may have very different resources, constraints, and needs than "de jure" female-headed households created through death or divorce. It is important, therefore, that agricultural researchers collect more data about the different types of female-headed households and test new strategies and approaches to alleviate existing constraints and develop the potential of women as a force in food production.

Closer inspection shows that training for women is not geared towards production. Home economics and nutrition are important subjects, but they are not of immediate use in the absence of adequate food supplies. Food is a major item of household expenditure despite the general productivity of these households. Presumably it is more profitable to sell the cash crops and buy food stuffs. If favorable prices were offered for the crops that women grow, the continued food shortages in Zambia might be eliminated.

Such policies would need to be supported by the extension services. The household production system, including subsistence should be viewed as a whole, and women given much more serious attention as subsistence producers. If the goals of development are increased productivity, more equitable distribution of resources, and more control by individuals over their own lives, then surely rural development must feature women's active participation rather than patronizing them. In general, women have shown their desire to be involved in development programs in many ways, and have in fact formed the requisite infrastructure for development. What they lack are access to credit and technical know-how. They are also too over-burdened with work to become actively involved in new programs separate from their traditional tasks.

The changes necessary to give women their rightful place in the food production system may seem revolutionary, yet in reality they merely acknowledge the actual state of affairs in food production in which women play a crucial role.

REFERENCES

ASSP-Agriculture Sector Support Programme
    1983   An Evaluation. Lusaka, Zambia: Government Printing Office.
Allen, J.M.S
    1984 Baseline Survey Report (1980-82) Chinsali. Integrated Rural Development Programme-Chief Mubanza's Area, Chinsali District Council, mimeographed.
Department of Agriculture
    1983   Northern Province Annual Report. Lusaka, Zambia: Government Printing Office.
Due, M.J. and Mudenda, T.
    1984   Women's Contributions Made Visible: of Farm and Market Women to Farming Systems and Household Incomes in Zambia. Urbana, IL: University of Illinois Agricultural Economics Staff Paper.
FAO-Food and Agriculture Organization
    1978 Regional Food Plan for Africa. Report of the 10th FAO Regional Conference for Africa. Rome: FAO.

Gaobepe, M.G. and Mwenda, A.
    1980   The Report on the Situation and Needs of Food
        Supplies: Women in Zambia. Lusaka, Zambia: FAO.
IRDP—Integrated Rural Development Programme
    1983   Serenje—Mpika—Chinsali Project. Occasional
        Paper No. 5
ILO—International Labor Organization
    1979   Activities in African Countries of Special
        Interest to Women Workers. ECA Regional Preparatory
        Meeting for the UNDP Conference on Women. Lusaka,
        Zambia: ILO.

Klepper, R.
    1979   Zambian Agricultural Structure and Performance.
        In B. Turok, ed., Development in Zambia. London:
        Zed Press.

Muntemba, M.S.
    1977   Women as Food Producers and Suppliers in the
        Twentieth Century: The Case of Zambia. Development
        Dialogue 1-2:29–50.
    1979   Expectations Unfulfilled: The Underdevelopment
        of Rural Zambia, Kabwe District 1960–1970. In
        Journal of Southern African Studies 5:1:59–85.

Richards, A.I.
    1939   Land, Labour and Diet in Northern Rhodesia.
        Oxford: Oxford University Press.
    1958   A Changing Pattern of Agriculture in East
        Africa. The Bemba of Northern Province.
        Geographical Journal.

Safilios-Rothschild, C.
    1985   The Implication of the Roles of Women in
        Agriculture in Zambia. Lusaka, Zambia: Government
        Printing Office.

Sklar, R.I.
    1975   Corporate Power in an African State.
        Berkely, CA: University of California Press.

Zambia, Republic of
    1980   Census of Population and Housing. Lusaka,
        Zambia: Government Printing Office.

26
A Diagnostic Survey of Female-Headed Households in the Central Province of Zambia

Robert E. Hudgens

An agricultural census by the Central Statistics Office which was conducted during the 1982/83 cropping season found that female headed households represented 15.3 percent of the traditional and small scale commercial farms in four districts of the Central Province of Zambia. Data from the as yet unpublished population census of 1980 revealed that the percentage of female-headed households in the Central Province varies greatly among districts and wards, and often exceeds 30 percent of the total households. While these figures are lower than those for other parts of the country, they have significant implications for agricultural development in the region. The importance of women in agriculture in Zambia is exemplified by the 1983 study in three provinces (including the Central Province) that showed that women contribute 53 percent of the total hours of labor in agriculture while men contribute 47 percent (Due and Mudenda 1983). The agricultural workload of women is substantially greater on low-income farms and in female-headed households where the resources for hiring labor are limited (Allen 1984).

The initial diagnostic studies of the Farming Systems Research and Extension Team (FSR/E), which will be referred to hereafter as the Central Province Adaptive Research Planning Team (ARPT), did not give special attention to the role of women in agricultural production, or more specifically to the situation of female-headed households. Recommendation Domains were based on the source of draft power, the type of starch staple, and the presence of cattle and cash crops in predominantly male-headed households.

Research priorities were subsequently established for each Recommendation Domain, on-farm experiments were conducted, and extension activities were initiated within the context of the descriptions of the farming systems that

374

evolved from informal and formal surveys. Special attention is now being given to female-headed households, which are isolated from the institutional infrastructure for agricultural credit and extension services, and have thus far been overlooked. These households may represent a specific target group within a larger group of traditional and small-scale commercial farmers. They comprise a subcategory like male hand hoe cultivators, subsistence farmers, and others for which we have very little information.

Consequently, a formal survey was undertaken in three Recommendation Domains in the Central Province to generate information on the resource base, constraints, cropping pattern, input usage, labor use pattern, income sources, and crop husbandry practices for female-headed households. The intention was to accumulate a sufficient information base to allow comparisons between female-headed households and male-headed households (as characterized in former surveys in each Recommendation Domain). This would then permit ARPT to develop a regional policy and relevant strategies to improve the agricultural productivity in female-headed households.

METHODOLOGY

This study was based on a five page questionnaire, similiar in many ways to that used in the ARPT formal surveys of male-headed households. Seven subheadings (general, maize husbandry, cropping pattern, input usage, labor, cash sources, and decision-making) contained questions about family composition, marital status, off-farm employment, kinship relations to the village headman, crop acreage, oxen ownership, livestock, labor hiring, collection of water and firewood, food preparation, labor calendar, use of fertilizer, yields, land clearing, and information sources. The questionnaire was coded in advance and revised after being pre-tested on ten farms. Each interview required an average of one hour.

The sampling frame was patterned after that used in the Early Warning and Crop Forecast Survey (1982-1983), the Early Warning and Agricultural Surveys (1983-1984 and 1984-1985) and the Crop Forecasting Survey (1985-1986), all done by the Central Statistics Office. A list of female-headed households was compiled from the above mentioned surveys for "Standard Enumeration Areas" and "Census Supervisory Areas" near agricultural camps where ARPT was working in the Central Province. The names of female farmers in nearby villages, that had been sampled in the

annual agricultural surveys, were then given to the ARP
enumerators who lived in those areas.

Four enumerators were used, and 135 female-headed
households were sampled in the vicinity of six "agricultural camps" (Rural Extension Service Offices). An attempt
was made to recruit female enumerators through the Ministry
of Agriculture and Water Development's (MAWD) main office
in Lusaka and through the University of Zambia's Rural
Development Studies Bureau, but the logistics of supplying
transport and lodging in the field, as well as separation
from family and jobs for extended periods, limited the
feasibility of that approach. Consequently, ARPT "Trials
Assistants", who were originally seconded from the Extension Branch of MAWD to supervise ARPT on-farm experiments
and demonstrations, were used as enumerators. This was not
their first experience in that capacity. Two had been
involved in formal surveys of male-headed households, and a
third had conducted a detailed labor use study, that
involved data collection on the same farms twice a week.

Since the Trials Assistants had lived and worked in or
near the Agricultural Camps for between one and four years,
they were known to local farmers and extension camp staff
and in turn were familiar with the immediate areas.
Although the enumerators were male government employees,
there were no indications that this influenced the information received. If anything, their association with ARPT,
which has established credibility through farmer field days
and other local extension activities, worked to their
advantage. The enumerators were instructed to seek out the
individuals on the lists and then to request assistance
from those interviewed in locating other female-headed
households in the vicinity. All interviews were conducted
in the local dialects.

The data in the survey matrix was tabulated and analyzed on an Apple IIe microcomputer using MSTAT statistical
analysis software. The data were sorted and analyzed in
five ways: (1) by oxen ownership; (2) by presence of adult
males on the farm; (3) by Recommendation Domain; (4) by the
amount of cultivated acreage per farm; and (5) by the
number of maize fields planted. Although it was hoped that
a minimum of 50 questionnaires could be completed for each
of three recommendation domains, delays that diverted the
attention of the Trials Assistants to the ARPT on-farm
experiments and demonstrations were experienced due to the
distances involved, a death in the family of one of the
enumerators, and the onset of the rainy season. Therefore, this report is based on 88 questionnaires from

Traditional Recommendation Domain (TRD) #3, 23
questionnaires from TRD#2, and 24 questionnaires from
TRD#5. As a result, the major focus for comparing female-
headed households with male-headed households will be data
from TRD#3, that comprises the largest information base on
female-headed households. Information from the other two
recommendation domains will be used in a comparative evaluation of female-headed households over three farming
systems.

CHARACTERISTICS OF THE CENTRAL PROVINCE OF ZAMBIA

The location of the Central Province in relation to the
urban markets in Lusaka and the Copperbelt has given it a
comparative advantage for commercial agricultural production, and in the last decade commercialization in the small
farm sector has accelerated. As a result, the Central Province ranks among the most agriculturally productive regions in the country in terms of total volume of maize produced and marketed (Central Statistics Office 1981).
Although maize is the dominant starch staple and cash
crop in Zambia, the Central Province also has large
acreages of sunflower, groundnuts, sorghum, and millet.
The province, at 13—15 degrees South Latitude, has a low
rural population density of about three persons per square
kilometer, plateau characteristics with an elevation
between 1,065—1,220 meters above sea level, and a rainfall
period from November to April, averaging 800—1000
millimeters. Most of the area under cultivation is flat or
gently rolling with sandy ("Sandveldt") soils. The
exception are two small pockets of heavier textured soils
and low lying drainage areas (dambos). Dambos are used for
dry season grazing. The Central Province is traversed by a
railway and a paved highway system leading from Lusaka to
the Copperbelt and beyond to Tanzania and Zaire. The input
supply and crop marketing infrastructure has undergone an
institutional evolution of late from the parastatal
National Agricultural Marketing Board (NAMBOARD) to the
Central Province Cooperative Marketing Union (CPCMU), and
most recently, back to NAMBOARD. The prices of agricultural inputs and produce are controlled by the government.

FEMALE-HEADED HOUSEHOLDS IN THE PROVINCE

The survey results indicated that female-headed households averaged 4.4 members, and that family composition was
skewed toward "adult" males and females older than 15 years

of age. Of the 135 farms sampled, 43.7 percent reported having family members in off-farm wage employment, much of which however was seasonal or piecework. Divorcees comprised the largest single group of female farmers, followed by widows. The six categories of marital status: widowed, divorced, single with children, single with no children, married with husband working elsewhere, and autonomous polygynous, (one of several wives with her own field plots) are based on those defined by Safilios-Rothschild in her 1985 review of the role of women in agriculture in Zambia. In Zambia, a "village" is defined by the Central Statistics Office as any group of houses. A question regarding kinship ties with the "head" of the village showed that many (41 percent) of the female farmers surveyed headed such villages themselves. Others lived in villages headed by their brother (16 percent), uncle (14 percent), mother (14 percent), grandfather (nine percent), or son (six percent). Among the Bembas, the dominant ethnic group in the Central Province of Zambia, inheritance (women's access to land) is matrilineal (Safilios-Rothschild 1985).

An average of 18 percent of the female-headed households in the total sample owned oxen, and 17 percent owned ox-drawn equipment, usually a plow. The histogram from a frequency distribution of responses showed that those who own oxen normally own two. Overall, 45 percent of the sample reported hiring oxen, primarily from non-relatives, but a smaller percentage (11 percent) hired ox-drawn equipment. Nearly all the farmers owned other livestock, predominantly chickens. Ducks and pigeons were also mentioned.

Given the low population density in rural Zambia, land is readily available and used in a shifting cultivation pattern. Local chiefs apparently allocate land according to the farmer's ability to cultivate it (i.e. available draft-power). Nine percent of those sampled had no idea of the total farm area available to them, implying that land is not limiting and not allocated in fixed quantities. Farm size averaged 91.7 acres, of which an average of 4.3 acres was cultivated (5 percent of the total acreage available). Most of the farms had been in operation for the last four to six years. Maize occupied the largest proportion (34 percent) of the cropped land, followed by sorghum on 24 percent of the cultivated area (Table 26.1). The average maize acreage was 3.06 acres per farm, with 65 percent of the maize growers planting hybrid maize varieties. However, some of this must have been second generation seed harvested from last year's production fields, because only 52 percent reported buying seed from the Central Province Cooperative Marketing Union.

The planting period for maize was spread evenly over the two month period from early November through late December. Hired oxen accounted for seedbed preparation in 43 percent of the farms, while 26 percent of the farmers prepared maize fields using a hand hoe. Maize was generally planted by hand with only 15 percent of the farmers reporting the practice of dribbling maize seed behind the ox plow. Of the total, 78 percent of the farmers applied fertilizer on their maize crop. Contrary to extension service recommendations, the basal fertilizer dressing was not applied at planting, but rather when the maize plants were an average of 22 centimeters in height.

The average height of maize at top dressing application (75 centimeters) also indicated that top dressings were made somewhat later than is recommended. Rates of application averaged 162.7 kilogram/hectare for basal and 193.0 kilograms/hectare for top dressing. The average distance

---

TABLE 26.1

MAIN CHARACTERISTICS OF MALE AND FEMALE HEADED
HOUSEHOLDS IN TRD #3 (MKUSHI DISTRICT)

Item	Male H.H.	Female H.H.
Family Size (#/Household)	6.5	3.7
Area Cultivated (Acres)	16.0	3.1
Maize Acreage (Acres)	5.0	2.6
Percentage Cultivated (%)	18.8	5.0
Own Oxen (%)	43.9	15.9
Own Ox Implements (%)	43.9	14.8
Hire Oxen (%)	27.0	38.6
Hire Labor (%)	50.0	48.9
Use Hired Labor for Plowing (%)*	5.0	29.6
Purchase Maize Seed (%)	81.0	33.0
Use Fertilizer on Maize (%)	88.0	56.8
Cash Source	Maize/Beer	Chickens/Mushrooms
Cash Crops	Cotton/Sunflower	Squash/Pumpkins/ Beans

Note: *Refers to subgroup of farmers who hire labor.

to the source of fertilizer was 4.9 kilometers. Maize
fields were weeded once in January using a hand hoe;
however a histogram showed that weeding often extended into
February suggesting that some or all of the maize fields
were weeded late. Yields, calculated by dividing the
reported maize acreage by the number of bags produced,
averaged 1,237 kilograms/hectare, but in some fields the
maize was eaten at a green stage as "corn on the cob".

Although the majority of the farmers interviewed had
access to cleared land, 64 percent clear land annually,
usually from March to May. Farmers reported the greatest
labor demand to be from October through January. Labor was
needed during this period for land preparation, planting,
and weeding most of their staple food crops including both
maize and sorghum. A second period of labor constraint was
reported in June and July for harvesting all crops and for
scaring birds in sorghum. Labor hiring was found to be a
common practice (47 percent hired labor) in December for
land preparation, in January for maize weeding, and in July
for maize harvest. Cash was used for paying hired labor in
60 percent of the cases, but payment in munkoyo (non-
alcoholic brew), mealie meal, clothes, and crop produce
were also recorded. Of the respondents with husbands, 12
percent of the husbands returned to help in land preparation and 19 percent returned to help with the harvest.

Maize was generally the first crop planted on newly
cleared land, because it is an important cash crop and
benefits from the higher fertility found on new land.
Maize nshima (a dough-like staple) was also preferred over
sorghum nshima for eating. Nevertheless, almost a third of
the total survey sample reported planting sorghum on new
land because it is less risky. Children helped in food
preparation in 63 percent of the households, but food
preparation still required an average of three hours per
day. Also with the help of her children, women spent
another 1.8 hours per day in collecting firewood and 0.5
hours collecting water. On the average, the water source
was 1.4 kilometers from the farmstead.

The sale of livestock, primarily chickens, provided the
greatest source of family income followed closely by the
sale of crops and munkoyo (Table 26.2). Groundnuts and
beans were listed most frequently of the crops sold, but
squashes, pumpkins, sweet potatoes, cassava, and vegetables
also were mentioned. Only two percent of the respondents
reported income from the sale of maize. Approximately one
fourth of the farmers received remittances from relatives.
Decisions as to what to sell and what area to allot to each
crop were generally made by the women, but in approximately

TABLE 26.2

COMPARISON OF FEMALE-HEADED HOUSEHOLDS IN TRD#3 (MKUSHI DISTRICT) BY
PRESENCE OF MALES IN HOUSEHOLD, OXEN OWNERSHIP, AND SIZE OF ACREAGE CULTIVATED

Item	Presence of Men		Oxen Ownership		Size of Acreage	
	Without Men	With Men	Without Oxen	Oxen Owners	<3.5 Acres	>3.5 Acres
Marital Status Percentages:						
Divorced	45.0	29.0				
Widowed	33.0	22.0				
Married w/ Husband Elsewhere	10.0	27.0				
Family Size (#/Household)	2.0	5.5	3.5	3.5	3.8	3.5
Area Cultivated (Acres)	3.0	3.2	2.8	4.1	2.0	5.5
Maize Acreage (Acres)	2.8	2.3	2.5	2.7	1.8	3.7
Percentage of Households That:						
Own Oxen	11.9	22.0	—	—	25.0	63.0
Hire Oxen	38.0	35.0	39.2	8.0	14.3	22.2
Borrow Oxen	19.0	13.0	—	—	11.0	8.0
Hire Labor	43.0	51.0	41.9	57.0	38.0	67.0
Purchase Maize Seed	41.9	41.0	29.7	36.4	34.0	52.0
Use Fertilizer	67.0	68.0	62.2	71.0	57.0	89.0
Listen to Radio	39.0	46.0	51.4	50.0	40.0	46.0
Cash Source	Chickens, Caterpillars, Mushrooms, Munkoyo	Chickens, Crops, Mushrooms, Munkoyo	Chickens, Caterpillars	Chickens, Crops	Chickens, Caterpillars	Chickens, Crops
Cash Crops	Squash, Beans, Pumpkins	Squash, Beans Pumpkins	Squash, Beans Pumpkins	Squash, Beans Pumpkins	Squash, Beans Pumpkins	Squash, Beans, Pumpkins

15 percent of the cases men (sons, brothers, uncles, husbands) helped decide. Of those surveyed, 37 percent knew of extension demonstrations in their area and 30 percent listened to government radio broadcasts.

COMPARISON OF MALE AND FEMALE-HEADED HOUSEHOLDS IN TRD#3

Certain similarities are to be expected in male and female-headed households existing within the same agro-ecological zone and institutional infrastructure of input supply and marketing. Maize was found to be the dominant starch staple food crop regardless of gender. Labor scarcity, particularly during the critical period of land preparation and planting, was the single most limiting production constraint on all farms. Labor hiring for weeding and harvesting maize was also widely practiced.

Given the small percentage of total farm area cultivated, both types of farmers followed an "extensive margin" production strategy to utilize their limited labor resources to plant the largest possible acreage of maize in the narrow planting period at the beginning of the rainy season. While the quantity of available land is not limiting, maize yields suffer in both cases due to the late planting, late weeding, and late fertilizer application that results from extensive management.

On both male and female-headed farms, maize acreage is a function of power source. Various regression analyses on the tabulated data from the survey of female-headed households showed that a greater number of adult males, adult females, and oxen on the farms was significantly correlated to larger acreages planted in maize. Similar relationships were established in the formal survey of male-headed households. However, sorghum is also important, because it is less risky than maize and because sorghum beer can be used as a payment in kind for hiring labor. Unfortunately, maize and sorghum compete for labor at the same time during the growing season.

In spite of the similarities between male-and female-headed households, there are certain discrepancies. For example, female farmers reported being busier during the months of May-July, that coincides with bird scaring in sorghum, maize harvest, and maize shelling (Figure 26.1). The degree of commercialization and oxen ownership also differ significantly between the two groups (Table 26.2). The fact that fewer female farmers owned oxen than their male counterparts meant that a larger percentage of women were forced to hire oxen and labor for land preparation. The cost of hiring oxen restricted the acreage cultivated

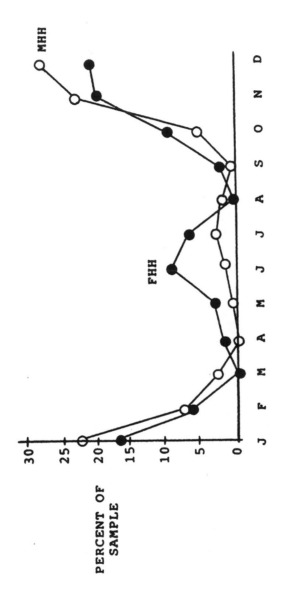

FIGURE 26.1

BUSIEST MONTHS FOR MALE AND
FEMALE HEADED HOUSEHOLDS

on female-headed households. Cash constraints were also
apparent on female-headed households in the lower percentages buying seed and fertilizer for their maize crops.
    However, the most significant difference between the
two groups are the sources of cash. The fact that crop and
beer sales figured predominantly in male-headed households,
whereas the sale of chickens and mushrooms were the main
cash source in female-headed households supports the argument that crop production on female-headed households is
primarily of a subsistence nature. None of the female
farmers interviewed mentioned making charcoal during land-
clearing, which is a secondary income source in many
male-headed households.
    To substantiate the hypothesis of labor scarcity the
data from female-headed households in TRD#3 was sorted and
analyzed by presence or absence of adult males and oxen, as
well as by the total land area cultivated. Although total
acreage was not found to be significantly higher on farms
where adult males were present, oxen ownership, labor
hiring, and the importance of crop sales increased. Oxen
owners also purchased more hybrid seed and fertilizer for
maize and cultivated larger acreages (Table 26.2).

COMPARISON OF FEMALE-HEADED HOUSEHOLDS
IN THREE FARMING SYSTEMS

    The farming systems of male-headed households identi-
fied in the three Recommendation Domains in question vary
in degree of agricultural commercialization. This same
gradient of commercialization dominates the comparison of
female-headed households in the same areas (Table 26.3).
The polar extremes in this comparison would be TRD#2
(Serenje District) at the subsistence end of the spectrum
and TRD#5 (Kabwe Rural District) at the commercial end.
The latter has more oxen owners and hirers, larger areas
cultivated, and utilizes more technical inputs (hybrid
seed, fertilizer, knapsack sprayers), that in part explains
the higher yields obtained. However, the labor constraint
is still evident in the fewer respondents reporting annual
land clearing and in the late top dressing application of
fertilizer on maize.
    The proximity of the urban center, Kabwe also has an
influence on the composition of the female-headed households in that area and the marital status of the farmers.
Lower percentages of both male and female working children
on farms in TRD#5 are probably due to better educational
and employment opportunities for adolescents in town where

TABLE 26.3

COMPARISON OF STATISTICS ON FEMALE HEADED HOUSEHOLDS
IN THREE RECOMMENDATION DOMAINS IN THE CENTRAL PROVINCE

Item	TRD#2	TRD#3	TRD#5
Average Family Size/Household)	7.0	3.7	5.0
Total Farm Holding Size (Acres)	36.4	96.6	127.1
Acreage Cultivated (Acres)	2.5	3.1	10.3
Firewood Collection (Hrs/Day)	2.0	2.4	1.4
Water Collection (Hrs/Day)	5.4	2.9	0.8
Maize Pl. Ht. at Basal Fert. Dressing (Cms.)	8.1	7.8	30.0
Bags of Basal Fertilizer/Acre	3.3	3.5	5.6
Maize Pl. Ht. at Fert. Top Dressing (cms)	62.1	72.4	90.0
Bags of Top Dressing Fertilizer/Acre	6.3	3.7	6.2
Total Maize Production/# Bags	12.9	10.4	32.1
Maize Yield (Kg/Ha)	1365.6	1284.4	1427.2
Percentage of Farmers:			
Own Knapsack Sprayers	–	7.1	30.4
Male Working Children	16.9	12.7	2.7
Female Working Children	18.1	12.7	–
Unmarried w/Children	–	5.7	20.8
Oxen Owners*	8.7	15.9	33.3
Oxen Hirers*	43.5	39.5	66.7
Oxen Borrowers*	8.7	10.5	4.2
Hire Oxen Equipment	–	6.9	39.1
Borrow Oxen Equipment	–	1.2	8.7
Use Hybrid Maize Seed	73.7	58.0	79.2
Prepare Land w/ Hand Hoe	44.4	29.9	–
Dribble Maize Seed Behind Ox Plow	31.6	16.2	–
Clear Land Annually	95.7	71.2	4.6
Buy Seed for Crops	82.6	38.0	100.0
Hire Labor	26.1	48.9	62.5
Receive Remittances	13.0	34.5	8.3

Note: *The sum of these percentages exceeds 100 percent
only because some oxen owners also hire and borrow
oxen.

they can live with relatives. The fact that many of these farmers were unmarried with children may also reflect a relaxation of traditional mores nearer urban areas. The conspicuous presence of commercial cash crops like cotton and soybeans in TRD#5 also reflects the existence of nearby urban markets.

IMPLICATIONS FOR FARMING SYSTEMS RESEARCH AND EXTENSION

Most of the assumptions held by ARPT regarding labor constraints and an emphasis on subsistence crops in female-headed households were supported by the findings of this study. The isolation of women from conventional government services such as agricultural education, credit, and extension, compounds the difficulty and designates female-headed households as a unique target group among agricultural producers. This study provides evidence that agricultural production on female-headed households is at a subsistence level because of a lack of exposure to new ideas and limited access to the means by which to obtain higher yields, rather than because of a lack of interest in commercialization. How then should ARPT address this problem?

It is obvious that extension workers within the Ministry of Agriculture and Water Development (MAWD), particularly at the district and camp levels, must be sensitized to the situation of female-headed households in the agricultural sector. This can be accomplished gradually through ARPT training activities. Female farmers should be encouraged to attend field days, and if women are intimidated by male dominance at the meetings, special tours of ARPT on-farm trials should be arranged for women. Some ARPT on-farm experiments and demonstrations should purposely be located on women's farms to involve them more directly in ARPT work. Future socioeconomic studies should place special emphasis on female-headed households.

Since it is not feasible for African governments to support and finance separate extension services for each sex, it follows that attitudes and methodologies of existing extension services must be modified to reach women as producers rather than simply as consumers. More attention to the extension messages for subsistence crops and chickens, especially over the radio, would be a humble beginning. General agricultural education themes such as land rotation techniques for sustaining yields on cleared land would also be timely. At the same time, ARPT should actively encourage women's groups and the training of

386

female extension workers in subjects beyond sewing and food
conservation.

It is reassuring to note that the results of this study
did verify that the research priorities established by ARPT
are applicable for all traditional and small scale farmers
in the province. Given the central theme of labor constraints, ARPT is justified in seeking mechanisms by
which to shift labor demand out of the critical planting
period (winterplowing, introducing late season cash crops
like sunflowers and soybeans, etc.). Labor saving technologies such as the use of herbicides for zero-tillage and
the mechanization of maize weeding with oxen cultivators
are also appropriate strategies under the circumstances, as
is the emphasis on improving sorghum, millet, and bean
production practices. Since much greater advances are
needed in directing extension services toward female-headed
households, studies of this nature insure FSR/E relevance
and at the same time increase the visibility of the silent
minority.

NOTES AND ACKNOWLEDGEMENTS

This paper is based on the work of the Adaptive
Research Planning Team (ARPT) in the Central Province of
Zambia, which is funded under USAID Contract 611-0201. The
author wishes to acknowledge contributions to the questionnaire design from A. Sutherland, ARPT Rural Sociologist,
and S. Kean, ARPT National Coordinator. The enumerators
for this study were G. Simwanza, H. Simuziya, J. Nshindano,
and M. Bwalya.

REFERENCES

Allen, J.M.S.
    1984   Baseline Survey Report (1980-82) Chinsali.
        Integrated Rural Development Programme - Chief
        Mubanza's Area, Chinsali District Council Zambia,
        mimeographed.
Central Statistics Office
    1981   National Commission for Development Planning:
        Economic Report. Lusaka, Zambia: Government of the
        Republic of Zambia.

Due, J.M., T. Mudenda, and P. Miller
    1983   How do Rural Women Perceive Development? A Case
        Study in Zambia. Illinois Agricultural Economics
        Staff Paper No. 83-E-265. Urbana, IL: University
        of Illinois Department of Agricultural Economics.
Due, J.M. and M. White
    1985   Differences in Earnings, Labor Input, Decision
        Making, and Perceptions of Development Between Farm
        and Market Women: A Case Study of Zambia. Urbana,
        IL: Illinois Agricultural Economics Staff Paper No.
        83-E-319.
Muntemba, M.S.
    1981   Women as Food Producers and Suppliers in the
        Twentieth Century: The Case of Zambia. Development
        Dialogue 1:2:29-50.
Safilios-Rothschild, C.
    1985   The Policy Implications of the Roles of Women in
        Agriculture in Zambia. New York, NY: The
        Population Council.

27
The Gender Factor and
Technology Options for Zambia's
Subsistence Farming Systems

Alistair J. Sutherland

Recent reviews of the role of women in farming systems have pointed to the variability of women's contributions across the developing world (Ferguson and Horn 1985; Lele 1985; Rockefeller and ISNAR 1985). In the Southern African context, some writers have noted that rural women do not comprise a uniform and undifferentiated group and that economic differentation between households may be a more significant variable than gender of household head in accounting for variation in farming (Brown 1981; Fortmann 1984; Keller 1984; Safilios-Rothschild 1985). In the case of Zambia, development efforts relating to women have moved increasingly from discussions at a general level to implementation in specific situations (Keller 1984; Safilios-Rothschild 1984; ZARD 1985). Effective implementation requires that empirical relations at the field level are properly understood in advance; such understanding implies detailed and thorough socio-economic studies of women's roles (Keller 1984).

In Zambia, along with a growing recognition of variation in gender roles, comes an awareness of the need for the fuller integration of women's interests within the larger context of rural development. This awareness is all important in view of the large number of projects (recently inventoried by Keller n.d.) that are supported by the government and a range of donors and that aim to increase women's participation in rural development. Agricultural projects targeted specifically at women's groups have proved relatively ineffective and have often failed to reach poorer households (Chilivumbo and Kanyengwa n.d.; Keller 1984).

Farming Systems Research (FSR) when institutionalised at the national level is an approach with the potential to integrate women's interests more fully into the agricultural development process (CIMMYT 1985). The time is ripe

for focused empirical studies that can provide the data
necessary to implement a research and development approach
that takes women's roles and their contribution to agriculture properly into account (Ferguson and Horn 1985).
Moreover, the treatment of gender as a social variable
should become a priority for social scientists working in
FSR programs (CIP 1985). Systematic country-wide comparisons of women's roles in agriculture should also be made a
priority. Such comparisons need to be systems specific and
to be done in conjuction with the identification of
recommendation domains and the subsequent setting of
research priorities. If a comparative and systematic
approach is not adopted, the danger is that a well
documented case from a particular country may be taken to
be typical of that country (or larger region in which that
country is included), when in fact it is not (Sharpe 1985).
FSR provides the framework for country-based comparative
studies that can be expanded to comparisons among
countries, subcontinents, and continents. This paper
initiates a country description for Zambia, beginning with
a short summary of Zambia'a farming systems research
program, and is followed by brief case studies that address
the issue of gender and its relationship to recommendation
domains and research priorities in three of Zambia's eight
provinces.

FARMING SYSTEMS RESEARCH IN ZAMBIA

Zambia's farming systems research program was begun in
1980 following a demonstration of farming systems methods
by CIMMYT's East Africa program in 1978/9 (Kean 1985). A
separate farming systems section known as the Adaptive
Research Planning Team (ARPT) was created within the
Research Branch of the Department of Agriculture, Ministry
of Agriculture and Water Development (MAWD). To date seven
provincial teams, comprised of a farming systems agronomist
and economist, have been established; eventually, each of
Zambia'a nine provinces will have its own ARPT team. A
rural sociologist as a consultant joined the program (Kean
and Sutherland 1984). Research extension liaison officers
and a nutritionist consultant also have been added to the
ARPT.
Gender issues received little attention in the initial
years of data collection in Zambia. ARPT team members were
trained through CIMMYT training programs that emphasized a
"user perspective," but did not specifically address the
importance of gender issues (CIMMYT 1985). Greater attention was brought to bear on the subject after the sociologist and the nutritionist joined the team.

THE CASE OF FEMALE-HEADED HOUSEHOLDS IN LUAPULA

Luapula Province lies in the northern high rainfall belt of Zambia. The Province is characterised by subsistence cropping throughout, and by fishing (commercial and subsistence) in areas close to the main rivers, swamps, and lakes. Cassava is the principal starch staple, supplemented with finger millet (the traditional staple) where chitemene shifting cultivation is still practiced and with maize where cultivation has become semi-permanent. Both rice and maize are grown on a limited scale as cash crops, largely in areas where institutional support and incentives are provided by donor projects.

Annual rainfall averages 950-1200 millimeters and is concentrated between the months of October and April. There is no frost. The soils are of two types, upland and dambo (low-lying areas prone to waterlogging). Nearly all crops apart from rice are grown on upland soils. Most soils have a sandy/gravelly texture and poor fertility due to a low Ph caused by extensive leaching.

Owing to disease problems and tsetse fly, cattle are very scarce and hardly ever used for draft puposes. The traditional method of chitemene using an axe is gradually giving way to the mounding and ridging of fields using a hand hoe. The population is concentrated in large villages along the major roads, rivers, and lakeshores of the Province. Population pressure on land adjacent to settlements is growing. Female-headed households (FHHs) comprise from 27 percent to 38 percent of all households by district, according to the 1980 census (Safilios-Rothschild 1984).

In view of the high proportion of female-headed households and the prominent role women play in subsistence farming, a study was conducted by the Zambian counterpart sociologist in 1983 to establish the role of women in farming in Luapula and to see if female-headed households comprised a separate target group for on-farm reseach (Haalubono 1984). Two different areas representing Traditional Recommendation Domains Seven and Four (TRD#7 & TRD#4) were chosen for the study: Mukunta Village, located in a fishing area of TRD#7 on the shores of Lake Mweru, and Mabumba Village, located on the plateau of TRD#4 that relied on cropping for subsistance.

In each village, a sample of ten female-headed households was compared with ten male-headed households with regard to division of labor in cropping, choice of crops, cash earning opportunities, land tenure, and decision-making. Data collected through unstructured interviews were supplemented by participant observation.

Certain features are common to both areas. The size of
semi-permanent land holdings is limited by the amount a
family (often with hired assistance) can prepare using a
hand hoe. Land tenure is based on customary rights allocated by the local chiefs and secured through clearing of
bush and continuous use. The kinship system is matrilineal
with uxurilocal residence giving way slowly to virilocal
residence. Along with this changing trend in residence
patterns, the custom of bride service is dying out and
being replaced with the payment of brideprice to facilitate
removal of a woman from her kin. This does not give the
husband's people custody over the children. Each household
is an independent production unit obligated to share with
relatives and neighbors. On death, the larger matrilineal
kinship group distributes the property; spouses and a man's
children are not allowed to inherit.

The division of labor in cultivation is organised and
allocated according to gender. In semi-permanent fields,
men are responsible for the preparation of mounds for
cassava and ridges for maize, while women are responsible
for planting, weeding, and harvesting. In chitemene
fields, men (preferably sons-in-law) are responsible for
lopping branches, while women are responsible for piling
and burning branches, sowing seed, and harvesting. In
terms of decision-making, women are primarily responsible
for food crops, while men tend to dominate cash cropping
decisions. However, as has been reported elsewhere in
Zambia, in most cases spouses consult each other before
making decisions (Due et al. 1984).

In Mukunta, the comparison of male-headed and female-
headed households revealed surprisingly few differences.
The two types of household enjoyed the same type of access
to land and the same choice of crops (except for one male
household which had a small pineapple and sugar cane plot).
The main difference was in the household size and composition; female-headed households were smaller with fewer male
adults, which made a difference in access to male labor in
land preparation. This obstacle was overcome by hiring
labor for cash as that was easily obtained by women in the
area who hired men from Zaire were as casual laborers. In
Mukunta, women participated in growing rice as a cash crop;
this was facilitated by the recruitment policy of the donor
project and by the difficulty in recruiting men for agricultural work due to the higher returns from fishing in the
area. Although women were excluded from fishing they were
the major fish traders in the area and both married and
unmarried women had access to this trade. A nutritutional
survey in Mukunta showed seasonal deficiencies in high

energy foods for children and in the calorie intake during the rainy season.

In Mabumba, there were more differences between male-headed and female-headed households than in Mukunta, but again these were not marked. No major crops were exclusive to one or other type of household. The most notable difference in cropping was that each female-headed household tended to have a groundnut field, but this was rare in male-headed households. Moreover, growing maize and rice as cash crops was confined mainly to male-headed households. This may have been due to a combination of two factors: a different donor was responsible for administration of credit and the area lacked fishing opportunities for men. This is supported by the fact that some female-headed households in the sample were active in fish trading as a source of cash income.

A nutritional survey carried out in Mabumba showed a similar pattern to Mukunta, except that intake of high energy foods and calories was lower during the land preparation season. This was due to the virtual absence of fish in the diet in Mabumba and the greater distance from residence to fields, giving mothers less time for preparing food for their young children.

With regard to recommendation domains, the conclusion drawn from the findings is that it is not necessary to treat female-headed households in TDR#4 or TDR#7 as separate target groups. However, research priorities have been adjusted. Ways to introduce more high-energy foods into the cropping system (such as groundnuts and sunflower) and to reduce the time and labor in cassava drying, storage, and processing to give women more time for cooking and childcare have been considered more seriously and technologies addressing some of these areas are being tested. While it does not make sense to treat female-headed households as a separate target group, it does make sense to ensure that at least one-third of cooperating farmers in the area are female heads of households and that wives are actively involved in trial planning, site selection, and trial assessment.

GENDER IN LUSAKA PROVINCE

Lusaka Province occupies the rural area adjacent to the national capital. It is characterised by a mix of cash cropping and subsistence farming. The Lusaka ARPT has divided the province into two traditional recommendation domains: TRD#1, the tsetse fly-infested area to the east and southeast of the capital where subsistence hand hoe cultivation prevails and TRD#2, the tsetse fly-free areas of communal lands closer to Lusaka where ox-cultivation and a mix of subsistence and cash cropping is common. The climate is much drier than in Luapula Province, with more extremes in temperature, rainfall, and altitude. The soils are highly variable, ranging from deep clay loams of high fertility to less fertile, shallow, gravelly soils on higher ground. Clay dambos are cultivated in the higher rainfall area and alluvial soils are important for cropping in the parts of TRD#1 adjacent to the Luangwa and Zambezi Rivers, that are prone to drought. Maize and sorghum are the main staples. Settlement is more scattered than in Luapula; villages are smaller, but more compact groups of households are found. Female-headed households are frequent, especially in the more remote district where they approach 45 percent of all households.

TRD#1 is inhabited by a mix of ethnic groups. The Senga and Kunda are matrilineal with a tradition of uxurilocal residence. Their Soli neighbours are also matrilineal, but with virilocal residence, while the Gova are partilineal with patrilocal residence. Household income in TRD#1 is supplemented by off-farm activities such as beer brewing, handicrafts, and gathering of wild produce for sale, and by money sent from relatives working in town.

In TRD#1, a survey was carried out covering eighty-eight households (sixty-seven male-headed and twenty-one female- headed households). The ARPT sociologist ensured that gender was included as a variable in the questions. Unfortunately the survey did not include the full proportion of female-headed households (only 24 percent in an area where the figure was about 40 percent) and the large size of households reported in some cases suggested that unmarried daughters with children were included in the households of their parents. While the results have yet to be fully analyzed, some observations are possible on the basis of preliminary data analysis.

In terms of access to land, there was very little difference between male-headed and female-headed households. In spite of some variation in field types, the two types of households had similar access to all kinds of fields. The two types of households had roughly the same proportion of children in school and in permanent employment. Male-headed households were on average about 20 percent larger than female households and their reported area cropped to staples was, on average, larger by a similar proportion. The methods of planting and land preparation and the main labor constraints in crop production were almost identical for the two types of household. The most significant difference was in the cropping pattern. Female-headed households were more than twice as likely as male-headed households to grow maize as a sole staple crop and less than half as likely to grow sorghum as a sole staple. The other difference was that no female-headed households grew cash crops, but 10 percent of male-headed households did so.

In TDR#2, the rainfall (800-900 millimeters per annum) is higher and more regular than TDR#1, and the higher altitude makes for lower temperatures and a greater incidence of frost. Soils are similar, but with the notable absence of alluvial sands. Staple crops are the same, but maize is much more important relative to sorghum; cash crops (cotton, sunflower, and soybeans) are quite commonly grown. Cash earning opportunities from off-farm activities are more numerous, with charcoal burning, casual labor, and petty trade being particularly important. Settlement tends to be scattered with family groups fragmented into more isolated independent households, particularly in areas closer to the city where more cash-cropping is found. The domain is comprised of a mix of Lenje, Soli, and Tonga peoples, with a considerable sprinkling of other incoming groups. Kinship tends towards matrilineal descent with virilocal residence.

In TRD#2, a less extensive rapid survey of thirty-four households (sixteen female-headed and eighteen male-headed households) for the purpose of farmer selection revealed large differences between the two types of household. Four out of five female-headed households grew maize only as their staple crop compared with only one out of three male-headed households in the sample. Male-headed households were more than twice as likely as female-headed households to grow a cash crop in addition to maize. The biggest difference was in access to draft power. Three in four male-headed households owned oxen compared to one in four female-headed households. There were also differences in household size, and size of land holdings: male-headed households had more members and tended to have larger holdings. Because of the concentration of oxen in male-headed households, it was difficult to separate oxen owner-ship from gender as a variable influencing cropping practices. However, male-headed households which lacked oxen had similar cropping patterns to female-headed house-holds, suggesting that access to draft power was the more critical of the two variables.

This comparison shows more differences between the recommendation domains within Lusaka Province than was the case in Luapula. In TRD#1, the initial findings suggest that the extent of the difference between households does not justify treating female-headed households as a separate target group. However, the difference is enough to support a shift in research priorities. Research on maize needs to be targeted more at female-headed households. Perhaps more importantly, women need more labor saving technologies for tasks such as scaring birds from sorghum, and weeding sorghum and maize (weeding was noted as the major labor bottleneck by all types of households). Fewer female-headed households now grow sorghum, in part because it requires more weeding and bird scaring than maize. Such technologies would make this useful drought resistant staple more available to female-headed households and consequently help reduce the risk of crop failure.

In TDR#2, big differences between the two household types, in combination with the similarities between female-headed households and cattleless male-headed households, suggest the need for a subdivision of the domain into two target groups. Although further research is required to verify findings, the most obvious suggestion is to sub-divide the domain into ox-owners and non-owners, using access to draft power rather than gender as the critical variable. As in TRD#2, women's activities need to receive higher priority in the development of labor saving devices, particularly those relating to weeding.

GENDER IN WESTERN PROVINCE

Western Province is inhabited by the Lozi and related peoples. Like Luapula, Western Province is characterised by subsistence agriculture and by fishing in areas adjacent to the major rivers and swamps. It differs in that live-stock, particularly cattle, are important for most house-holds in the Province. Other differences include much greater variation in ecology, more ethnic groups and resulting farming systems, and a greater dependence on imported staple foods. The last feature is due to the com-bined effects of a more risk-prone system of cropping susceptible to both drought and to flooding on less fertile soils.

The case reported below arises from research in Senanga West District in the southwest of the province during the planting season of 1983 (Sutherland 1984). The research did not set out specifically to compare female-headed and male-headed households, but included type of household as a variable in a study of cropping patterns, land tenure, and access to draft power. The study was the first attempt in the province to look systematically at gender in relation to recommendation domains and research priorities. While the definition of recommendation domains in the Province had included gender of household head as a variable, responses from extension workers indicated that only 10 percent of household heads were females. It was only after the study was complete that a much higher proportion (38 percent) was found in the study area and examination of the 1980 national census data revealed a much higher figure across the Province (31 percent to 29 percent in different districts).

Senanga West is a drought prone area with a subsistence economy that combines crops with cattle keeping and has some fishing, hunting, and gathering. Its 18,000 square kilometers are predominantly sandy flood plain, studded with forested ridges and termite mounds. However, along the banks of the main rivers of the district, especially the upper Zambezi, more extensive belts of forest with sandy loam soils are favored for settlement and cultivation. A population density of 3.2 per square kilometer means that grazing is relatively abundant. Cattle are numerous and are kept primarily for local subsistence uses: draft power, milk, manure, and sale in emergency. They also have a social role in local institutions such as brideprice, debt payments, and funeral rites. In the flood plain areas particularly, arable land is quite limited. Land tenure is administered through the office of the Lozi paramount chief by local indunas and village headmen; rights are secured by residence and by use. The division of labor in agriculture makes males reponsible for livestock, ox plowing, and manuring. Adults of both sexes plant, while women weed and harvest.

Ecological variation gives rise to three main soil and field types: mutemwa -- fields on sandy loams found on forested ridges; mazulu -- fields on wooded termite mounds with sandy clay loam soils; and sitapa -- fields on seepage soils adjacent to rivers. In addition, many hamlets have manured gardens close by where cattle are penned at night. Settlement is scattered, with people living in hamlets containing from three to fifteen households, a household being defined as an owner (plus dependents) of a house and granary, who has independent rights to arable land. A hamlet is a discrete cluster of households related by kinship and usually sharing a cattle pen. A collection of hamlets of different sizes makes up a locality or "neighborhood". This area is inhabited by three main ethnic groups: Lozi, Shanjo, and Mbunda. Different ethnic groups do not share hamlets, but often share the same neighborhood. Kinship varies somewhat between groups. Lozi and

Shanjo have cognatic kinship with virilocal residence, while Mbunda have matrilineal kinship with a mix of uxurilocal and virilocal residence.

Farming in Senanga West is typically subsistence. Cash cropping is almost unknown, and the district has been the recipient of famine relief due to three years of drought and influx of refugees from neighbouring Angola and Namibia. Farmers reduce the risk of crop failure by using a range of drought resistant staples spread over different field types and planted at different times. While all households have independent rights to arable land and its products, very few are truly independent in production. Rights in land and assistance in production are secured by virtue of membership in a hamlet and a neighborhood. Nearly everyone is active in farming in spite of unfavorable soils and rainfall, frequent crop failure, and often pitiful yields. Production is largely uncommercialized and the substitution of cash for other scarce factors is very rare.

The survey of 190 households revealed differences in farming practices and resource endowments that corresponded with two social variables in particular: ethnic group and sex of household head. Polygynous households comprised nine percent of the sample and were classified as male-headed). There was no significant difference between the three main ethnic groups in the proportions of femaleheaded households (38-40 percent for all groups).

Lozi and Shanjo had, on average, significantly more fields and more types of field, scattered over a wider geographical area than Mbunda. In Lozi and Shanjo settlements, female-headed households had significantly fewer fields per household and also fewer types of field. With Lozi, female-headed households relied more on mutemwa fields while male-headed households relied more on mazulu, sitapa, and manured fields. Shanjo female-headed households relied more on mazulu fields, while Shanjo male-headed households relied more on manured fields and mutemwa fields. The differences between household types suggest that female-headed households are less likely to spread risks in cropping, and this places them more in a position of dependence on male-headed households, especially in bad seasons.

With crop preferences and planting priorities, there were more differences among the three ethnic groups than between types of household in relation to preferences and planting priorities for the three staple cereals of maize, pearl millet, and sorghum. Overall there was little difference between types of households and staple crop preferences. However, with Shanjo and Lozi, female-headed households using mazulu and sitapa were significantly more likely than male-headed households to plant pearl millet as the first crop, while Mbunda and Lozi female-headed households using mutemwa fields were more likely to plant maize. There was also some variation between household types in relation to crop preferences and cropping patterns. Female-headed households were proportionately less likely to have planted early than male-headed households. This tendency was most pronounced in Lozi and Shanjo female-headed households depending on mazulu and sitapa fields. In fact, on mutemwa fields female-headed households were proportionately more likely to have planted an early crop than male-headed households. This is because most Mbunda female-headed households relying on mutemwa fields only, depend largely on hand hoe cultivation and so do not have to wait their turn for oxen.

Gender of household head was very significant for the ownership of oxen, plows, and harnesses. Male-headed households were six times as likely as female-headed households to own oxen. On average, only one household in five owned oxen, but with significant differences among ethnic groups. Ox ownership was confined to one in three Shanjo, one in five Lozi, and one in seventeen Mbunda households. With male-headed households, polygynists were more likely to own oxen (47 percent) and more likely to own more than two pairs of oxen (29 percent) than were monogamists (29 percent and 6 percent respectively). This strongly suggests that in cattle-owning hamlets, cattle ownership and polygyny are indicators of status and influence. The ownership pattern of plows and harnesses was similar to that of oxen; male-headed households being much more likely to owners. Of the households that plowed, on average only one in seven (one in five in Lozi, one in seven in Shanjo, and none in Mbunda) was independent in terms of oxen and equipment. Only one in ten households hired oxen for cash; the majority (72 percent) cooperated on the basis of kinship and neighborly relationships. No female-headed households were independent in plowing. In Lozi and Shanjo hamlets, most female-headed households depended on assistance from male relatives and neighbors, while in Mbunda villages they mostly hired for cash or relied on hand hoe cultivation. In Lozi and Shanjo villages, two in seven male-headed households were independent in plowing; no Mbunda male-headed households were independent.

The findings in Senaga West raise a question regarding the applicability of the concept of recommendation domains

for Western Province. The high degree of variation over a
small area and population make the notion of recommendation
domains unworkable. Thus while gender of household is a
significant variable in accounting for difference in farming practices in Senanga West, ethnic and ecological variation overlay that of gender, making for a complex pattern.
In place of recommendation domains, it is appropriate to
look for technical options that will benefit the majority
of households in the area, particularly the resource poor
households, including female-headed households. In this
context, there is a need for improved drought-resistant
varieties of pearl millet and maize, for more labor saving
weeding technologies, and for ways to generate a cash
income to give cattleless households access to oxen as
required. In the testing of new technologies, cooperating
farmers need to be sub-divided into ox-owners and non-
owners, and about 40 percent of cooperators should be
female-headed households from all ethnic groups.

CONCLUSION

The comparison above provides an initial indication of
the variability of women's roles in Zambia's subsistence
farming systems as evaluated from the point of view of
identifying recommendation domains and setting research
priorities. Gender roles were sigificantly influenced by a
range of factors: cultural, economic, political, and ecological. In order to begin a basis for comparison, these
influences are summarized briefly below.

The type of kinship system influenced the degree of
autonomy of women: matrilineal systems gave women greater
autonomy from spouses and male relatives than did cognatic
systems. But kinship was overlayed by residence at marriage and by the presence or absence of cattle and ox
plowing. Perhaps more important than the type of kinship
system was the type of residence at marriage: uxurilocal
residence being associated with both greater female autonomy and more submission of younger male-headed households
to older female-headed households than was the case with
virilocal residence, where married women and female-headed
households were subject to more male authority.

The presence or absence of draft power was also a
highly significant factor. Where draft power was important, women generally, and female-headed households in
particular, were far more dependent on men and male-headed
households for successful crop production. Thus, when
ecological conditions make cropping a high risk activity
and tillage is by ox plowing, female-headed households and

wives are more dependent on males for food supply than
where ecological conditions favor a reliable cropping
system that does not depend on ox-plowing.

The availability of scarce types of arable land was
also a factor related to gender differences. This was most
pronounced in Western Province where access to draft power
and animal manure imposes limitations on the kinds of land
that can be easily cultivated. Significantly, in parts of
TRD#1 in Lusaka Province, where alluvial soils are highly
valued but do not depend on draft power for cultivation,
available data suggested little difference between household types in degree of access.

The local availability of cash and labor/draft power
also influenced gender differences. Where cash earning
opportunities for women were plentiful and labor/draft
power was easy to hire, female-headed households were in a
less disadvantaged position than where these were in short
supply.

These tentative observations relate to farming systems
that are subsistence oriented. As a subsistence farming
system moves towards a cash cropping system, as is beginning to happen in parts of Luapula and Lusaka Provinces,
other factors more closely connected with the larger political economy influence gender differences (Chilivumbo
1984). The reliance on credit and extension advice for
cash cropping tends to favor males and the better off in
spite of donor efforts to reach the rural poor (Due et. al.
1984; Mungate 1983). Moreover, the fragmentation of settlement and weakening of kinship and neighborly ties that
accompanies the growth of cash cropping, tends to place
greater labor burdens on women in the households of cash
croppers and makes it more difficult for female-headed
households to secure assistance from kin and neighbors in
critical farming operations.

At a more immediate level of application, this paper
concludes with some tentative recommendations regarding the
treatment of gender issues in the identification of recommendation domains and setting of research priorities for
FSR in Zambia.

The gender issue needs to be broadened beyond the consideration of whether or not to treat female-headed households as a separate target group. From the data available
in Zambia and elsewhere, female-headed households are often
an internally differentiated group, while married women
often face constraints similar to those confronted by single women in the management of their farms and households
(Fortman 1984). This suggests that when setting research

priorities it should be the interests of women as a group, whether they are married or not, that should be considered.

The introduction of oxen into subsistence farming as well as cash earning opportunities are mainly restricted to males and this tends to create larger wealth differences between male-headed households and female-headed households. Most female-headed households become impoverished and dependent on male-headed households; a "feminization of poverty" occurs (Brown 1981:17). However, female-headed households are not the only households affected by this impoverishment and it probably makes more sense to identify recommendation domains on the basis of a household's access to draft power than on the basis of the gender of its head.

When setting research priorities and conducting trials on farmers' fields in Zambia, there is a tendency not to consider the gender aspect. Instead, returns to cash invested or to labor input (measured as a genderless input) are the main yardsticks of evaluation. There is, moreover, insufficient technical expertise in the development and testing of appropriate technology that will reduce women's burden of household work not directly connected with crop production (Keller 1984). The solution to this problem is not simply to improve gender sensitivity in setting research priorities and testing new technologies. It will also require lobbying for a national commitment to developing and testing appropriate household technology as one of the activities of the Research Branch of the Ministry of Agriculture.

While there is a place for thorough studies of gender roles, it is quicker and cheaper simply to include gender as a variable during survey work. The results of such surveys often do not tell the whole story, however, and too much time can be spent on detailed studies of gender roles in agriculture, with little visible benefit to research. Consequently it is necessary to continue exploring the importance of gender in particular systems by ensuring that women are actively involved in problem diagnosis, pre-screening, testing, and evaluation of new technologies. This will require sensitizing FSR practitioners to the importance of involving women farmers at all stages of the farming systems research and extension process, and in setting out clear guidelines as to how this can be achieved.

In establishing the significance of gender through application of the diagnostic sequence put forward by CIMMYT (Collinson 1981), great care is required during the delineation of recommendation domains. The experience from Zambia is that extension workers and local leaders (and

often local social scientists as well) underestimate the
proportion of female-headed households. Data from reliable
sources such as census data should be used to supplement
information and a sociologist may be needed to carry out
rapid estimates in the field. When carrying out surveys,
social scientists on the FSR team must ensure that a representative proportion of female-headed households are
included in the interview sample. Moreover, steps must be
taken to ensure that women are involved in the interview,
if necessary by identifying specific questions to be
answered by women only. The opinion of women farmers
should be sought when setting research priorities and
designing on-farm trials. Moreover, a representative
proportion of farmer cooperators should be female-headed
households.

The above suggestions should serve to make FSR in
Zambia more sensitive to the importance of gender and more
effective in incorporating a gender perspective into the
research program. In order to facilitate this, rural
sociologists working with ARPT have a gender component
written into their job descriptions.

REFERENCES

Brown, B.
   1981   The Impact of Male Labor Migration on Women in
      Botswana. Presented at the Annual Meeting of the
      African Studies Association, Bloomington, Indiana.
Chilivumbo, A.
   1984   Small-Scale Farmer Resource Endowment;
      Evaluation, Research and Development Issues.
      Presented at ARPT/CIMMYT Workshop on Role of Rural
      Sociology (and Anthropology) in Technology
      Generation and Adoption, Ridgeway Hotel, Lusaka,
      November 27-29.
Chilivumbo, A. and J. Kanyengwa
   n.d.   Women's Participation in Rural Development
      Programmes: the case of SIDA-LIMA Programme. Rural
      Development Studies Bureau, University of Zambia,
      Mimeograph.

CIMMYT-International Maize and Wheat Improvement Center
   1985   CIMMYT'S Experience with the User's Perspective
          in Technology Development. In Women and Agricultural
          Technology: Relevance for Research Vol. II -
          Experiences in International and National Research.
          Rockefeller Foundation and ISNAR, eds., Pp. 13-26.
          Netherlands: The Hague.
CIP-International Potato Center
   1985   Women and Potatoes in Developing Country Food
          Systems: The CIP Experience. In Women and
          Agricultural Technology. Rockefeller Foundation and
          ISNAR, eds., pp. 27-34, The Hague, Netherlands:
          ISNAR.
Collinson, M.
   1981   A Low cost Approach to Understanding Small
          Farmers. Agricultural Administration 8:433-450.
Due, J. and T. Mudenda with P. Miller and M. White
   1984   Women's Contribution Made Visible: of Farm and
          Market Women to Farming Systems and Household Income
          in Zambia, 1982. Illinois Agricultural Economics
          Staff Paper No. 84 E-285: University of Illinois at
          Urbana-Champaign.
Ferguson, A. and N. Horn
   1985   Situating Agricultural Research in a Class and
          Gender Context: the Bean/Cowpea Collaborative
          Research Support Program. Culture and Agriculture
          26:1-10.
Fortmann, L.
   1984   Economic Status and Women's Participation in
          Agriculture: A Botswana Case Study. Rural Sociology
          49:3:452-464.
Haalubono, A.
   1984   Causes and Characteristics of Female Farming in
          Mansa and Nchelenge Districts, Luapula Province.
          Adaptive Research Planning Team Special Study
          Report, (Draft). Lusaka, Zambia: Ministry of
          Agriculture and Water Development.
Kean, S.
   1985   Institutionalising Zambia's Farming Systems
          Research Programme: the Case of the Adaptive
          Research Planning Team. Mimeograph. Lusaka,
          Zambia: Research Branch, Ministry of Agriculture
          and Water Development.
Kean, S. and A. Sutherland
   1984   Institutionalising Rural Sociology into
          Agricultural Research - The Zambian Case Study.
          Presented at ARPT/CIMMYT Workshop on Role of Rural
          Sociology (Anthropology) in Technology Generation
          and Adoption, Ridgeway Hotel, Lusaka, November

Kean, S. and A. Sutherland
    1984   Institutionalising Rural Sociology into
        Agricultural Research – The Zambian Case Study.
        Presented at ARPT/CIMMYT Workshop on Role of Rural
        Sociology (Anthropology) in Technology Generation
        and Adoption, Ridgeway Hotel, Lusaka, November
        27–29.
Keller, B.
    1984   Report on Current Efforts to Integrate Zambian
        Women in Development. Mimeograph. University of
        Zambia, Lusaka: Department of African Development
        Studies.
Lele, U.
    1985   Women and Structural Transformation. In Women
        and Agricultural Technology. Rockefeller Foundation
        and ISNAR, eds., pp. 243–265, The Hague, ISNAR.
Mungate, D.
    1983   Women, The Silent Farm Managers in the Small
        Scale Commercial Areas of Zimbabwe. Zimbabwe
        Agriculture Journal 80:6:245–249.
Rockefeller Foundation and International Service for
        National Agricultural Research (ISNAR), eds.,
    1985   Women and Agricultural Technology: Relevance
        for Agricultural Research. Volume 1 – Analyses and
        Conclusions. The Hague: Netherlands.
Safilios-Rothschild, C.
    1984   The Policy Implications of the Roles of Women in
        Agricultural Research in Zambia. (Preliminary
        Draft). Mimeograph. Lusaka, Zambia: The
        Population Council and National Council for
        Development Planning.
    1985   The Persistence of Women's Invisibility in
        Agriculture: Theoretical and Policy Lessons from
        Lesotho and Sierra Leone. Economic Development and
        Cultural Change 33:2:299–317
Sharpe, B.
    1985   Social Knowledge and Farming Systems Research:
        Ethnicity, Power and the Invisible Farmers of
        North-Central Nigeria. African Social Research 38
        Sutherland and Chileya, eds., Special Issue on
        African Farming Systems: Contributions of
        Anthropology and Sociology.
Sutherland, A.
    1984   Draft Power and Other Socio-Economic Aspects of
        Farming Systems in Senanga West District, Western
        Province: A Preliminary Report. Mimeograph,
        Adaptive Research Planning Team. Lusaka, Zambia:
        Ministry of Agriculture and Water Development.

406

ZARD–Zambia Association for Research and Development
1985   Womens Rights in Zambia.   Proceedings for the
Second National Women's Rights Conference, Mindolo
Ecumenical Foundation, Kitwe, March 22–24, 1984.

28
Using Male Research and Extension Personnel to Target Women Farmers

Anita Spring

To many Farming Systems Research and Extension (FSR/E) proponents, commodity-oriented scientists do not focus on the whole farming system and therefore they cannot appreciate the complexities of small farm management and smallholder needs and problems. This is analogous to the way those involved in research on gender issues feel about the lack of appreciation and consideration of gender issues and intra-household dynamics displayed by many FSR/E practitioners. Farming systems researchers did not invent the fact that farmers have to deal with a multitude of environmental, familial, infrastructural and other factors, so that a focus on a single commodity might not remedy the problems of the farming system. So too, researchers who consider women's role in agriculture did not invent the sexual division of labor, the semi-autonomous nature of different family members, the differential access to land, labor and capital, or the fact that women are becoming more involved in the smallholder sector in some developing countries because of extensive male migration (Chaney and Lewis 1980; Gladwin et al. 1984).

Evidence is accumulating that technology transfer is frequently hindered when intra-household dynamics are not taken into account (see for example Rogers 1979; McKee 1984). Often, technologies are ill-suited or only partially adopted because the resource base -- personnel, capital, land and equipment -- is inappropriate or inadequately understood. A consideration of intra-household labor allocations and decision-making shows that in many places female family members will have to provide the labor and will either make or be involved in the decision as to whether or not to adopt the technology. A failure to look at who does what farm operations, who makes which decisions, and who receives the remuneration and makes further

investments, will affect the practice of FSR/E. For
example, a higher yielding cereal variety might require
more labor in managing, harvesting, processing and storing,
especially in synchronously maturing varieties (Ferguson
and Horn 1985; McKee 1984) or a livestock intervention
might target one group of producers at the expense of
another. For example, in a case from Senegal, men made
decisions about the planting of cereal crops, but women
contributed much of the labor for the crops' weeding, harvesting and processing. Women made decisions about legume,
vegetable, and condiment crops. If women did the extra
work for the new variety of cereal crop, they had less time
for the crops that they managed. In livestock production,
male farmers favored livstock interventions that "would
increase live-weight and quality of stock" because size and
number of the herd were determinants of wealth. But women
controlled the milk allocation and sale of milk products
and "would gain most from interventions which increased
calf survival or ... permitted an increase in the number of
animals under current land or labor constraints" (McKee
1984: 598-599).

There are specific methodologies needed to understand
intra-household variables within the FSR/E process (McKee
1984). In the pre-diagnostic stage, the ethnographic
literature that provides information on the household's
division of labor, decision-making, and allocation of
resources must be reviewed for specific recommendation
domains. In the diagnostic stage, the types of households
and the types of representative farmers need to be considered. For example in areas where there are many households
headed by women, as in the case of much of Africa and the
Caribbean, it is necessary to include such households in
the sample and to ask if their resources and needs are the
same as or different from the households headed by men.
Socioeconomic and agronomic variables have to be assessed
in terms of various household members in the different
types of households. The interventions have to be geared
to the needs of the types of households and the constituent
members. In the technology design stage, it is necessary
to make sure that the researchers do not use incorrect
assumptions about gender; McKee suggests the input of
female scientists and field workers, but this is not always
possible or even a guarantee that gender issues will be
considered. There is no reason why both male and female
scientists cannot work on the problem. In the testing
stage Mckee says that one must monitor "how the farm
household actually copes with the reallocation of resources
required by the new requirements" (McKee 1984: 602). In
the final extension stage, McKee argues that it is

important "to involve women farmers and farm workers, as
well as female extension agents, in diffusing technologies
for crops and tasks in which women predominate" (McKee
1984: 602).

The major thrust of this paper is that men as well as
women agricultural researchers and extensionists have to
become involved and to target farmers of both genders. The
argument here first considers the gender-related characteristics of extension services and how these characteristics
affect reaching a variety of farmers, especially women.
Then, a case study from Malawi shows that women are important in agriculture but neglected in extension services and
in the practice of FSR/E. In order to study and correct
the problem, the results of two sets of trials are presented. In one, the analysis shows the results of using different groups of farmers in the sample. In the other,
mechanisms by which the male staff can work with women
farmers are described. Based on the lessons learned, the
paper concludes with a recommendation that the mandate of
male extensionists can be extended legitimately to include
female farmers.

CHARACTERISTICS OF EXTENSION SERVICES

Researchers tend to use the extension and research
services as they already exist in the host country. This
increases the difficulty in including women farmers in
FSR/E because it is usually male extension personnel who
work with researchers to locate, interview, select trial
cooperators, and target disseminators. The number of male
extension workers far exceeds the number of women who
receive training and who are employed as extensionists in
most places.

Many writers comment on the paucity of female extension
workers compared with male ones (Jiggins 1986; Berger et
al. 1984; Staudt 1975-76). Worldwide, including North
America and Europe, only 19 percent of the agricultural
extension staff members are women. The average number of
female extensionists for Africa is 3 percent; for Latin
America and the Caribbean it is 14 percent and for Asia and
Oceania the figure is 23 percent. Only in the Philippines
are 40 percent of the staff members female. Berger et al.
(1984) estimate that of extensionists specially designated
as agriculturalists, 41 percent do home economics rather
than agriculture.

Table 28.1 shows the number of men and women trained in
Malawi, a country where women are critical in agricultural
production. This table shows that women extensionists are
found in the bottom education tier and that their training

TABLE 28.1

GRADUATES OF MALAWI AGRICULTURAL TRAINING INSTITUTIONS:
BUNDA COLLEGE OF AGRICULTURE, COLBY COLLEGE OF AGRICULTURE,
AND THUCHILA FARM INSTITUTE (FARM HOME ASSISTANTS)
BY YEAR AND SEX

	Male	Female	Total Percentage Female
Bunda			
Degree (5 year)	273	24	8
Diploma (3 year)	797	124	13
Colby			
(2 year)	2255	45	2
Thuchila*			
(1 year)	–	291	100
Average per year	–	22	

Note: *Course for women only.

is much shorter than that for men. As a consequence, women
extensionists often are not regarded as professionally com-
petent as men in their knowledge of field crops and of
livestock. Female workers are often pressured to work in
home economics programs rather than to work in the agricultural programs for which they were trained. The contacts
of female extension workers with male farmers tend to be
limited concomitantly; the male extensionists tend to deal
with male farmers rather than with all farmers (Jiggins
1986). Since only a small proportion of farmers usually
are reached by extension services, there is even less
reason to further restrict extension to only male farmers.
For example, Table 28.1 shows the low number of women
trained in Malawi, the country that is the subject of this
paper and where women are critical in agricultural
production.

In the extension services themselves, male personnel
hold a variety of positions, including decision-making ones
that affect programs and policies. The female extensionists, with the exception of a few supervisors, usually are
concentrated in the lower ranks. Often male workers are
given the tasks of offering practical agricultural services

either through the training and visit system or through other regimes, while the female workers are supposed to form women's groups for small scale income generating activities. Most extension services in developing countries were modeled after the systems in North America and Western Europe during the last century with men providing agricultural information to male farmers and women providing home economics and nutrition information to women (Mead 1976; Berger et al. 1984). Ironically, home economics programs in the developed countries have changed a great deal since the 1930s and have become relevant to the needs of American farm women today, focusing on such topics as human development, consumer education, household finances, and marketing. By contrast, the teaching of domestic science in Africa is mostly focused on sewing, embroidery, cooking and basic hygiene/nutrition.

Coupled with this restriction of women to home economics programs, is the notion that there is better communication between members of the same sex than between members of the opposite sex. Sometimes these notions are strongly stated in terms of tradition or cultural constraints and operationalized so that only women are permitted to work with women and only men are permitted to work with men. However, the paucity of women in agricultural services assures that rural women will remain uncontacted and unassisted. Although it is probably true that many people prefer to learn or to work with people of their same sex, coeducational programs have worked in a large portion of the world. Berger et al. remark that "since very little empirical work has been done in this area, there is really no basis on which to judge the relative effectiveness of men and women agents in assisting women farmers" (1984:54).

The polarization of extension services by gender contributes to the inability of FSR/E to assess intra-household dynamics and differential access and control over resources by household members. This failure in turn detrimentally affects the design of technology testing and dissemination. The question to be asked is what would happen if the equation were changed and if extension and research programs in practice were geared to all farmers regardless of sex. This might entail new procedures to target and reach the neglected farmers rather than the standard procedure of assuming that one method works for all. A case study from Malawi examines the problem of relying on male extensionists in FSR/E and reports on some methods that were undertaken to change extension and FSR/E procedures to reach female as well as male farmers.

412

CASE STUDY FROM MALAWI

The Women in Agricultural Development Project (WIADP) was funded by the Office for Women in Development, USAID, and operated in Malawi from 1981 to 1983. It was of national scope and its aims were multifaceted: to research women's and men's roles in smallholder farming; to use farming systems research to ascertain smallholder, and especially women's needs; to disaggregate agricultural data by sex; to work with extension and research units to target women as well as men farmers; to evaluate women's programs; and to orient policy makers to consider women farmers in agricultural programs.

Primary and secondary research by the WIADP showed the contributions by gender for various commodities (Clark 1975; Spring et al. 1983b). Women formed the bulk of the agriculturalists in the rural areas. They spent as much time on their farm work as on their domestic work. Approximately one-third of the households in the country were headed by women, but in some areas as many as 45 percent of the households were female-headed. Women were assuming management of more family farms, not only in households they headed, but also in married households because of male out-migration. Farm operations were differentiated by sex in some areas and in some households, while in other places and households they were not. The so-called standard sexual division of labor where men prepared the land and women planted, weeded, and harvested had given way to expediency in many places (Spring et al. 1983b). The adult who was home on the farm did the operations and in many cases this meant that the women were doing the work and making the farm decisions. Women were involved in all aspects of farming including land clearing, plowing, applying fertilizer, crop protection, etc., either routinely or when male labor was unavailable. Women were involved in a variety of cropping patterns from mixed sub-sistence to cash crops, and grew maize, groundnuts, rice, cassava, tobacco, cotton, coffee, and tea. They performed many operations, such as spraying cotton and planting tobacco seedlings, that were commonly believed to be done by men only (Clark 1975). Women in many areas were involved in the care of livestock, especially of small ruminants and poultry. Free ranging cattle were mostly owned by men and cared for by boys and men, but when the animals were brought into the village for fattening in stall feeding projects, their care fell to women (Spring 1986a).

Agricultural development projects had increased the
amount of time in hours per day and in days per month that
both men and women had to work (Clark 1975). The agricultural services provided by integrated development and local
projects such as training, inputs, and credit programs, and
agricultural extension, mostly by-passed women. As a
result, the farming efficiency of many households was
reduced. However, some women were able to participate in
development programs to increase their productivity. Some
male extensionists included women farmers with the male
farmers they targeted for training, credit and visits
(Spring et al. 1983b).

The WIADP documented the delivery of agricultural
extension services to men and to women in a variety of
ways. First, the WIADP analyzed the extension survey that
was part of a large national multi-instrument survey conducted by the Ministry of Agriculture and financed by the
World Bank. Second, the WIADP interviewed and observed
extension personnel in the field in terms of the way they
worked with clients. Third, the WIADP conducted FSR/E
surveys and trials and studied the ways the extension personnel were utilized to identify and to work with farmers.
Fourth, meetings and interviews were held with the staff
and managers of agricultural projects who supervised extension and research efforts to examine their procedures.

Extension Survey

The results from the national survey (The National
Sample Survey of Agriculture or NSSA) showed that farmers'
contacts with extension workers, including personal and
field visits, attendance at group meetings and demonstrations, and participation in training courses, differed by
sex (Table 28.2) (Spring et al. 1983b). Contact with
extension workers was the major source of advice for both
men and women farmers, but men received more personal
visits and more advice than women. Group meetings tended
to reach more farmers than personal visits, but men were
the primary participants. Relatively few farmers of either
sex viewed extension demonstrations, but more men than
women learned from this method. Field visits reached even
fewer women and the WIADP observed that many male
extensionists simply dismissed the women working in the
fields while they concentrated on the men.

The WIADP disaggregated the NSSA data into three
categories: male household heads, female household heads,
and wives of the male household heads. The data showed
that men received more services than women and wives

TABLE 28.2

TYPE OF EXTENSION CONTACTS FOR MALE HOUSEHOLD HEADS (MHH), FEMALE HOUSEHOLD HEADS (FHH), AND WIVES FROM THE NSSA EXTENSION SURVEY, MALAWI (IN PERCENTAGES), 1980–81

Type of Contact	Lilongwe			Ngabu			Karonga	
	MHH (N=147)	FHH (N=35)	WIVES (N=35)	MHH (N=95)	FHH (N=31)	WIVES (N=95)	MHH (N=70)	WOMEN* (N=73)
Personal Visit	41	28	23	28	12	4	44	29
Group Meeting	66	44	49	43	12	8	61	34
Demonstration	13	6	6	5	1	0	16	10
Field Visit	13	9	6	15	5	2	16	4

Note: *Female Household Heads and Wives tabulated together.

Source: Spring, et al. 1983b.

received more services than female household heads. Very few wives received agricultural information from their husbands. The presumed transfer of technology from husbands to wives and from men to women in the household did not take place. The assumption that if men are trained or assisted that other family members learned or were assisted was not confirmed by the data (Spring et al. 1983b).

FSR/E Surveys

Analysis of FSR/E surveys and trials indicated that women farmers were contacted infrequently by reconnaissance or survey teams and were not often part of the recommendation domains delineated. Host country and expatriate researchers tended to ignore the women in the fields during rapid reconnaisance surveys. When production and social scientists relied on the extension workers, which they often did, the extension workers usually took them to interview and work with the men. Only male cooperators were selected for on-farm farmer managed trials. Sometimes the male cooperators carried out trial work themselves. Other times their wives and female relatives assisted or did much of the work, sometimes producing errors in the way the trials were conducted. This may have happened because these women had not received the instruction directly and the male cooperators did not pass on the information. To understand the problem, the WIADP asked trial cooperators who actually performed each operation. The information obtained showed that wives and female relatives carried out many of these tasks (Spring's notes from Kawinga and Phalombe FSR surveys).

The WIADP participated in several attempts to change the way in which surveys and trials were conducted. The WIADP conducted its own FSR surveys in three different regions of the country and worked with a German team in Central Region (Spring 1982; Spring, et al. 1982; Spring et al. 1983a). Each time, male extension workers who accompanied the teams tended to direct the teams to the better male farmers. To remedy the situation, it was explained to staff and team members that it was necessary to examine a range of environments, family types, and economic situations.

The WIADP prepared guide sheets that detailed the types of households and families that needed to be considered and requested that the following categories of farmers be sampled by teams doing the diagnostic survey:

(1)   A diversity of economic situations: low
resource farmers, including those who must work
for others; subsistence farmers; and wealthy
farmers who grow cash crops and hire laborers.
(2)   A diversity of household types: families
composed of (a) a wife, a husband, and children;
(b) a husband, two or more wives, and children;
(c) a married woman with children, but the husband
was away; and (d) an unmarried woman and children.
(3)   A diversity of ages and life-cycle situations: older people and recent widows and
widowers; young couples just beginning to farm;
long time farmers; and women recently divorced or
on their own (Spring 1982).

The WIADP also helped to place women researchers and
extensionists on the teams, and subsequently it became
fairly standard practice to have women on FSR/E teams.

On-Farm Experimentation

The WIADP conducted its own on-farm experiments and
worked with another USAID project on its trials (Hansen
1986).   Two examples of trials that included female farmers
illustrate the problems in obtaining women cooperators as
well as the lessons learned by considering women.   The
first example concerns trials held in a low resource area,
where there is land shortage, a drought-prone climate and
37 percent of the households are headed by women.   Average
holding size was one hectare, but more than 60 percent of
the households cultivated less than a hectare and almost a
third cultivated less than half a hectare.   Male out-
migration was pronounced and women and children remained to
work family farms.   The trials consisted of comparing an
improved cultivar with a local variety using "a simple
nonreplicated 2 x 2 factorial arrangement with two maize
varieties and two levels of fertilizer (0 and 30 kilograms
per hectare)" (Hildebrand and Poey 1985: 127-8).   Since
area farmers intercrop, all the treatments had a mix of
maize, cowpeas, and sunflowers (Hansen 1986).
The village headmen and the extension workers who
selected the farmers were specifically asked that half of
the cooperating farmers be women.   However, only 40 percent
in one village and 30 percent selected in the second
village were women.   In some cases the women and the men
farmers selected in a village were not comparable as
farmers.   The men tended to be vigorous individuals in

their middle years and many were high resource farmers who
owned cattle. Most of the women tended to be low resource,
older individuals at the end of their life cycle. They
were probably selected because age is revered and it was
considered an honor to be selected. Therefore, comparisons
between male and female farmers in terms of management and
yields would not be valid to show gender differences in
farming skills (Table 28.3). However, the data show
differences between high and low resource farmers and more
women are in the latter category.

TABLE 28.3

MAIZE YIELDS FROM FARMER-MANAGED ON-FARM TRIALS, PHALOMBE, MALAWI, 1981-82

| Four Treatments | \multicolumn{8}{c}{Farmers in First Village} | Treatment Mean for Village |
| | 1 | 2 | 3 | 4* | 5 | 6* | 7* | 8 | |
	\multicolumn{8}{c}{----tons/hectare----}	
Local Maize	2.2	2.2	1.9	1.2	1.3	0.9	1.0	0.5	1.4
Fert. Local	3.6	3.7	4.3	3.2	2.3	2.3	3.1	2.8	3.2
CCA Maize	3.5	2.0	2.9	0.4	0.6	0.5	0.6	0.3	1.3
Fert. CCA	5.0	4.7	4.3	3.5	2.4	1.7	3.0	2.8	3.4
Mean for Farmer	3.6	3.2	3.3	2.1	1.7	1.3	1.9	1.6	2.3

| Four Treatments | \multicolumn{6}{c}{Farmers in Second Village} | Treatment Mean for Village |
| | 1 | 2 | 3 | 4* | 5 | 6* | |
	\multicolumn{6}{c}{----tons/hectare----}	
Local Maize	1.8	1.1	1.6	1.0	1.6	0.6	1.3
Fert. Local	3.2	2.5	2.9	1.2	1.9	0.8	2.1
CCA Maize	2.2	0.7	0.9	0.3	1.1	0.3	0.9
Fert. CCA	2.9	2.5	2.1	1.1	0.8	0.4	1.6
Mean for Farmer	2.5	1.7	1.9	0.9	1.4	0.5	1.5

Note: *Female Farmer (Designation added by Spring).

Source: Hansen (1986); Hildebrand and Poey (1985).

A modified stability analysis was carried out on these data by Hildebrand (Hildebrand and Poey 1985: 126- 134) and by Hansen (1986). Because of the inclusion of a range of farmers, young and old and male and female, an evaluation of the types of environments could be made where "environment...becomes a continuous quantifiable variable whose range is the range of yields from the trial" (Hildebrand and Poey 1985: 126). In the same area, there were a range of environments in terms of farmer management, soils, rainfall and the like, and the cultivars responded differently. The local cultivar was superior in "poor" maize environments while the improved maize was superior in "good" maize environments (Figures 28.1 and 28.2). Both cultivars responded "favorably to fertilizer in both good and poor environments" (Hildebrand and Poey 1985:129). The analysis demonstrated that there were two different recommendation domains. Although there were both men and women in each domain, there was a tendency for the women farmers to be in the poorer environment, probably because more of them were low resource farmers. Further analysis using confidence levels allowed the high and low environment farms to be compared. The results showed that only farmers in the better environments should choose the improved variety (the composite) and that they should fertilize the crop (Hildebrand and Poey 1985: 133). In the poorer environments, the local variety was better (Hildebrand and Poey 1985: 134). Fertilizer helped but only for the farmers who could afford it. Farmers who owned cattle and used the manure on their fields were in the better environments. Women did not own cattle too frequently, although the one high resource farmer in the sample did. The final recommendation was "to fertilize the local maize variety in the poorer environment and to use the composite maize with fertilizer in the better environment" (Hildebrand and Poey 1985: 132).

By comparing people at different ends of the spectrum, two recommendation domains were defined. All but one of the farmers in the better environment were men. Most of the farmers in the poorer environments were women, although there were some men. Recognizing different segments of the population, including those at particular risk, resulted in the delineation of multiple domains and technology solutions within a fairly homogeneous area. The female-headed households constrained by labor and cash would find it difficult to use fertilizer, coupled with their smaller holdings and lack of extension advice this would make their use of the improved cultivar disastrous. Another study on these households documented that they were being ignored by extension and credit programs (Evans

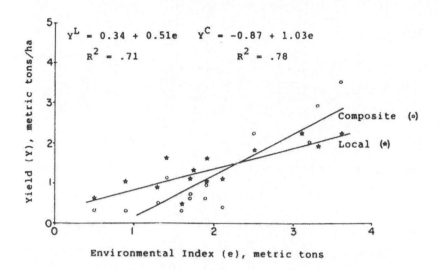

FIGURE 28.1

GRAIN YIELD RESPONSE FOR LOCAL MAIZE (L) AND
CCA COMPOSITE (C) TO ENVIRONMENT, WITHOUT FERTILIZER,
PHALOMBE PROJECT, MALAWI

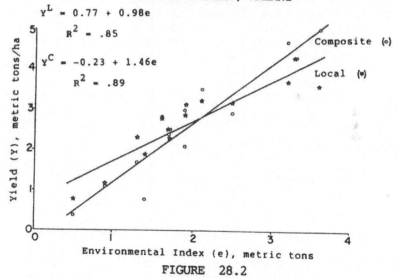

FIGURE 28.2

GRAIN YIELD RESPONSE OF LOCAL MAIZE (L) AND
CCA COMPOSITE (C) TO ENVIRONMENT, WITH FERTILIZER,
PHALOMBE PROJECT, MALAWI

Source: For Figures 28.1 and 28.2 Hildebrand and Poey,
1985: 130.

1981). A British researcher devised a multi-step method to remedy the situation. First, the cooperation of male extensionists and male village leaders was sought to bring women into the extension arena and to enable them to articulate their problems. Second the notions of "credit worthiness" had to be changed. Finally, the actual credit packages had to be modified.

Male village leaders were asked to designate women farmers for leadership training. The women were taught leadership skills by both male and female extension personnel. Then they were able to articulate their problems in farming. In general, they noted that extension services by-passed them in terms of credit and training. To solve the problem of credit services by-passing women, the male staff members were told to target women. But a different method of determining "credit worthiness" had to be devised for these women and a way for them to pay back the cash for inputs had to be found. Since these women were not members of farmers' clubs organized by the male workers and they lacked collateral, their "credit worthiness" was nil by the usual standards. A new method was devised in which they could be vouched for by male village headmen. It was not known how these women would pay back the loans since these households were thought to have no cash sales; the inputs supplied were to help them attain food self-sufficiency. The women, much to everyone's surprise, began paying back the loan from the sale of beer and crafts even prior to the harvest (Evans 1983).

Another difficulty was that the standard credit packages of improved seed and fertilizers (in multiples of one acre) were too large. The solution to the actual credit package itself was the creation of mini-technical (for 1/2 acre) packages of fertilizer and seed. With the assistance of the male extension staff, the number of women getting credit in the project increased from 5 percent to 20 percent of the credit recipients in a single year. These households went from food deficient to food selfsufficient households. However, non-standard techniques had to be used and the male extension staff members' contacts with women farmers were important to the success of the endeavor.

The second example of using women as trial cooperators concerns demonstrations and trials with soybeans (Spring 1985, 1986b). The Ministry of Agriculture determined that soybeans would improve the rural diet deficient in fats and proteins. Female extensionists were therefore instructed in the preparation of recipes for soybeans in their annual refresher course. However, the female extensionists were not taught the appropriate cultivation techniques for

soybeans. In particular they lacked training on rhizobium
inoculation and the use of fertilizers. In a test of
whether or not the male extension staff could work with
women farmers, the WIADP held demonstrations one year and
gave inputs and instruction to fifty-nine female cooperators. There were a range of environments and it was possible to compare the performance of women farmers. The
better farmers had better management and viable inoculum.
The poorer farmers had problems with pests and unviable
inoculum because they failed to reinoculate after late
rains and delayed planting. As a result of the demonstrations and surveys of both men and women farmers involved in
soybean production in a number of areas, the problem of how
to get viable inoculum to the rural areas was identified as
a general problem affecting both men and women. There were
also gender-specific smallholder problems such as the lack
of training and the limited seed given to women. Trials
were held the following year with twenty female cooperators
selected by the male extension staff. In addition to trying to solve the technical problems, two other questions
were asked. Could women do on-farm research with precision? Could the male extension staff work with women and
what were the methods that worked best? The answers to
both of the questions were affirmative. The women were
able to learn to conduct the trials correctly. Second the
male extension staff had no difficulty in identifying,
instructing, and monitoring the women cooperators (see
Spring 1986b for a full discussion of the trials).

    To capitalize on the discovery that women could be part
of trials and and that the male extension staff could work
with women the WIADP prepared an extension circular. The
circular, entitled "Reaching Female Farmers Through Male
Extension Workers," was published in August 1983 and issued
by the Ministry of Agriculture (MOA) rather than by the
WIADP (MOA 1983). Extension circulars from the MOA are
regarded as technical recommendations for the entire extension staff and must be heeded by them. As a MOA publication, the circular legitimized and advertized the fact that
male extensionists could work with women as well as with
men farmers in their areas and that working with women
farmers was not only the concern of the few female extensionists. It was distributed to all grassroots workers and
to agricultural project managers.

    The circular drew attention to the fact that extension
services need to reach women because of women's involvement
in Malawi's smallholder agricultural sector. It pointed
out that where women have been offered agricultural programs, they have learned new technologies and increased

their production. The circular used photographs taken from the MOA's own collection and depicted women in various farming operations, attending extension demonstrations led by male extensionists, attending village meetings with male farmers, receiving credit inputs along with men, and receiving a certificate of recognition for excellence in farming.

The circular presented methods for improving the delivery of extension services to women and for getting women into extension and research activities. Techniques were given that explained how to use leadership training to encourage women to attend village meetings and agricultural training courses and how to increase women's participation in credit and soil conservation programs and farmers' clubs. The male extensionists were directed to include women at their demonstrations, trials, and field days. They were also told to keep records of contacts and program involvement in terms of the number of women and men participating. The circular noted that there are a variety of household types and that women both as wives and women heads of households needed to be targeted.

CONCLUSIONS

Those involved in FSR/E must consider gender issues and intra-household dynamics at all stages of their work. In the pre-diagnostic stage, they must consider primary and secondary sources that detail the sexual division of labor and the changing roles of various household members. Sometimes this type of information is available, but where it is not, researchers may have to disaggregate agricultural data sets in order to ascertain gender differences or to collect their own data from local women and men farmers.

Researchers must confront extension workers with the need to include in their surveys a diversity of farmers in terms of resources and households and to consider people at different points of the life cycle in the diagnostic stage. Strict instructions need to be given to extension personnel for them to include in their surveys (1) low as well as high resource farmers, (2) women farmers with both low and high resources, and (3) women as both household heads and wives. The sexual division of labor and differential management strategies will have to be described. If questions about the allocation of labor and resources, problems and needs, and remuneration and investments, cannot be answered for different categories of farmers, the work is incomplete. If only men provide the answers about women, the data are most likely biased.

In the design of trials, intra-household dynamics and
the needs of various household members must be considered;
a range of farmers and of environments need to be included.
In the actual trials, women as well as men have to participate as cooperators. In some trials it may be necessary to
restrict the cooperators to the sex that actually is responsible for a particular commodity, e.g., groundnuts are
often grown by women in some areas. In others, recognition
of the fact that women and men do different farm operations
means that both male and female household members will have
to be considered as trial participants and that both will
have to be instructed accordingly. The extension and
research workers who help select and monitor the trials
will require strict instructions as to how to choose and to
work with these farmers. Researchers should not be fearful
about including a range of environments, but they need to
be careful about how they select too many farmers in certain categories and in comparing farmers at different
resource levels. Recommendation domains and technologies
tested may or may not be gender specific. In the dissemination of information, the male research and extension staff
members will be important to the success of adopting a
technology. The WIADP recognized that it was often difficult for individual extension workers on their own to make
special attempts to deal with neglected segments of the
population.

Although these techniques might not work in every cor-
ner of the world, the general argument that states that men
cannot work with women farmers needs to be reexamined. New
methods and techniques have to be devised that are feasible
and that consider cultural traditions. Female extensionists must not be left out. The number being trained in
agricultural subjects must be increased and incentives and
promotions need to be recommended for them. The curriculum
of the home economics courses for rural women must be
modified to include materials that are directly related to
women's productive roles. Usually the FSR/E personnel will
have the clout to influence policy and sometimes to provide
motivation and incentives for the work that extension is
doing. FSR/E personnel can therefore attempt to set the
tone and to require that women as well as men be targeted.
They can assist extension workers in discovering the
techniques that will work in an area.

424

REFERENCES

Berger, M., V. Delancey, and A. Mellencamp
    1984   Bridging the Gender Gap in Agricultural
        Extension. Washington D.C.: International Center
        for Research on Women.
Chaney, E. and M. Lewis
    1980   Women, Migration and the Decline of Smallholder
        Agriculture. Washington, D.C.: Office of Women in
        Development, USAID.
Clark, B.
    1975   The Work Done by Rural Women in Malawi. Eastern
        Journal of Rural Development 8:2:80-91.
Evans, J.
    1981   Phalombe Rural Development Project, Women's
        Programmes Programme Plan. Malawi: Blantyre
        Agricultural Development Division.
    1983   Women's Involvement in the Seasonal Credit
        Programmes in the Phalombe Rural Development
        Project. Mimeo. Chintheche, Malawi: National
        Credit Seminar.
Ferguson, A. and N. Horn
    1985   Situating Agricultural Research: Class and
        Gender Issues in Project Advisement. In Women
        Creating Wealth: Transforming Economic Development.
        R. Gallin and A. Spring, eds., pp. 85-90.
        Washington, D.C.: Association for Women in
        Development.
Flora, C.
    1982   Farming Systems Research and the Land Grant
        System: Transforming Assumptions Overseas. Farming
        Systems Research Paper Series, No. 3. Manhattan, KS:
        Kansas State University.
Gladwin, C., K. Staudt, and D. McMillan
    1984   Reaffirming the Agricultural Role of African
        Women: One Solution to the Food Crisis. In
        Proceedings of the Association of Facilities of
        Agriculture in Africa. Manzini, Swaziland: Fifth
        General Conference on Food Security.
Hansen, A.
    1986   Farming Systems Research in Phalombe, Malawi:
        The Limited Utility of High Yielding Varieties. In
        Social Sciences and Farming Systems Research:
        Methodological Perspectives on Agricultural
        Development. J. Jones and B. Wallace, eds., pp.
        145-169. Boulder, CO: Westview Press.

Hildebrand, P. and F. Poey
    1985   On-Farm Agronomic Trials in Farming Systems
        Research and Extension. Boulder, CO: Lynne Rienner
        Press.
Jiggins, J.
    1986 Gender-Related Impact, and The Work of the Inter-
        national Agricultural Research Centers. Consult to
        the group on International Agricultural Research,
        Study paper #17. Washington, D.C.: World Bank.
Mead, M.
    1976   A Comment on the Role of Women in Agriculture.
        In Women and World Development. I. Tinker, M. Bo
        Bramsen and M. Buvinic, eds., New York: Praeger.
McKee, C.
    1984   Methodological Challenges in Analyzing the
        Household in Farming Systems Research:
        Intra-Household Resource Allocation. In Proceedings
        of the Kansas State University's 1983 Farming
        Systems Research Symposium. C. Flora, ed.
        Manhattan, Kansas: Kansas State University.
MOA-Ministry of Agriculture
    1983   Reaching Female Farmers through Male Extension
        Workers. Extension Aids Circular 2/83. Lilongwe,
        Malawi: MOA.
Rogers, B.
    1980   The Domestication of Women. New York:
        St. Martin's Press.
Spring, A.
    1982   Adopting CIMMYT Farming Systems Survey
        Guidelines to the Malawian Situation. Report No. 5.
        Mimeo. Lilongwe, Malawi: The Women in Agricultural
        Development Project.
    1985   The Women in Agricultural Development Project in
        Malawi: Making Gender Free Development Work. In
        Women Creating Wealth: Transforming Economic
        Development. R. Gallin and A. Spring, eds., pp.
        71-76. Washington, D.C.: Association for Women and
        Development.
    1986a   Men and Women Participants in a Stall Feeder
        Livestock Program in Malawi. Human Organization
        45:1: 154-162.
    1986b Trials and Errors: Using Farming Systems
        Research in Agricultural Programs for Women. In
        Social Sciences and Farming Systems Research
        Methodological Perspectives on Agricultural
        Development. J. Jones and B. Wallace, eds., pp.
        123-144. Boulder, CO: Westview Press.

426

1986c   Reaching Female Farmers Through Male Extension
         Workers in Malawi. Training for Agriculture and
    ,   Rural Development, In 1985. pp. 11-22. Rome:
         FAO/UNESCO/ILO.
Spring, A. C. Smith, and F. Kayuni
    1982   Karonga Farmer Survey. Report No. 12. Mimeo.
         Lilongwe, Malawi: The Women in Agricultural
         Development Project.
    1983a   Studies of Agricultural Constraints Facing Women
         Farmers in Phalombe Rural Development Project.
         Report No. 19. Mimeo. Lilongwe, Malawi: The
         Women in Agricultural Development Project.
    1983b   Women Farmers in Malawi, Their Contribution to
         Agriculture and Participation in Development
         Projects. Washington D.C.: Women in Development
         Office and Ministry of Agriculture.
Staudt, Kathleen
    1975-76 Women Farmers and Inequities in Agricultural
         Services. Rural Africana 29:81-94.

29

The Role of Women Farmers
in Choosing Species for
Agroforestry Farming Systems
in Rural Areas of Ghana

Kofi Owusu-Bempah

Agroforestry as a farming system involves organizing available resources to produce forest tree and agricultural crop/livestock products on the same piece of land either simultaneously or sequentially for the well-being of the farmer (Combe and Budowski 1979; Douglas and Hart 1976; ICRAF 1979; Wiersum 1981). There are three basic types of agroforestry systems: (1) agrosilviculture systems (integration of forest trees with agricultural crops); (2) silvopastorial systems (integration of forest trees and livestock); and (3) agrosilvopastorial systems (integration of forest trees, agricultural crops, and livestock). When species and technology are well chosen and efficiently managed, agroforestry provides productive, protective, and other socioeconomic benefits to the farmer on a sustainable basis. Compared to other systems, agroforestry is productive on marginal lands, and meets rural dwellers' diverse food and forest products demands. It minimizes farmer pressures or encroachment on national forest reserves.

Out of Ghana's 23,854 million hectares of total land area, forest now occupies only 1,7658 million hectares (Forestry Dept. of Ghana 1985). An estimated annual deforestation rate of 22,000 hectares was recorded for 1981–1985 (FAO/UNDP 1981). The annual rate of loss of productive and unproductive woodlands (land degradation) was estimated at 20,000 hectares and 30,000 hectares, respectively, for 1980–85 (FAO/UNDP 1981). In 1980, an estimated annual rate of forest renewal was only 3,000 hectares (FAO 1985). Total annual planting rate of fuelwood and pole plantations between 1981–1985 was estimated at 700 hectares (Forestry Dept. of Ghana 1978). These figures clearly indicate that the largely public afforestation efforts are insignificant compared to the high rate of deforestation. The adverse effects of deforestation and land degradation are concentrated in the interior and coastal savannah. Both processes are taking place at an alarming rate in the Forest Savannah Transitional Zone (FSTZ).

The agroforestry system (taungya) has been practiced in the constituted forest reserve since 1929 (Taylor 1960). The tradition of conserving certain useful forest trees on farms outside the forest reserve (traditional agroforestry) is as old as rural farming activities in Ghana. Agroforestry outside forest reserves within the moist tropical zones of West and Central Africa as a substitute for shifting cultivation within bush fallow has been anticipated (Grinnel 1975). Increased income for the farmers through the sale of the planted tree crops in addition to improved food crops due to more site rejuvenation under the tree crops has also been anticipated (Grinnel 1975). Experience has shown that agroforestry should be based on locally productive, ecologically sound, and conservation-oriented activities which can be done most efficiently by local people. It cannot be done by government project interventions or by punitive enforcement of law (Weber and Hoskins 1983). To save the forest savannah transitional zone of Ghana from the environmental hazards of deforestation, land degradation, and desert encroachment, successful agroforestry systems implemented by the local people on their marginal lands are required. These systems should utilize local species, be flexible, sustainable, resilient, economically attractive, and acceptable to the local population. They should be administered with a true interdisciplinary spirit and by an efficient program that considers males and females of all socioeconomic levels equally.

Women farmers, who must produce subsistence crops on increasingly degraded soils, walk further for water, fuelwood and fodder; they identify and gather herbs to treat family health problems and are the principal sufferers from environmental degradation (Fortmann and Rocheleau 1984; Fouad 1982). Severe disruption of the hydrology balance through a high deforestation rate, widespread soil degradation, and erosion (Santos 1981; Antonini et al. 1975) dries up streams making the responsibility for fetching water more tedious and time consuming for women. Involvement of women in agroforestry project definition, design, and implementation is highly desirable but in most cases they are excluded (Fortmann and Rocheleau 1984). Reluctance on the part of male forest extension officials to work with females is well documented (Spring 1983). In Ghana, the role of women in promoting forestry development is not recognized. Their significant contribution to agricultural development is underestimated. Despite their deeper knowledge and experience in traditional agroforestry, women farmers' involvement in formal and informal

agroforestry, women farmers' involvement in formal and
informal agrofroestry surveys is insignificant.

A major constraint for conducting feasibility studies
on new agroforestry projects in Ghana is lack of data on
the traditional agroforestry areas. The Rural Agroforestry
and Forestry for Development Programs (RAFDP) is a response
to this development constraint. Agroforestry Farming Systems Research and Development (AFSR&D), a program of RAFDP,
embarked on data collection in 1981. The main objective of
this survey conducted by an AFSR&D team was to identify
useful local tree species to choose some for designing on-
farm research trials for areas in the Forest Savannah Transitional Zone of Ghana. Using an interdisciplinary teach
approach, joint projects of women in agroforestry and
"barefoot" agroforesters were conducted to come out with
the best species alternatives for particular areas of the
Forest Savannah Transitional Zone.

METHODOLOGY

The study covered the Forest Savannah Transitional Zone
of Ghana. Women and men farmers were distinguished, and
each gender was given equal attention during the home and
on-farm interviews conducted. Information was collected
randomly from 1,200 men farmers and 1,200 women farmers
from 240 villages of the study area. Two hundred and forty
members (mainly students and teachers) of the Agroforestry
Club and Conservation Society, native to the randomly
selected villages, collected the data during the Christmas
holiday of 1981 and the summer holidays of 1982. It was
envisaged that interviewing women farmers directly with
well-formulated specific questions during the formal and
infomal surveys would enable a researcher to obtain
accurate information (Shaner et al. 1982). Methods for
gathering accurate information about farm women were
utilized for this study (Martius-von-Harder 1979; McSweeney
1979; Mencher et al. 1979; Safai 1979; Smock 1979).

Differentiation of shifting cultivation from other
types of farming was a problem during the analysis of data
for this study. A ratio R (percentage) expressed by the
length of the cultivation period (C) times 100, divided by
the length of fallow period (F) was applied. Thus, the
formula:
$$R = \frac{C}{C + F} \times 100$$
adapted from previous studies (Josten 1962, Ruthenberg
1980) was used. The system is shifting if R is less than
33 percent (long fallow), semi-shifting or semi-permanent
when R is between 33 and 66, and permanent cultivation when

R is greater than 66. Names of tree species were collected in local dialects (Brong, Asanti, Ewe) and the botanical names were identified with the help of a botanist and relevant textbooks (Taylor 1960; Irvine 1961). Herbal practitioners and craftsmen were consulted to confirm the farmers' experience of the use of the named trees and shrubs. The names of agricultural crops that combine well with the trees and the names of livestock (including domesticated wildlife) which rely on forage of the trees and shrubs during dry season feeding periods were also recorded.

An approximately one hectare farm of a woman farmer from Cheremfaso village besides the Cheremfa River was selected and the species conserved were studied. Slide pictures of some of the identified species were taken. A diagrammatic presentation of the farm was drawn (Figure 29.1). A model design of an on-farm research trial plot was proposed for a typical Forest Savannah Transitional Zone village farmer. Some aspects of diagnosis and design methodology developed in a previous study (ICRAF 1979) were used.

RESULTS AND DISCUSSIONS

Farmers' Family Structure, Goals and Farm Types

For the 2,400 households included in the study (Table 29.1), 25.3 percent were found to be de facto headed by women. This figure falls within the 25-33 percent range estimated for all households in the world (Buvinic and Youssef 1978). Male out-migration, death of husbands, divorce, and desertion/abandonment were the main reasons for farming offered by the households headed by women. On average, women farmers listed 57 forest products that could be harvested or produced from the nearest forest reserve, while the men farmers listed an average of only 14.

Individual farmers gave complex and varied responses in the survey. Seven important goals were recorded (Table 29.2). While food, nutrition/shelter, and domestic fuelwood/energy were the top priorities of the women farmers, higher income and better clothes/social status were the most important for the men farmers. Both groups of farmers complained that the higher the education their children receive, the more their children neglect them and shun the rural life. The male farmers failed to realize that nutritional status of the family has direct effect on productivity and hence income. Malnutrition leads to low physical development resulting in low productivity and that in turn leads to low income (Berg 1973; Hicks 1980; Myrdal 1972; Wheeler 1980). But agroforestry as a farming system

TABLE 29.1

FARMERS' HOUSEHOLD TYPE AND THE AVERAGE NUMBER OF FOREST PRODUCTS THEY KNOW BY AGE GROUP

Age Group	Total Number of Households[a]		Households de facto Headed by Women			Listing of Forest Products by Farmers[b]	
			No.	%	Main Reasons	Men	Women
Below 21	161	(86)	10	6.2	Male Out-Migration	6	27
22-35	391	(204)	121	30.9	Male-Out-Migration and Divorce	10	39
36-45	920	(510)	237	25.8	Divorce and Desertion/Abandonment	16	62
46-60	928	(400)	240	25.9	Divorce and Death of Husband	25	98
TOTAL	2400	(1200)	608	25.3	Average:	14	57

Note: [a]Number of women farmers in brackets.

[b]Average number of forest products that can be harvested from the nearest natural forest reserve (nearest whole number).

Source: Compiled by the author.

FIGURE 29.1

DIAGRAMMATIC PRESENTATION OF A WOMAN FARMER'S TRADITIONAL AGROFORESTRY FARM IN THE FSTZ ZONE OF GHANA

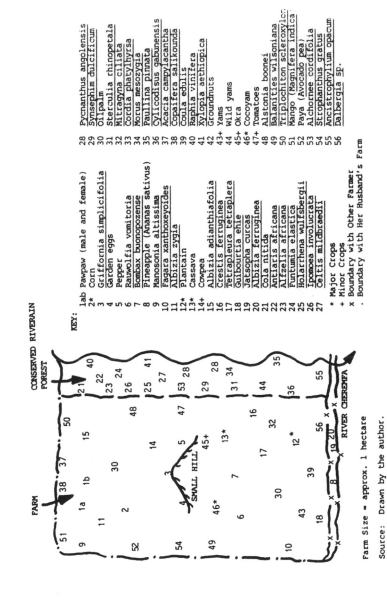

KEY:

1ab	Pawpaw (male and female)
2*	Corn
3	Griffornia simplicifolia
4	Garden eggs
5	Pepper
6	Rauwolfia vomitoria
7	Bombax buonopozense
8	Pineapple (Ananas sativus)
9	Manosonia altissima
10	Fagara xanthoxeyoides
11	Albizia zygia
12*	Plantain
13*	Cassava
14*	Cowpea
15	Albizia adianthiafolia
16	Crestis ferruginea
17	Tetrapleura tetraplera
18	Guibourtia ehle
19	Jatsopha curcas
20	Albizia ferruginea
21	Cola nitida
22	Antiaris africana
23	Alfzelia africana
24	Funtumia elastica
25	Holarrhena wulfsbergii
26	Ipomoea involucrata
27	Celtis mildbraedii
28	Pycnanthus angolensis
29	Synsephim dulcificum
30	Oilpalm
31	Sterculia rhinopetala
32	Mitragyna ciliata
33	Cordia phatylhyrsa
34	Morus mesozygia
35	Paullina pinnata
36	Cylicodisus gabunensis
37	Acacia campylacantha
38	Copaifera salikounda
39	Coula edulis
40	Raphia vinifera
41	Xylopia aethiopica
42	Groundnuts
43+	Yams
44	Wild yams
45+	Okra
46*	Cocoyam
47+	Tomatoes
48	Alstonia boonei
49	Balanites wilsoniana
50	Triplochiton scleroxylon
51	Mango (Magnifera indica)
52	Paya (Avocado pea)
53	Alchornee cordifolia
54	Strophanthus gratus
55	Ancistrophyllium opacum
56	Dalbergia sp.

* Major Crops
+ Minor Crops
x Boundary with Other Farmer
. Boundary with Her Husband's Farm

Farm Size = approx. 1 hectare

Source: Drawn by the author.

TABLE 29.2

GOALS OF FARMERS IN THE FOREST SAVANNAH
TRANSITIONAL ZONE OF GHANA

Needs	Role of Forest Products	Men Order	Percent	Women Order	Percent
Adequate Food, Nutrition, and Shelter	High	3	89.6	1	100.0
Domestic Fuelwood/ Energy	High	6	82.9	2	98.6
Health/Medicinal Plants	High	4	73.4	3	95.2
Higher Income	Low	1	99.8	4	97.4
Better Clothing and Social Status	Low	2	92.3	5	88.9
Better Children's Education	Low	5	93.1	6	79.8
Cultural Values	Moderate	7	83.7	7	75.4

Source: Compiled by the author.

could be programmed to provide the adequate food, nutrition, fuelwood, and medicinal plants requirements of a farmer's family (Dykstra 1980).

Eight types of farmers were identified in the study area (Table 29.3). A majority of the farmers were found to be shifting cultivators (46.9 percent) and permanent subsistence farmers (43.5 percent). Almost all the permanent subsistence farmers are women. Extent of tree species conservation, incidence of nutrient cycling, and consideration of household food, nutrition and domestic energy demand (fuelwood) were found to be relatively high for the subsistence systems of farming. However, the option of varying capital, income generating, and labor

TABLE 29.3

AGRICULTURAL SYSTEMS (FARM TYPES) IN THE FOREST TRANSITIONAL ZONE OF GHANA

FARM TYPES	Percent of the Farmers Involved[a]	Percent of Sale of Food Products[b]	Extent of Forest Tree Species Conservation	Mono-Cropping/Inter-Cropping	Incidence of Nutrient Cycling	Direct Contribution of Farming to Family Food, Nutrition and Energy	Option of Varying Capital	Productivity
SHIFTING CULTIVATION								
Combined Hunting-Gathering and Shifting Cultivation	1.5 (0)	Less than 2% (Low)	Moderate	No/Yes	Low	Moderate	Low	Low
Shifting Cultivation (Long Fallow)	17.0 (25.0)	Less than 4% (Low)	Low	No/Yes	Moderate	Moderate	Low	Low
Semi-Shifting Cultivation (Short Fallow)	28.4 (21.4)	Less than 4% (Low)	Low	No/Yes	Low	Moderate	Low	Low
Total	46.9							
PERMANENT CULTIVATION								
SUBSISTENCE								
Small-Scale/Unproductive Subsistence	24.3 (90.1)	Less than 10% (Low)	High	Yes/Yes	High	High	Low	Moderate
Large-Scale/Productive Subsistence	19.2 (86.3)	Between 10-50% (Moderate)	High	Yes/Yes	High	High	Moderate	Moderate
Subtotal	43.5							
COMMERCIAL								
Small-Scale Commercial (less than 100 ha)	4.6 (18.2)	Over 50% (High)	Low	Yes/No	Low	Low	High	High
Large-Scale Commercial (over 100 ha)	2.8 (10.8)	Over 50% (High)	Low	Yes/No	Low	Low	High	High
Cooperative Commercial	1.2 (15.4)	Over 50% (High)	Low	Yes/No	Low	Low	High	High
Subtotal	8.6							
TOTAL	52.1							

Note: [a]Percent women farmers in brackets.

[b]Income generating capacity in brackets.

Source: Compiled by author.

productivity were found to be lower for permanent subsistence farming compared with commercial farming. Areas where permanent subsistence farmers were selling 40-50 percent of their farm surplus and were cultivating relatively small-sized, highly productive land areas had better living conditions.

The highly productive subsistence farmers were found to be those practicing traditional agroforestry. They could list several trees that combine well and improve the yield of crops around them. According to this group of farmers, when the weather is favorable, the yields are high but the produce fetches a low price in the market. However, they are better off than their counterparts who are commercial farmers since they make additional income from their forest tree products to add to their income. They keep more domestic livestock than others. Thus, their ability to survive adverse financial periods is significantly higher than any of the other types of farmers. Though their production rate is lower compared with the commercial farmers, they are better risk avoiders.

The commercial farm group (8.6 percent) has the least percentage of women farmers. Their labor productivity, rate of capital use, and income were found to be relatively high. However, their consideration for household food, nutrition, and domestic energy (fuelwood) was found to be low. They are poor conservators of forest trees and very exploitative, especially those engaged in intensive cattle ranching. A commercial farmer whose labor productivity is rather high but purchases all his family food was found to have far lower living standard than a less productive subsistence farmer who raises all or most of his or her family food and sells little surpluses (Harwood 1979).

Deforestation, Suggested Species and Conservation

Eleven important deforestation activities were identified during the survey (Table 29.4). Men farmers were found to engage in these activities more than the women farmers. The rural farmers were particularly concerned with the loss of medicinal plants, that they attributed to the clearing of the riverine sacred forest formerly reserved by traditional taboos close to villages. To them even lesser known species are potential sources for healing unknown diseases. Almost all useful species named by the farmers have a role in coping with their health problems. Since women farmers are more concerned with household health problems, they know more about the uses and some

TABLE 29.4

TYPES OF DEFORESTATION AND LAND DEGRADATION ACTIVITIES IN FOREST SAVANNAH TRANSITIONAL ZONE OF GHANA

Types of Activity	Contribution to Deforestation Rate	Contribution to Land Degradation Rate	Extent of Men Farmers' Involvement	Extent of Women Farmers' Involvement
Charcoal Making	High	Low	High	Low
Tobacco Curing	High	Low	Moderate	Moderate
Mass Fuelwood Collection for Sale	High	Moderate	High	Low
Collection of Fuelwood for Domestic Use	Low	Low	Low	High
Mass Collection of Medicinal Plants for Sale	Moderate	Low	High	Low
Collection of Medicinal Plants for Family Use	Low	Low	Moderate	High
Shifting Cultivation	High	Moderate	High	Low
Intensive Ranching	Moderate	High	High	Low
Clearing of Sacred/Riverine/Watershed Forest for Farming	High	Moderate	High	Low
Burning as a Means of Hunting	High	High	High	Low
Timber Exploitation	Moderate	Low	High	Low

Source: Compiled by the author.

still conserve them against deforestation activities through traditional agroforestry practices (Figure 29.1). Forest guards of the Forestry Department complained of encroachment of the national reserved forest by shifting cultivators, charcoal makers, medicinal plant gatherers, and sometimes by firewood gatherers.

Farmers named 658 plant species and suggested that those with multiple uses should be integrated with agricultural crops. Since poor soil fertility was one of their major problems, they asked for nitrogen fixing trees which are also capable of providing dry season feeding for their livestock. The subsistence farmers interviewed were mostly keeping sheep, goats, poultry (chickens and ducks), rabbits, and domesticated wildlife like grasscutters (Thryonomys swinderianus). Commercial farmers were more interested in cattle and pigs. The study showed that an agroforestry package that integrates these local livestock with forest trees to ensure adequate forage for sheep, goats, and cattle during dry seasons would be readily adopted by the farmers in forest savannah transitional zone of Ghana.

Conservation and sustainable exploitation of natural forest species are still practiced by some women farmers, despite the numerous deforestation activities (Table 29.4) going on. Figure 29.2 is a diagrammatic presentation of a woman farmer's farm that was studied during the survey. She was able to identify 82 different plant species. She was able to recite their uses and to provide some historical background for some of them. She lamented that none of her children and grandchildren are interested in the conserved plants. The possibility of losing them through deforestation activities is high when she becomes too old to farm. This was her major concern, and she repeatedly attributed the young generation's lack of conservation and sustainable exploitation consciousness to education.

The study indicated that the farmers lack the tradition of raising tree seedlings and cultivating them. Considering their interest in these species, there is a need to embark on a program aimed at examining the most important species for food, health, social, and economic reasons (Ayensu 1983) and for including their cultivation in the national land use system (Levingston and Zamora 1983).

FIGURE 29.2

PROPOSED FIELD LAYOUT OF ON-FARM AGROFORESTRY
RESEARCH PLOTS FOR A TYPICAL VILLAGE IN FSTZ OF GHANA

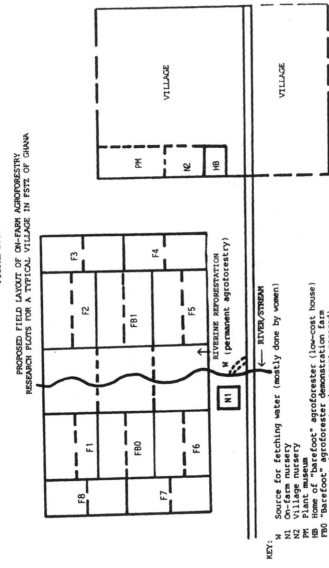

KEY:

W   Source for fetching water (mostly done by women)
N1   On-farm nursery
N2   Village nursery
PM   Plant museum
HB   Home of "barefoot" agroforester (low-cost house)
FB0 "Barefoot" agroforester demonstration farm
     (research-farmer-extension agent managed)
F1 & F5   Farmer-managed farm but supervised by "barefoot" agroforester
F2 & F6   Researcher (student) and former partnership farm (research and
          farmer managed)
F3 & F7   Farmer-managed farm but supervised by student and "barefoot"
          agroforester
F4 & F8   Farmer and "barefoot" agroforester partnership farm
FB1 Community farm supervised by barefoot agroforester

Source: Designed by the author

On-Farm Research

The Agroforestry Farming System Research and Development (AFSR&D) Program adopting some survey, planning, designing, and implementation strategies outlined by farming systems specialists (Harwood 1979; Shaner et al. 1982), has initiated a project to utilize some of the useful ideas and tree species identified during the survey. An interdisciplinary collaboration with a coordinator who has brought the disparate insights and skills of various scientific specialties together is undertaking this on-farm research project. The field layout of typical new agroforestry on-farm research for the Forest Savannah Transitional Zone is illustrated in Figure 29.2. It is a joint project of barefoot agroforesters and women. A barefoot agroforester is an agriculturist/forester with basic training in agroforestry who has settled within a rural farming community as a full-time farmer, on-farm researcher, and an agent for extension services to farmers using his/her farm as a demonstration farm to promote transfer of agroforestry technology. A revolving fund that will engage two students (Women in Agroforesty Project) and one barefoot agroforester (Farmer-Research-Extension Agent) and eight local collaborating farmers per site is being solicited. Researcher, farmer, and joint researcher/farmer-managed trials are considered in the design. The field design adopted involves paired treatments per farm, analysis of variance calculations, and t test for significance (Shaner et al. 1982). A re-establishment of "sacred" forest forms part of the joint project. Local plants species as well as foreign species (Leucaena leucocephala and Gliciadia speium) are being considered.

Planning, design/field layout, and implementation strategy were formulated with both men and women farmers' involvement. The farmers offered the following ideas for the project: local and foreign species with multiple uses that are relevant to rural well-being should be included; and reestablishment of riverine/sacred forest and integration of medicinal plants in agroforestry should be considered in the planning. A long-term project when efficiently implemented would improve site rejuvenation, increase income for the farmers (Grinnel 1975), and also improve the hydrology (Santos 1981; Antonini et al. 1975) that could be observed by the farmers, from where they fetch water (W in Fig. 29.2).

FINDINGS, RECOMMENDATIONS AND CONCLUSION

The following are some of the findings from farm surveys conducted in the rural Forest Savannah Transitional Zone of Ghana:

(1) Food, nutrition, health, and energy (fuelwood) are more important to women farmers than income, clothes, and social status which are comparably more important to men farmers.

(2) Traditional subsistence agroforestry farmers (mostly women) capable of selling between 45-50 percent surplus products acknowledged the benefits of forest trees and proposed the most useful trees for the new agroforestry systems.

(3) Multiple purpose local trees like Rauwolfia vomitoria, Alstonia boonei, Tetrapleura tetraplera, and Fagara xanthoxeyoides as well as foreign species like Leucaena leucocephala and Gliriadia speium have high potential of being accepted by farmers in the Forest Savannah Transitional Zone.

(4) Women farmers were found to be better conservators and more resourceful than their men counterparts.

(5) An agroforestry package that integrates local livestock with forest trees to ensure adequate forage, especially for sheep and goats during the dry season, would be readily adopted by the subsistence farmers. A similar package favoring cattle would be favored by commercial farmers.

(6) Plants with medicinal or healing properties had the highest probability of being accepted for both agroforestry and conservation purposes.

In order to utilize the above findings to promote rural development through agroforestry and conservation projects, researchers should consider: the direct involvement of women in formal/informal surveys and the planning, designing, and implementation of projects, as well as, and further research on species recommended by the farmers. National and international agencies should consider community efforts like RAFD and promote attempts to involve women in agroforestry and forestry development. Rural farmers' participation in the planning and designing of agroforestry onfarm research projects would enable researchers to incorporate their useful ideas into solutions for rural problems. Since women farmers and subsistence farming are vital components of total agroforestry systems, women's roles in choosing appropriate species for projects must be recognized.

REFERENCES

Antonini, G., et al.
1975   Population and Energy: A Systems Analysis of
        Resource Utilization in the Dominican Republic.
        Gainesville, FL: University of Florida Press.
Ayensu, E. S.
1983   The Healing Plants. Unasylva 35:140:2-6.
Berg, A.
. 1973   The Nutrition Factor. Washington, D.C.:
        Brookings Institute.
Buvinic, M. and N.H. Youssef
1978   Women-Headed Households: The Ignored Factor in
        Development Planning. Washington, D.C.:
        International Center for Research on Women.
Combe, J. and G. Budowski
1979   Classification of Agroforestry Techniques.
        Proccedings of Symposium on Agroforestry Systems in
        Latin America, pp. 17-47, Turrialba, Costa Rica:
        CATIE.
Douglas, J. S. and R.A. de J. Hart
1976   Forest Farming: Towards a Solution to Problems
        of World Hunger and Conservation. London: Watkins.
Dykstra, D. P.
1980   Food and Fuelwood: A Preliminary Mathematical
        Programming Analysis for an Ujamaa Village in
        Tanzania. Prepared for National Seminar
        Agroforestry, Nairobi, Kenya: University of
        Nairobi/ICRAF.
FAO
1985   Forest Resources of 1980. International Year of
        the Forest. Rome: FAO.
FAO/UNDP
1981   Tropical Forest Resources Assessment Project (in
        the Framework of GEMS). Forest Resources of
        Tropical Africa. II. Country Briefs. FAO: Rome,
        UN 32/6, 13d-78-04.
Forestry Department of Ghana
1978   National Progress Report on Forestry — Ghana
        1976-1977. Prepared for the 5th session of the
        African Forestry Commission, Accra, Ghana.
1985   National Progress Report 1980-1984. Prepared
        for the 12th Commonwealth Forestry Conference,
        Victoria, B.C., Canada, September.

442

Fortmann, L. and D. Rocheleau
    1984   Why Agroforestry Needs Women: Four Myths and a
            Case Study. Unasylva 36:146:2-11.
Fouad, I.
    1982   The Role of Women Peasants in the Process of
            Desertification in Western Sudan. Geojournal
            6:1:25-30.
Grinnel, H. R.
    1975   Agri-Silviculture: A Suggested Research
            Programme for West and Central Africa. Ottawa,
            Canada: IDRC.
Harwood, R. R.
    1979   Small Farm Development: Understanding and
            Improving Farming Systems in the Humid Tropics.
            Boulder, CO: Westview Press.
Hicks, W.
    1980   Economic Growth and Human Resources. World Bank
            Staff Working Paper No. 403.
ICRAF-International Council for Research in Agroforestry
    1961   Newsletter 1:1. Nairobi: ICRAF.
Irvine, F. R.
    1976   Woody Plants of Ghana with Special Reference to
            Their Uses. Oxford, United Kingdom: Oxford
            University Press.
Josten, J. H. L.
    1962   Wirtschaftliche und agrarpolitische Aspekte
            tropischer Landbausysteme. Gottingen, Federal
            Republic of Germany: Institut fur landwirtschaft-
            liche Betriebslehre.
Levingston, R. and R. Zamora
    1983   Medicine Trees of the Tropics. Unasylva
            35:140:7-10.
Martius-von-Harder, G.
    1979   How and What Rural Women Know: Experiences in
            Bangladesh. Studies in Family Planning 10:405.
McSweeney, B. G.
    1979   Collection and Analysis of Data on Rural Women's
            Time Use. Studies in Family Planning 10:379.
Mencher, J. P. et al.
    1979   Women in Rice Cultivation: Some Research Tools.
            Studies in Family Planning 10:408.
Myrdal, G.
    1972   Economics, Theory and Underdeveloped Regions.
            London: Methuen.
Ruthenberg, H.
    1980   Farming Systems in the Tropics. 3rd Edition.
            Oxford: Clarendon Press.

Safai, M.
    1979   Circumventing Problems of Accessibility to
            Rural Muslim Women. Studies in Family Planning
            10:405.
Santos, B.
    1981   El Plan Sierra: Una Experiencia de Desarrollo
            Rural en las Montanas de la Republica Dominicana.
            In Agricultura de Ladera en America Tropical. A.
            Novoa B. and J. L. Posner, eds. Turrialba, Costa
            Rica: CATIE.
Shaner, W. W. et al.
    1982   Farming Systems Research and Development.
            Boulder, CO: Westview Press.
Smock, A. C.
    1979   Measuring Rural Women's Economic Roles and
            Contributions in Kenya. Studies in Family Planning
            10:385.
Spring, A.
    1983   Women in Agricultural Development in Malawi.
            Presented at the 11th International Congress of
            Anthropological and Ethnographic Sciences,
            Vancouver, Canada.
Taylor, C.J.
    1960   Synecology and Silviculture in Ghana. London:
            Thomas, Nelson and Sons, Ltd.
Weber, F. and M. Hoskins
    1983   Agroforestry in the Sahel. Blacksburg, VA:
            Virginia Polytechnic Institute and State University,
            Department of Sociology.
Wiersum, K. F.
    1981   Outline of Agroforestry Concept. In Viewpoints
            on Agroforestry, K. F. Wiersum, ed., Wageningen,
            Netherlands: Agricultural University of Wageningen,
            Department of Forestry.
Wheeler, D.
    1980   Human Resources Development and Economic Growth
            in Developing Countries, A Simulation Model. World
            Bank Staff Working Paper No. 407. Washington, D.C.:
            World Bank.

Contributors

Amalia M. Alberti received her Ph.D. from Stanford University in 1986. She has lived and worked in Honduras, Ecuador, and the Philippines. Her special interests include non-formal education, ethnicity, stratification, and the relationship between women and poverty.

Susan A. Beckerson received her M.S. from the University of Guelph, in Ontario. She collected the data on which her paper is based in 1981-82 while on the staff of Bunda College of Agriculture, Malawi. Currently she is teaching family studies at Elmira High School in Ontario.

Vasantha Chase is a social economist with the Caribbean Agricultural Research and Development Institute (CARDI) in Trinidad. She received her Ph.D. in economic anthropology from Cornell University. She is currently coordinating a project with the Food and Agricultural Organization of the United Nations on women and agricultural development in the English Speaking Caribbean.

Jean M. Due is a professor of agricultural economics at the University of Illinois at Urbana-Champaign. She has spent twenty years in Africa doing research on many development topics including allocation of credit to small farmers, land tenure changes in Kenya and Zimbabwe, farming systems in Tanzania and Zambia, and the role of women in development. As a member of the Bean/Cowpea CRSP, she has been active in farming systems research in Tanzania since 1980.

Lila E. Engberg is an associate professor in the Department of Family Studies, University of Guelph. She worked in Africa on three occasions: 1963-68 in Malawi with the Food and Agricultural Organization of the United Nations; 1970-75 as Head of the Home Science Department, Faculty of

Agriculture, University of Ghana; and 1982–85 as Senior
Lecturer in Home Economics, University of Malawi, Zomba.

Patricio Espinosa is an agricultural economist and an
agronomist. He is former director of Smallholder
Programming and current head of Planning and Agricultural
Economics at the Instituto Nacional de Investigacion
Agropecuarias, Ecuador.

Alison Evans is an economist working as a research officer
at the Institute of Development Studies, University of
Sussex, United Kingdom. Her research interests include
rural household economics and agricultural development in
Central and Southern Africa, especially the gender implications of agricultural development and on-farm technological change. She is currently working in Zambia on gender
issues in farm household labor allocation.

Anne Ferguson is the women in development specialist with
the Bean/Cowpea Collaborative Research Support Program
(CRSP) at Michigan State University. She has a Ph.D. in
anthropology from Michigan State University and has
conducted research in both Africa and Latin America.

Maria E. Fernandez is the field coordinator for Grupo
Yanapai in Huancayo, Peru and a Research Associate of the
Department of Sociology, University of Missouri, Columbia,
in the Small Ruminant Collaborative Research Support
Program (SR-CRSP). She has an M.A. in rural social
development from the University of Reading, England, and is
presently working on a Ph.D. in agricultural systems at the
same institution.

Mary Futrell is a professor of human nutrition at
Mississippi State University. She has been a USAID
consultant in nutrition and home economics to Ahmadu Bello
University in Nigeria and has conducted nutrition studies
in India, Sudan, Burkina Faso, Haiti, Honduras, Mexico and
the Dominican Republic. Currently, she is working with
Eunice McCulloch on a Health and Human Services research
project dealing with ethnic comparisons of aging.

Rita S. Gallin is an associate professor of sociology and
director of the Women in International Development Program
at Michigan State University. She has done extensive

research in Taiwan and the United States on women and socioeconomic change.

Patricia Garrett is a sociologist and was the former co-principal investigator on the Bean/Cowpea Collaborative Research Support Project, Instituto Nacional de Investigaciones Agropecuarias, Ecuador and Cornell University. She is currently working with the Bush Institute in Child and Family Policy in Chapel Hill, North Carolina.

Jane E. Gleason received her B.A. at Illinois University and her Ph.D. at the University of Illinois in agricultural economics. Her current research interest is women in development in Asia.

Bahnisikha Ghosh is a lecturer in economics at the Center for Human Resource Development, University of Kalyani, West Bengal, India. His specializations include economics of population, agricultural, rural, and human resource development, and related interdisciplinary studies. He has studied, worked and published in India and several other countries.

Art Hansen is associate professor of anthropology and chairs the Steering Committee of the Food in Africa Program at the University of Florida. He has conducted research in Zambia, established an adaptive research program in Malawi, and consulted and worked in development in Botswana, Tanzania, Bolivia, and the Dominican Republic.

Jeanne Koopman Henn is currently an assistant professor of economics at Northeastern University in Boston. She has lived and taught in Africa for nine years, including four years in Cameroon and three in Tanzania.

Peter E. Hildebrand is a professor of food and resource economics and is involved with the Farming Systems Program at the University of Florida. He established the graduate level FSR/E course in 1980 and helped formulate the farming systems minor for graduate students. He is Senior Counsel to the Farming Systems Support Project and has authored or co-authored three farming systems books.

Robert E. Hudgens received his B.S. degree from the University of California at Davis, and his M.S. and Ph.D. degrees in agronomy from the University of Florida. He has

over 13 years of experience in agricultural development in
South America, Central America, Africa, and South Asia.

Janice Jiggins is currently working as an independent rural
development consultant. She has extensive experience in
South Asia and anglophone Sub-Saharan Africa. Her
interests include organization and management of local
producer services, agricultural extension, and advocacy on
behalf of "women in development."

Jonice Louden is currently a graduate student-tutor in the
Department of Sociology at the University of the West
Indies at Mona, Jamaica. She has lived and worked in
England and Nigeria in health related fields. She is
primarily interested in the study of women's work, the
relationship between production and reproduction, and in
evaluation methodology.

Melanie Marlett is working on her Ph.D. in nutrition at
Tufts University. She worked in The Gambia as a Peace
Corps volunteer for two years and was involved with the
Mixed Farming Project (Colorado State University/USAID)
where she managed the Women's Program. She intends to
return to The Gambia to do her dissertation research.

Eunice R. McCulloch is a research sociologist with the
Social Science Research Center at Mississippi State
University. Her research interests are varied, dealing
with such diverse topics as women and development, aging,
and child development. Most of her studies are conducted
using a cross-cultural perspective.

Manasse Timmy Mtoi is a lecturer in agricultural economics
at the Sokoine University of Agriculture in Tanzania. He
has conducted extensive socio-economic studies on farming
systems in Tanzania and is currently a Ph. D. candidate at
Giessen University in West Germany.

Sudhin K. Mukhopadhyay is a professor of economics and
director of the Center for Human Resource Development,
University of Kalyani, West Bengal, India. His areas of
specialization include economics of population,
agricultural, rural, and human resource development, and
related interdisciplinary studies. He has studied and
worked in India and in several other countries.

Margaret Norem is working for the University of Arizona's Environmental Research Laboratory. She received her Ph.D. in agronomy from the University of Arizona. She went to The Gambia in 1985 on a Women in Development Fellowship and began working for the Mixed Farming Project and its Women's Program.

Rosalie Huisinga Norem is associate professor of family environment and teaches in the Technology and Social Change Program at Iowa State University. She has written in the areas of training in farming systems, families and international development, women in development, and communications and farming systems. She is a member of the advisory committee, Intra-Household Dynamics and Farming Systems Case Study Project (Population Council/FSSP).

Kofi Owusu-Bempah received a B.S. in agriculture from the University of Science and Technology, Kumasi, Ghana. The research for his paper was conducted during his enrollment in the International School for Agricultural and Research Development, Colorado University as a Hubert H. Humphrey fellow.

Thelma R. Paris is an assistant scientist in the Department of Agricultural Economics of the International Rice Research Institute. She has a M.S. in agricultural economics from the University of the Philippines, Los Banos. She is involved in action research in a farming systems project and has conducted intensive field studies on incorporating nutrition and women in farming systems research.

Elizabeth Chola Phiri received a post graduate diploma in nutrition planning and food resources related to community development and has a M.A. in agricultural extension. She has worked for twelve years as the head of the female extension section in Zambia and served as national coordinator of three donor-funded women's agricultural programs.

Susan V. Poats is the associate director of the Farming Systems Support Project (University of Florida/USAID) and served as the acting director of the Women in Agricultural Development Program from December 1986 to September 1987. She received her Ph.D. in anthropology at the University of Florida and did a post-doctorate at the International Potato Center, Lima, Peru. She is currently responsible for FSSP training and networking activities in Africa.

Andrée Rassam is a research associate in the Farm Resource Management Program at the International Centre for Agricultural Research in Dry Area. She holds a B.S. in economic science from Aleppo University and a M.A. in sociology from the University of Western Ontario. A Middle East Research Award sponsored by the Ford Foundation, IRDC, and Population Council (MEAward), supported part of the research reported in her paper. Agricultural labor is the main focus of her current research.

Dianne E. Rocheleau M. is a program officer with the Ford Foundation in Nairobi, Kenya. She has a Ph.D. in Geography from the University of Florida. She has conducted ecological research on hillslope farming systems in the Dominican Republic (1979-81) and has worked as a Rocke- feller Foundation Fellow and Senior Scientist at the International Council for Research in Agroforestry (1983- 86), focusing on environmental and equity issues and community-based research in agroforestry.

Sandra Russo is employed as assistant research scientist for the University of Florida. She received her Ph.D. in agronomy from the University of Florida. She has worked in Africa on several projects and was employed by the Mixed Farming project in The Gambia as a Forage Agronomist.

Jean H. Sabry is professor of applied human nutrition and coordinator of graduate studies in the Department of Family Studies, University of Guelph. She has served on the faculties of the University of Manitoba, University of Toronto, Pennsylvania State University, and Beirut College for Women in Lebanon.

Marie Sambou is working for the Gambian Ministry of Agriculture on the Women's Program of the Mixed Farming Project. She completed her degree in agronomy in Nigeria.

Marianne Schmink is associate professor of Latin American Studies at the University of Florida. From 1983-87 she served as Co-Director of the Women in Agricultural Development Program, University of Florida and as Organizer of the 1986 conference on "Gender Issues in Farming Systems Research and Extension." Since 1980, she has been Executive Director of the Amazon Research and Training program at the Center for Latin American Studies, University of Florida.

Anita Spring is associate dean of the College of Liberal Arts and Sciences and associate professor of anthropology at the University of Florida. She was director of the Women in Agricultural Program at the University of Florida from 1984-86. From 1981-83 she directed the Women in Agricultural Development Project in Malawi. She received her Ph.D. from Cornell University and has worked in Zimbabwe and Botswana.

Alistair J. Sutherland received his Ph.D. in social anthropology from the University of Manchester. He has been the senior rural sociologist in the Research Branch of Zambia's Ministry of Agriculture and Water Development since 1983. Before that he carried out applied social research in Botswana and the United Kingdom. His current research interests include research/extension linkages and the role of social scientists in farming systems research nand extension.

Mary N. Tembo received a diploma in nutrition and has seven years of experience in nutrition and agricultural activities in the Ministry of Agriculture and Water Development, Zambia.

Dennis Tully is coordinator of the Agricultural Labor and Technological Change Project in the Farm Resource Management Program, ICARDA, Syria. He received his B.S. in chemistry from Massachusettes Institute of Technology and his Ph.D. in anthropology from the University of Washington. He has done research in Ethiopia, Sudan, and Syria.

Eva Wollenberg is a Ph.D. candidate in the Department of Forestry and Resource Management, University of California, Berkeley. She has spent nearly two years in the Philippines and three months in India doing work in rural development. Her special interests are resource socioeconomics and tropical ecology.

Peter Wotowiec, Jr. holds a B.S. degree specializing in Horticulture from the Ohio State University and a M.S. Degree in Agricultural Education from the University of Florida. Current research interests include the economics of subsistence farm households and research/extension linkages in applied agricultural development.

Printed and bound by CPI Group (UK) Ltd, Croydon, CR0 4YY

23/10/2024

01778240-0007